Penguin Books

A CERTAIN GRANDEUR

Graham Freudenberg was born in Brisbane in 1934. He worked as a journalist and press secretary to Arthur Calwell before joining Gough Whitlam's staff as press secretary in 1967. Between 1972–5 he was special adviser to the Prime Minister. Since 1976 he has been consultant to New South Wales Premier Neville Wran and his successor Barrie Unsworth and since 1983 to Prime Minister Bob Hawke.

A CERTAIN GRANDEUR

Gough Whitlam in Politics

GRAHAM FREUDENBERG

PENGUIN BOOKS

Penguin Books Australia Ltd,
487 Maroondah Highway, PO Box 257
Ringwood, Victoria 3134, Australia
Penguin Books Ltd,
Harmondsworth, Middlesex, England
Viking Penguin Inc.,
40 West 23rd Street, New York, NY 10010, USA
Penguin Books (Canada) Limited,
2801 John Street, Markham, Ontario, Canada L3R 1B4
Penguin Books (NZ) Ltd,
182-190 Wairau Road, Auckland 10, New Zealand

First published 1977 by Macmillan
Published in Sun Books, 1978
Published in Penguin, 1987

Copyright © Graham Freudenberg 1977

All Rights Reserved. Without limiting the rights under copyright reserved above, no part of this publication may be reproduced, stored in or introduced into a retrieval system, or transmitted, in any form or by any means (electronic, mechanical, photocopying, recording or otherwise), without the prior written permission of both the copyright owner and the above publisher of this book.

Offset from the Sun Books edition
Made and printed in Australia by
The Book Printer, Maryborough, Victoria

CIP

Freudenberg, Graham, 1934-
A certain grandeur: Gough Whitlam in politics.

Rev. ed.
Includes index.
ISBN 0 14 010049 0.

1. Whitlam, Gough, 1916- . 2. Australia — Politics and government — 1965- . 3. Australia — Politics and government — 20th century — Biography. I. Title.

320.994

To my parents

Contents

Introduction to Penguin edition ix
Author's Preface xix
Introduction xxi
1. The Menzies Inheritance 3
2. 'Is the Tumbril Ready?' 24
3. How Australia Got Into Vietnam 39
4. The Leader 63
5. 'The Party, the Policy, the People' 85
6. The Decline of Holt 106
7. The Coming of John Grey Gorton 118
8. Resignation 129
9. 'I did it my way' 138
10. Phoenix 151
11. Reconstruction 166
12. The Destruction of Gorton 179
13. Whirlwinds of Change 189
14. 'Tiberius with a telephone' 215
15. 'It's Time' 224
16. The Duumvirate and After 239
17. Opposition 267
18. The Economy v. the Programme 276
19. Double Dissolution 288
20. A Question of Legitimacy 302
21. 'Scandal' 317
22. The Last Casualty 327
23. Loans 342
24. Ambush 366
Afterword 414

Notes and Sources 418
Appendix 424
Index 430

Introduction to Penguin edition

By the tenth anniversary of the dismissal of the Whitlam Government, it was easy enough to believe that the wounds had healed. The Australian Labor Party had reached a political ascendancy throughout Australia unmatched since the 1940s, holding government in the states of NSW, Victoria, South Australia, Western Australia and the national government itself. The events of 1975 had so far passed into legend as to have received already the ultimate acceptance as past history, the television miniseries.

The series had been scheduled by the Channel Ten Network for screening during the last week of February 1983. There was a hurried rescheduling when, on 3 February, Malcolm Fraser procured a double dissolution and an election for 5 March. It was a day of drama almost comparable with 11 November 1975. Bill Hayden, the man to whom Whitlam had proffered the succession amid the ruins of 13 December 1975, resigned the leadership he had secured in 1977. Bob Hawke, Whitlam's next choice in those desperate hours, became leader-designate. The Ten Network agreed with Labor Party representations that it would be inappropriate to screen so provocative a program in the height and heat of an election campaign. By the time *The Dismissal* was seen by a huge television audience, Fraser, devastated by his election defeat, was not even a member of parliament.

Such was the healing effect of time and events that by November 1985, Hawke was able to name Fraser as Australia's 'eminent person' on a committee set up by the Commonwealth Prime Ministers to negotiate with the South African Government on apartheid. The reaction from the Labor Party was surprisingly muted.

Nor was there much evidence of Labor *schadenfreude* when Fraser's mission came to its foredoomed failure.

Kerr's successors as Governor-General of Australia, Sir Zelman Cowan and Sir Ninian Stephen both worked diligently to sound the 'healing' theme, and restored a certain respect, if not acceptance, to the office. Kerr himself was spending most of his time overseas, but he was never more an exile than whenever he was at home.

Whitlam returned in triumph for the tenth anniversary. He was now living in Paris where he was Australia's Ambassador to UNESCO. This was the very post accepted by Kerr, and then abandoned in the face of bitter hostility in 1978. On 11 November 1985, Kerr happened to be in Paris. He called the Australian Embassy. He received no reply. Next day, he asked an Australian official why the Embassay had been shut. Was it, he inquired, something to do with 'Whitlam and that business'? He had to be reminded that, for France – and even for Australia – the eleventh day of the eleventh month has other meanings.

Whitlam himself, in his triumphant travels to five capitals – Canberra, Sydney, Melbourne, Adelaide and Perth – was careful to emphasise the change in circumstances: his own, the Party's the nation's.

He stressed the longer term meaning of his famous invocation on the steps of Parliament House. Its shorthand version had passed into history as merely 'Maintain the rage'. His actual words had been: 'Maintain your rage and your enthusiasm through the campaign for the election now to be held and until polling day'. Now, ten years and many polling days later, it had been the enthusiasm, Whitlam insisted, which had been maintained and which had brought the Australian Labor Party to its present splendid position.

The main event of the 1985 festivities was the launching by Prime Minister Hawke of Whitlam's book, *The Whitlam Government 1972-75*, at the National Press Club, Canberra, on 11 November. Of the book itself, Hawke said, 'There is no false modesty about it – and why should there be'. And of their personal relationship he said,

> Gough, in 1942 Roosevelt wrote to Churchill that it was fun to be in

the same decade with him. I can truly say that for all our ups and downs – and there were a few – it was fun being with you in the same decade that we were together as the leaders of the political and industrial wings of our Movement. It was more than fun. It was great.

In his book, Whitlam concentrated almost exclusively, and certainly exhaustively, on the development and implementation of policy and his personal involvement in that work. Sir Paul Hasluck described it in a review as 'perhaps the longest trumpet voluntary in history'. But Whitlam had a purpose beyond self justification or self glorification.

He was anxious that he and his government would not be remembered, like King Charles I, chiefly for a spectacular demise. For all that had been written about the Whitlam Government there was a considerable imbalance in the various analyses of what happened. All the books – my own no exception – highlighted the drama at the expense of the substance. He wrote:

> The manner of my Government's destruction, the extent of the ALP defeat at the elections on 13 December and the extreme nature of Fraser's conservative stance on most issues throughout the subsequent seven years have all contributed to another misconception about our work and its results. It is a misconception regrettably and all too widely shared by ALP supporters.
>
> It is a misconception amounting to myth that our Government's achievements were largely dismantled and our work was largely nullified in the seven years which followed. It has been central to the purpose of this book to remove that misconception and to dispel that myth.

The record of substantial achievement was impressive. Whitlam pointed out that the basic thrust of his new approach to Australia's international relations had never been reversed. He paid tribute to the Fraser Government which had 'continued and built upon our pioneering work in China' and pointed to the continuity of policy in Australia's relations with Papua New Guinea, ASEAN, Japan, the Middle East and, surprisingly, southern Africa. In large areas of domestic policy, many basic structures had remained sufficiently intact to permit the restoration of key Whitlam programs by the Hawke Government, especially in education, health, welfare and the status and rights of women, migrants and aborigines. Whitlam wrote:

> The great law reforms endure: the Federal Court, the abolition of appeals to the Privy Council, family law, legal aid, trade practices law, the Law Reform Commission and the Ombudsman. In electoral reform the hard-won principle of one vote one value is now firmly embodied and regular redistributions of electorates, according to this principle, now guaranteed.
>
> The quality of the electorate and the responsiveness of parties to the electorate have been permanently enhanced by the granting of votes to 18-year-olds. Public funding of elections and public disclosure of political donations have been achieved by the Hawke Government... The application of environmental impact statement conditions to State grants trammels even the most wanton and profligate of State Governments. The Australian Heritage Commission remains the watchdog of the national estate.
>
> The most important of our initiatives in the arts and media continue: the Australia Council, the Australian Film and Television School, the Australian Film Commission, the Public Lending Right and the introduction of FM radio. There is a new and internationally acclaimed dimension to community participation in public broadcasting. Blue Poles has not been sold.

The book disappointed those who were looking for some 'inside story'. But, as Whitlam claimed at the Press Club launch, 'there really is no inside story, new or unrevealed. With us, what you saw was very much what you got.' Its other feature was the absence of any lengthy re-examination of the events of 1975, including the dismissal itself. This was itself a reflection of Whitlam's chief purpose in his voluminous testament, to redress the balance by stressing the record. Indeed, by 1985, he felt he could be almost dismissive of the dismissal. But the relatively brief treatment he gave was also due to the fact that he had already written his full account of these events in *The Truth of the Matter*, his reply to Kerr's own account *Matters for Judgment*, published within weeks of each other in the summer of 1978–79.

An important by-product of the tenth anniversary was the public endorsement by former Liberal Senators of Whitlam's central assertion about the nature of the 1975 crisis – that it was essentially a political crisis capable of a political solution; and that in acting as he did, when he did, Kerr had made a political assessment which was both improper and incorrect.

In *The Truth of the Matter*, Whitlam wrote:

> The entire operation which my colleagues and I conducted in October and November 1975 had a single and simple objective – to force the Senate to do its constitutional duty and vote directly and explicitly on the Budget, either to pass it or to reject it. The Opposition tactic depended for its success on avoiding such a vote at all costs. Even in that 'tainted Senate' of 1975, which two Premiers had perverted from the composition decided by the electors in 1974, no motion to reject the Budget or refuse Supply outright could have succeeded. The numbers were just not there. That is why the Opposition had to adopt the stalling tactic of refusing to bring the matter to a direct vote. That tactic, and the political nerve of its authors, were on the verge of exhaustion on 11 November 1975.

By 11 November 1985, ex-Senators Bonner and Jessup were prepared to state publicly that they and at least two other of their colleagues were within 24 or 48 hours of changing their vote when Kerr intervened. In other words, the Budget would have passed by 12 or 13 November. It was a remarkable retrospective vindication of the Whitlam strategy. It was equally remarkable as a retrospective rebuke to Kerr – more than a rebuke, a complete rebuttal of Kerr's defence. For if, as he has always claimed, the sole purpose of his action was to break the deadlock between the Houses, then it was simply unnecessary.

On what basis, then, did Kerr make his assessment that the deadlock could not be broken – an assessment we now know to have been wrong? The only conclusion to be drawn is that he accepted Fraser's word that there was no possibility of any change of heart or mind by any Opposition Senators. In other words, he chose to accept the false advice of the Leader of the Opposition while rejecting the correct advice of his Prime Minister. It adds a lurid dimension to the tragedy of 11 November 1975 when it is established as a matter of fact – not mere speculation, but something placed on the permanent public record – that it need never have happened.

Thus, ten years on, it was tempting to accept that time and events had done their healing work. And yet . . . and yet . . . grave doubts, a deep uneasiness remained among many who pondered their country's future. It was impossible not to see that a political vengefulness and personal vindictiveness had entered Australian public life to an extent we had not experienced since the '20s and early '30s. This deterioration was abetted massively by the media, which had learned the full extent of its new power in 1974 and longed for opportunities to deploy it.

In the final analysis, so complex a system as the Australian democracy rests upon something more than its rules and conventions, written and unwritten; it rests on something more even than basic respect for its key institutions. All these things had suffered seriously but not irretrievably from the events of November 1975. But there is something even more fundamental needed for the proper functioning of our complex and often contradictory system – fundamental, indeed, to the spirit of the nation itself. It is the essential element of goodwill. It is that underlying spirit of goodwill which legitimises any government whose election will have been opposed, by the very nature of the party system, by nearly half the people.

I have argued in this book that virtually from the beginning, and specifically, from April 1973, the opponents of the Whitlam Government attempted to deny its legitimacy by denying it the essential element of goodwill.

In the first months, this denial was usually expressed by the assertion that the election result of December 1972 was merely an aberration. When Labor was confirmed in power in May 1974, the Opposition's mood – and its approach – changed; it became altogether a more deadly affair. It was this change of mood which ultimately enabled Fraser to depose Snedden. Not only the Opposition's leadership changed, but its whole approach. A new destructiveness, a personal vindictiveness, had entered Australian political life.

The events of 1975 – not just the crisis and the coup but the war of attrition throughout the year – transformed a transitory power play into something systematic, endemic and deep-seated, in the attitudes and methods acceptable to both sides. Indecencies never before contemplated became routine. The massive psychological shock of 11 November dealt a devastating blow to that essential element of goodwill which provided a core of cohesion in a society by no means rich in other sources of unity provided by history, culture, religion or tradition; and to a nation without the stability given the United States by its official piety and patriotism or to Britain by its institutionalised hypocrisy. It is a factor in the national equation more important to us than most. Its loss, therefore, is more damaging to us than most.

We can now see that first great breach of constitutional convention over the Senatorial replacement of Lionel Murphy was a portent – immensely significant, immensely sinister. It not only set in motion a course of events which led to the dismissal. It sowed the seeds of a political and personal persecution which was to lead to the tragedy of Lionel Murphy's destruction in 1986.

Yet if Lionel Murphy was to be a future victim, Fraser himself was the first great victim of the new regimen created by 1975. Despite his massive majorities of 1975 and 1977, his remained a government utterly without goodwill. It neither showed goodwill nor received goodwill. It was always a government with no threshold of tolerance for mistakes or personal misjudgements. As a direct consequence, throughout its seven years, the Fraser Government was an extraordinarily unstable government. Don Chipp, dropped after the caretaker ministry, broke away and formed a new party, the Australian Democrats. Phillip Lynch was forced to resign as Treasurer in the middle of the 1977 campaign. Andrew Peacock resigned in April 1981, taunting Fraser with the same words he had used against John Gorton almost exactly ten years earlier: 'This man is unfit to be Prime Minister'.

Four ministers, including Reg Withers, the architect of the Senate obstruction, were removed on the grounds of improper conduct. Three ministers, including Robert Ellicott, the draftsman of the plan of action for Kerr and Barwick, resigned in bitterness and recrimination. Altogether sixteen ministers were dropped, suspended or removed themselves between November 1975 and February 1983. This is a record of instability unprecedented since Federation. The Fraser Government never escaped the taint of its origins.

The importance of restoring the spirit of goodwill as the key element in Australian national life lay at the heart of Hawke's appeal to the electorate in the election campaign which brought Labor back to power on 5 March 1983. Hawke focussed that appeal on the issue of industrial relations, with the promise of co-operation instead of confrontation. The centrepiece of Labor's campaign was an agreement with the ACTU on the basis for a wages policy, later to be enshrined as the Prices and Incomes Accord. The spirit of consensus had its apotheosis at the National Economic Summit in April 1983.

With Australia facing its worst economic crisis since the Depression, there did seem to be a widespread desire for the restoration of some degree of national co-operation and consensus. Hawke sought to apply this to the solution of the immediate industrial and economic problems. But in a deeper sense he was trying to reach back before 1975; he was seeking to exorcise the baleful ghost of 1975. For at least two years it seemed he had been remarkably successful. But when, in July 1985, he attempted to revive the spirit of the Economic Summit to the question of tax reform, the limits of consensus became clear. So did its limitations.

By 1986 very serious difficulties had emerged in the Australian economy, largely due to the collapse in the terms of international trade. Hawke again sought the path of national consensus and the revival of the spirit of goodwill.

The severity of Australia's problems in the last years of the 1980s is such that his success or failure will not be known for many years yet. It will, of course, be first and last a test of the Labor Government's skill and foresight in economic management. But in the deepest possible sense it will be the test – and the measure – of what really happened to Australia and the Australian people in 1975. The deepest of all the meanings of the events of 1975 is yet to unfold. Australia is yet to discover whether the terrible wound inflicted on the spirit of goodwill which must underlie the cohesion of this nation was temporary or permanent; whether the damage finally passed with the passing of the principal participants from the great stage; or whether the wound was mortal. In that sense, the full history of 1975 is yet to be written and the people of Australia will be their own historians.

Graham Freudenberg
1987

Author's Preface

From August 1961 to February 1966, I was Press Secretary to the Leader of the Opposition, Hon. A. A. Calwell; from 8 February 1967 to 5 December 1972, I was Press Secretary to Hon. E. G. Whitlam, and then Special Adviser to the Prime Minister until 11 November 1975. I was, therefore, privileged to be a close observer and occasionally a participant in the events I attempt to describe and explain. On none of these events do I pretend neutrality or impartiality; but if holding strong views were a disqualification for writing contributions to history, very little worthwhile would ever be written. As a description of my approach, if one may avoid impertinence or pomposity, I can do no better than quote Winston Churchill in his preface to his account of World War I, *The World Crisis*: 'It is a contribution to history strung upon a fairly strong thread of personal reminiscence. It does not pretend to be a comprehensive record, but it aims at helping to disentangle from an immense mass of material the crucial issues and cardinal decisions. Throughout I have set myself to explain faithfully and to the best of my ability what happened and why.'

I wish to record my deep appreciation for the unfailing assistance of '*The Women*', Whitlam's associates, all of whom played a real, if unacknowledged part, in the events herein described: Barbara Stuart who did most of the typing for the book and gave valuable advice; Joy O'Brien who typed the earlier part of the manuscript and gave encouragement when it was most needed; Carol Summerhayes, Lorraine Dwyer, Denise Darlow, and Norma Thompson, who has been Whitlam's electorate secretary for nearly a quarter of a century, all of whom shared the burden but precious little of the glory; and Margaret Whitlam, always 'first lady' in our hearts.

Beverley Lloyd prepared the index. Clem Lloyd, father of Frances, read the manuscript.

I am deeply grateful to the Premier of New South Wales, Hon. N. K. Wran Q.C., who provided me with part-time employment which enabled me to eat as well as write. His staff freely assisted in many chores, in their own time.

E. G. Whitlam offered his services as a research assistant and proved quite satisfactory on the few occasions I thought it proper to avail myself of them. He had done the real work in the years before.

Graham Freudenberg
1977

Introduction

It was the flight of the three Prime Ministers: the blue, white and silver BAC-III, pride of Squadron 34, Royal Australian Air Force, ex Melbourne, ETA Canberra, 1.15 a.m., 11 November 1975. Aboard were the Prime Minister of Australia, Edward Gough Whitlam; William McMahon, whom Whitlam had defeated in December 1972, to form the first Labor Government of Australia for 23 years; and John Malcolm Fraser, Leader of the Opposition since March 1975. At Government House, Canberra, the documents by which the Governor-General, Sir John Kerr, would dismiss the Prime Minister and install the Leader of the Opposition in his place had been drawn up and copied ready for release to the press. Aboard Flight 2, only the Leader of the Opposition was aware of the likelihood that within twelve hours he, and not his host, would be Prime Minister of Australia.

*

For the Prime Minister, it had been an exhilarating day. He was convinced that the constitutional crisis which had gripped Australia for the past month would end this week and end decisively in his favour. His confidence seemed well-based. For a month, the Australian Senate had refused to pass the Government's financial measures, the Budget. Whitlam's strategy was simple and had a single purpose: through Parliamentary and public pressure to force the Senate to a direct vote on the Budget, either to pass it outright or reject it outright. Every sign pointed to this being the week when the Senate would crack. In Sydney, at the office of the *Bulletin*, a weekly journal hostile to the Government, the plates were ready for the printing-run early next morning. The cover would show a photograph of Malcolm Fraser, battered and dazed, a halo of stars spinning around his head with the caption: 'Fraser—the man in the muddle.' Before the run could be completed, Fraser was Prime Minister. Opinion polls, press and public reaction around Australia and, above all, the hardening of belief in Parliament House, compounded of rumour, gossip, shrewd speculation and exact information, all pointed towards a tremendous victory, not only a personal victory for Whitlam, but an historic, irreversible triumph of the House of Representatives over the Senate.

On that night, the feeling had pervaded the banquet of the Lord Mayor of Melbourne, the annual show of strength and solidarity of the Melbourne Establishment, where a Labor Prime Minister was an exhibit and an enemy. These were the men (even in 1975 it was still almost exclusively a male gathering) who, more than any other group in Australia, had put pressure on the Opposition to use their numbers in the Senate to block the Budget and bring down the Labor Government. As Whitlam, in his speech, gently taunted them, they listened half-hating, half-admiring the nerve and style of this impossible man and were certain they had lost.

Whitlam had asked Fraser and his deputy, Phillip Lynch, to travel to Canberra with him so that they could meet early in the morning in his office in Parliament House. He proposed to tell them that if the Opposition persisted in blocking the Budget in the Senate, he would ask the Governor-General to hold an election for half the Senate; it was to be the last turn of the screw.

Two other members of Parliament, one Liberal, one Labor, travelled in the rear section of the aircraft—Andrew Peacock, soon to be Foreign Minister, and Ted Innes, Labor MP for Melbourne. Fraser, unable to sleep, strode down and said amiably to Innes, 'I thought this flight was for VIPs only.'

At the Air Force Base at Fairbairn, Canberra, the farewells were polite and perfunctory. The principals would meet in the morning. At the Prime Minister's Lodge, the author showed Whitlam the draft of a speech prepared as the last of a carefully planned series on the constitutional issue which had divided the nation. 'I think it's probably being a bit rough on Malcolm, but let's see what happens in the morning,' Whitlam commented.

At Government House twelve hours later, Sir John Kerr, with no preliminary discussion, handed the Prime Minister a letter dismissing him. It was all over in half a minute, and the work of ten years lay in ruins; ended, literally, by the stroke of a pen.

This is a story of those ten years. It is only marginally about the constitutional crisis of 1975 and the actions of Sir John Kerr. Yet those few hours, imprinted upon the political memory of Australia, and still a raw wound on the spirit of those whose lives have been irreversibly changed by them, go to the heart of any understanding of the career of Edward Gough Whitlam. That career makes sense only in the context of his extraordinary exertions to make the Australian parliamentary system an instrument for social change and to convince the Labor Party that national parliamentary power was the only way to achieve social change. That is what all his conflicts within the Labor Party — conflicts which brought him to the edge of expulsion and political destruction — have been all about. The opposing view is that Parliament is a sham and social democracy a sell-out. When Sir John Kerr dismissed the Whitlam Government and forced an election which the Labor Party must lose, he did more than reward Labor's enemies; he powerfully vindicated those on the Australian 'left' who despise the system and detest Whitlam as its champion. The means by which this was done, the man by whom it was done and the man to whom it was done, powerfully reinforce all those who argued against Whitlam throughout the 1960s that even if Labor secured a majority in the House of Representatives and formed a government, it would never be allowed to govern and would be destroyed by any means. If the Labor Party has not now succumbed to that argument (and that it has not is shown by its remarkable achievement in electing the Wran Labor Government in New South Wales only six months after Whitlam's dismissal) it is largely due to the continuing hold of the Whitlam philosophy on the members and supporters of the Labor Party. But it cannot be denied that Sir John Kerr exposed the flaws of that philosophy, with its commitment to conserving the parliamentary system, in a way that Whitlam's old opponents within the Labor Party or the new Marxist critics could not.

This then is the story of ten years of struggle for the idea of the Australian Parliament as the principal instrument of change and reform in Australia; how that struggle was briefly crowned with shining success; and how the parliamentary system itself was used to destroy a government led by the Australian who has fought hardest and longest for the meaningful survival of the system.

It is a story in the high and enduring Australian tradition — tragedy so mixed with comedy that the sin of despair becomes impossible; failure so mixed with farce that the unendurable becomes after all laughable. It is a story about the Australian paradox — effort carried on intensely in the expectation of failure. The classic Australian doctrine, 'She'll be right', is in fact the Australian's taunt against a fate which has forced him to great effort but dooms him in the end to failure.

Almost on the eve of his greatest success, Gough Whitlam, the central figure of this story, hinted at the inevitability of ultimate failure. Writing in 1971 about another Australian overthrown at the peak of fame and power, E. G. Theodore, Australian Treasurer in the Scullin Labor Government of 1929–31, Whitlam said:

> Theodore's career and fall is a powerful reminder of a persistent feature in Australia's political history: our chief men and our chief efforts have been singularly associated with failure and frustration . . . There is a deep poignancy in the fate of a remarkably long list of our chief figures from the very beginning: Phillip embittered and exhausted; Bligh disgraced; Macquarie despised here and discredited at home; Macarthur mad; Wentworth rejecting the meaning of his own achievements; Parkes bankrupt; Deakin outliving his superb faculties in a long twilight of senility; Fisher forgotten; Bruce living in self-chosen exile; Scullin heartbroken; Lyons dying in the midst of relentless intrigue against him; Curtin driven to desperation and to the point of resignation by some of his own colleagues at the worst period in the war. Theodore's particular tragedy was that of a supremely able man suddenly struck powerless at the very time when his power and his ability were at their peak and most needed.

Only one Australian Prime Minister seemed to have retired wholly exempt from the Australian doom. Yet Robert Gordon Menzies, the only real success story of Australian politics, had ended his first test as Prime Minister in 1941 as a classic example of Australian failure. His first Prime Ministership from 1939 to 1941 began brilliantly and ended disastrously as a result of a conspiracy against him by some of his own colleagues. Luck and guts won him a second chance. Thirty years after Menzies' first career collapsed, another Australian Prime Minister, Gough Whitlam, would write to him: 'Your remarkable achievement in rebuilding your own party and bringing it so triumphantly to power within six years has been an abiding inspiration to me.'

This book covers the decade from Menzies' retirement at the beginning of 1966 to the fall of the Whitlam Government at the end of 1975. It tells how Whitlam was able to follow that part of Menzies' political achievement to which he paid tribute—bringing his party from disaster to power within six years, and attempts to explain why Whitlam failed to secure as long a term for his government as Menzies had for his.

These were, above all, the Vietnam years. Vietnam is not just the name of a country or a war. It is a name for an epoch. The link between the Kennedy Presidency and the Carter Presidency is Vietnam. In his Inaugural, John F. Kennedy had said: 'Let every nation know, whether it wishes us well or ill, that we shall pay any price, bear any burden, meet any hardship, support any friend, oppose any foe to assure the survival and the success of liberty.' That was the promise and half the world fell in love with the man who made it; Vietnam was the fulfilment of that promise. By 1966, Imperial America was at its zenith. Harold Holt's 'All the way with LBJ' was more than a slogan; it signalled the final transfer of Australia's colonial allegiance from Britain to the United States. For the next ten years, Vietnam and its consequences—its economic consequences even more than its shattering political consequences—dominated the societies, the economies and the politics of the United States and Australia. The world economic disruption of the seventies springs from Vietnam. Every Australian Prime Minister involved with Vietnam was damaged by it. All save one—the Prime Minister who first locked Australia into the war, Robert Gordon Menzies. In a profound sense, the Whitlam Government was one of the last casualties of the war in Vietnam.

'These flaws have to be balanced carefully against its notable record of policy innovation and political stimulation. In the generally undistinguished, and often tawdry, atmosphere of Australian national politics, it is impossible to deny the Whitlam Government a certain grandeur.'

> C. J. Lloyd and G. S. Reid on the first term of the Whitlam Government (December 1972–May 1974) in *Out of the Wilderness*.

I
The Menzies Inheritance

Sir Robert Menzies' remarkable political achievement—power unbroken for sixteen years—was based on his unrivalled ability to express and exploit the conservatism of the Australian people. He exploited not only the conservatism of the Australian electorate at large, but also the conservatism of his opponents—the Australian Labor Party itself. His most humiliating electoral defeat, the rejection of his referendum to ban the Communist Party in 1951, came from a gross misjudgement about the strength of one expression of conservatism, in the form of anti-communism, and its expression in the form of reluctance to tamper with the Australian Constitution.

Menzies never made the mistake of raising Australian expectations. It was a key to his success. His great skill lay in never despising the obvious, and in refining his technique of getting the obvious across. He made platitudes seem eternal verities. He boasted not just that he simplified but oversimplified. The surest sign that Menzies was about to put forward the most doubtful proposition was his characteristic preface: 'It is abundantly clear that ...' He was the great simplifier. His recovery from near defeat in the 1961 election was an exercise in seizing tenaciously one great simple fact—the fact that he was in possession and the power that came from possession. He had after all retained a majority—of two. But that was enough. And his confident grasp of that simple fact was the foundation of his complete dominance from then until his retirement four years later.

Menzies did not seek to impose his successor on the Liberal Party. Harold Edward Holt was chosen by the Parliamentary

Liberal Party simply because his claims were beyond dispute. His own delighted reaction in his victory telephone call to his wife Zara was 'I got there without stepping over any dead bodies'. That was true enough; but the outcome had never been in question. Menzies left Holt a rich inheritance—a healthy parliamentary majority, a united coalition, a steady economy and, above all, a popular war.

The new Prime Minister's vigour, touched with a certain glamour, was neatly balanced by the dour maturity of the Leader of the Country Party, John McEwen, a supremely political animal, half lion, half fox. The economy had fully recovered from the stresses of the early sixties which had so nearly destroyed both the Menzies Government and the Menzies legend at the 1961 election. The stresses imposed on the international economy by the Vietnam war would not show up for years. Australians were again sharing fully in the prosperity of the industrialized West. Growth was king. To add spice to life on the quiet continent, the people of the 'lucky country' had the vicarious excitement of an important war, fought overwhelmingly by others but officially for Australia's own national survival.

Yet the crust of stability and security was developing fissures. Menzies never understood the forces he had let loose by locking Australia into Vietnam. In particular, he did not understand its impact on the uncommitted Australian middle class, those Australians who were not necessarily bound, professionally or economically, to the fortunes of corporate enterprise, whose interests the Australian Liberal Party upholds. The dominant presence of Menzies deadened self-questioning. In his own time, his sweeping simplifications were enough to suppress the truth about the war. His successors could not get off so lightly. Over the next decade, four Prime Ministers, Holt, Gorton, McMahon and Whitlam, would all be damaged, in very different ways, by Vietnam. The only Prime Minister associated with Vietnam to emerge politically unscathed by it was Menzies, the man responsible for the original commitment. In his last active years, he had become a living institution. Like most institutions, he was perceived as vaguely unsatisfactory but acceptable for want of an alternative. With Menzies gone, the doubts grew. The self-questioning became increasingly insistent.

Years later, the best man of them all expressed the view of his generation and type, if so rare a spirit can be said to be a type. Going public politically for the first time, the writer Patrick White, holder of the Nobel Prize, would say in 1974:

> Some of you to whom I am speaking may be in a quandary over how to cast your vote, as I, too, found myself in a quandary at a point in the post-Menzies era. Brought up in a Liberal tradition, I realized we had reached a stage where a change had to be made if we were to cure ourselves of mentally constipated attitudes and heave ourselves out of that dreadful stagnation which has driven so many creative Australians to live in other parts of the world.

For most of those for whom White spoke, Vietnam was the emetic.

Nor did Menzies ever grasp the new attitudes emerging among the new urban class, even though his policies on housing, education and immigration had helped create it. It had always been true that most Australians lived in cities, but the growth over which Menzies had presided spawned a new breed of urban Australians and cast them into the new urban wildernesses of the capitals, especially Sydney and Melbourne. As much as their parents, this generation sought the fulfilment of the Australian dream of home-ownership. Forced into the new suburbs, they found the reality was often costly, bleak and lonely. They began to wonder if this was what life in this rich and happy nation was really all about, and was this all that Australia had to offer?

By 1966, expectations were beginning to rise. For the conservatives, this could be dangerous. Nevertheless, in January 1966, there was no reason at all to believe that the coalition would not be well able to meet any test. Menzies had left them a powerful legacy, the assumption of the coalition's fundamental right to rule.

This confident coalition faced a demoralized opposition. One of the world's oldest organized parties and the world's first social democratic party to form a government, the Australian Labor Party in 1966 was in one of the periods of turmoil inseparable from the life of a great political party. For many years it has been fashionable for political commentators to ascribe to the Australian Labor Party a 'death wish'. On the contrary, its outstanding characteristic has been its capacity for survival and self-renewal. It has survived three great splits—in 1916, over conscription for World War I; in 1932, over economic policy to combat the Great Depression; and in 1955, over allegations by the Federal Leader, Dr H. V. Evatt, that Catholic-based groups, under the leadership of Mr B. A. Santamaria's National Civic Council, had infiltrated large sections of the Labor movement, particularly in Victoria. From this Split emerged the Democratic Labor Party which for seventeen years skilfully, singlemindedly and successfully pursued its aim of preventing the election of a Federal Labor Government. For the Labor Party, the most serious result of the growth of the DLP was to hasten the leaching of its traditional Catholic support. The DLP provided a bridge over which middle-class Catholics with a Labor background could cross to the Liberals.

All attempts to explain the 1955 Split have concentrated on the personal and political aspects. Yet even the best account, Robert Murray's *The Split*, is forced to the conclusion that there was no real necessity for the split, except that the chief actors lacked the will to prevent it. But perhaps the basic reason was that after 1949, after Labor's loss of power, the new Catholic middle class needed a party to provide a bridge to enable them to vote against Labor with a clear conscience. The historic function of the DLP was to provide this bridge. The crossing-over would have occurred inevitably. The DLP, acting within the preferential voting system, eased the Catholic passage and eased the Catholic conscience. The DLP was the political expression of a change of class by Australian Catholics, or more accurately, a change in their own perception of their new class by a new generation of Catholics born of Labor-voting parents.

By 1966, the DLP was at the peak of its influence. In the 1963 elections for the House of Representatives, its preferences had been cast decisively to the coalition in a score of crucial seats; in the Senate, the Government was dependent on DLP Senators for its majority. The strength of the DLP gave the coalition a further reason for confidence in its future.

In one respect, the departure of Menzies immediately strengthened the Government against its official Opposition, the Australian Labor Party. It was to the Government's immediate advantage to have a younger leader. Only one man seriously believed that Menzies' removal from the board redressed in any way the political balance in Labor's favour. That was the Leader of the Australian Parliamentary Labor Party, Arthur Augustus Calwell, then nearing 70, only a year younger than Menzies, 13 years older than Holt and 21 years older than his deputy, Gough Whitlam. Menzies had barely concealed his benign contempt of Calwell and had never hesitated to exploit it. In 1965, Calwell speculated on the possibility of an early election following Labor's decision to oppose the sending of combat troops to Vietnam. Menzies replied, 'I had never heard of the idea until I read the paper this morning and I almost fell out of bed. Then I said to myself, "Poor old Arthur, he's at it again."' It struck no one as ironic that Menzies, almost on the eve of his retirement, should refer with such condescension to his slightly junior contemporary. Menzies' retirement confirmed Calwell in his determination to continue as Leader of the Opposition until the general elections which had to be held by the end of 1966. But in fact he needed no external encouragement to stay on. Events in his own party and his relations with his colleagues worked upon everything in his nature and temperament to cling on. By January 1966, he was a man burning with an intense sense of grievance which supplied any want of real political purpose. Neither fame nor ambition was the spur; the accumulated frustrations and resentments of two decades drove him on.

The consequences of Calwell remaining leader after Menzies' retirement had been foreshadowed a year before by the Deputy Leader of the Parliamentary Labor Party, Edward Gough Whitlam. In one of a long series of explanations of one or other of his actions he had to make to the National Executive of the Australian Labor Party, Whitlam wrote in January 1965:

> As to the next elections, both major parties are in a state of flux and their fortunes over the next ten years could be decided in the next two years. Their leaders are the oldest in Australian history and their attitudes are old-fashioned. When they both went, the

> advantage would be with Labor for our front-bench men are all younger and more modern than the Liberals. If Mr Calwell retired before Sir Robert Menzies, the latter's attitudes would appear more dated and, despite his big start, we might catch up. If, however, Sir Robert retired before Mr Calwell, then the Liberals believe and most Labor men fear that Sir Robert's successor would appear more modern than Mr Calwell; and it might then be still more difficult to persuade the people to choose him as Prime Minister for the first time at 70.

Not for the last time, Whitlam discovered that the deepest insult to an old man is to call him old. Old men do not forget.

In one way, all of Calwell's long political life had been a preparation for 1966. Increasingly he had fallen out of sympathy with the kind of Australian society then emerging. He had long sought that one great issue which he could relate to his own concept of Australia, to his own past and to his own concept of the Australian Labor Party. He found it, not in the Vietnam war, but in conscription of soldiers to fight in the war in Vietnam. His opposition to Vietnam was secondary, and began reluctantly. His opposition to conscription was basic, immediate and instinctive. Conscription for Vietnam gave him a specific and familiar issue in which he could sink all other issues facing the Labor Party and the nation. All his present, pressing difficulties might be resolved in a contest he was renewing for the third time in half a century. The old enemies and allies of earlier conscription fights were gone — Hughes, Curtin, Mannix, Ward. He had begun his climb through the hierarchy of the Labor Party machine when it had split over conscription in 1916. In 1943, a member of Parliament for Melbourne, he and Ward had reduced Labor Prime Minister Curtin to tears and resignation by their methods in opposing Curtin's efforts to get approval of conscription for overseas service in war time. Now, more than twenty years later, he alone was left.

For how could these newcomers, upstarts like Whitlam or even Cairns, really understand what it was all about? Everything was changing, and all for the worse — the Party, the Church, the society, the nation, the world. He at least had stood firm. So nothing had changed. Besides, conscription would again serve the purpose fulfilled by his confrontation with Curtin in 1943. In 1941, Curtin had worked to exclude him from the place in Cabinet his service to the Party deserved. Harassment won what patient merit had not; Curtin was obliged to place both Calwell and Ward in the Ministry after his landslide election victory in 1943. Now in 1966, it was Calwell who was the leader, beleaguered within his own party. Once again, he would break through with the well-tempered weapon which he alone really understood how to use.

Calwell's three great loves and loyalties were his country, his Party and his Church. With each, his relations were ambiguous and unsatisfactory. His patriotism was uncluttered and unembarrassed. He had a standard line for political rallies, superb in its sweep, heroic in its meaninglessness: 'We are Labor because we are Australian and Australian because we are Labor.' It never failed to raise a cheer. He was never embarrassed saying it and his audience was never embarrassed hearing it—from him. Yet his response to electoral failure and political setbacks was invariably abuse of the majority of Australians. He did not like or understand the new Australian society. And, for all his deeply felt loyalty to the Labor Party, he damaged every leader, not excepting himself, from Curtin to Whitlam.

He was a man of unshakeable faith but he came to see the Catholic Church, its Bishops, its priesthood, as an enemy. His autobiography, *Be Just And Fear Not*, is an extended denunciation of the Catholic Church in Australia. Almost at the end of his life he wrote his considered judgement of the Catholic Church in Australia, her political motivation in the great Labor Split of 1955, her relations with his Party, and with him: 'The Catholic Church in Australia had its theology perverted by the National Civic Council and the DLP. Calumny is no longer a sin and detraction is now a major virtue. You would almost think there is a move to ex-communicate the Pope because he preaches Christian charity and exhorts everyone to work for universal peace and world disarmament.' He had said of his boyhood: 'Even as a boy I was in sympathy with Savonarola, the heretic, and I have always been something of an iconoclast.'

The trouble was that everybody had let him down—the

Australian voters, the Labor Party from Curtin to Whitlam, the Church ... even Daniel Mannix, his hero from the first great fight against conscription in 1916, the turbulent Irish prelate whose claims to the cardinal's red hat he had championed in 1945 against those of Archbishop Gilroy of Sydney. Yet even Mannix had turned against him and against the Labor Party.

Archbishop Mannix was the hero of his youth and middle-age. He always attributed to Mannix advice which had led to a marked and increasingly remarked upon change in his personal habits: 'A man who drinks before he is 40 is a fool and a man who doesn't drink after he's 40 is a fool.' The relationship between them is best told in his own words:

> On the occasion of his ninetieth birthday in 1954, I sent him a telegram wishing him many years of good health and added: THE OBJECTIVE IS ONE HUNDRED. He replied thanking me for my greetings and said: I ACCEPT THE CHALLENGE. This was typical of his wit and his humour. He told a friend that he received 1,000 telegrams on his birthday. 'But', he said, 'I only replied to three of them. I replied to the Governor of Victoria, to Mr De Valera (Prime Minister of Ireland) and to Mr Calwell.'

This was the high point. Within six months, the great Split in the Labor Party ended that friendship, as it broke up friendships and families throughout Victoria. Love and reverence on Calwell's side and courtly respect on the other ended in deep bitterness. Yet not quite to the end. Less than half an hour before death finally came to Mannix in 1963, Arthur Calwell was at his bedside. For the first time since the death of his 16-year-old son in 1948, he wept.

The bitterness of 1955, within the Catholic Church even more than the Labor Party, is a hard thing to recreate after a generation. A casual reference Calwell made to the Democratic Labor Party being 'like the mule, without pride of ancestry or hope of posterity' called forth the charge from a country priest that he was insulting the celibacy of the clergy. The award of a Papal Knighthood in 1964 caused a fresh outburst of unbridled vituperation against Calwell from the Catholic press.

Calwell's boyhood hero was Daniel O'Connell—Irish passionate patriot, parliamentarian, loser. In 1973, Whitlam speaking in Parliament in his valedictory, was to use this description of O'Connell and Calwell:

> He was a thorough Celt. He represented all the impulsiveness, the quick-changing emotions, the passionate exaggerated loves and hatreds, the heedlessness of statement, the tendency to confuse impressions with facts, the ebullient humour and all the other qualities that are especially characteristic of the Celt.

Arthur Calwell was the last, as he had been the greatest and the most articulate, of the red-blooded Australian racists in the early tradition of the Australian Labor Party. He was racist in the tradition of Deakin, Watson, the Australian Workers' Union, Henry Lawson and the *Bulletin*. 'Red-blooded' was his chosen word — a fascinating choice, describing exquisitely the whole nature of racism; it is not after all the colour of the *blood* which is the problem.

In the end, race became his last obsession, his last weapon to hit out at everyone who had hurt him — the Labor Party under Whitlam, the new middle class, the Catholic Bishops, even those migrants who had come to Australia through the post-war immigration which he had started. In May 1972, the Ecumenical Council of Australia issued a joint statement on behalf of the World Council of Churches and the Catholic Bishops deploring the measure of race as the basis for the selection of migrants to Australia. By then, such a statement was scarcely ahead of overt public opinion. In a coda of fury and frustration against all who had fought or thwarted him for the past fifty years, he had put up on the notice board in the Parliament House Press Gallery a last testament to the people of Australia:

> I am neither terrorized nor influenced by the pious outpourings of those prelates who seek to stigmatize all red-blooded Australians who want to keep this country as our pioneers, those who were born here or came here before Federation, made it. Why can't the Bishops develop a new theology on war? When will most of our Christian clergy stop pandering to the middle class in our supposedly Christian community and stop posturing when it comes to opening the flood-gates to unwanted and unnecessary

coloured migrants? The silence of Dean Maitland was eloquent alongside the mammoth silence of most of our Church leaders on everything associated with the brutal, filthy, immoral, unwinnable, criminal, genocidal, civil war in Vietnam, Laos and Cambodia.

There is a curious and instructive comparison to be made between the contrasting experience of the two most significant Australian Catholic laymen of their time — Arthur Calwell and Bartholomew Santamaria. It is curious because by utterly opposite paths they came to share the same feeling — disillusion and a sense of alienation from the Church they both loved so passionately. It is instructive because their alienation, opposite in its causes, is a measure of the social and political changes which have taken place with the Church in the last twenty years. Calwell was bitter because, under his old friend and hero, Mannix, the Church had supported the right-wing DLP. At the end of his life he was contemptuous of the new liberal attitudes expressed within the Church. Santamaria, one of the architects of the Split, became bitter (though with none of Calwell's vituperation) because the Church, under the successors of his friend and hero, Mannix, seemed to him to be lurching to the left.

Santamaria was unique in Australian history as her only political intellectual in the high European tradition. In that sense, his enemies were correct — he was an alien force. By sheer force of intellect and persistence of will, he created and sustained a political idea which influenced a generation and influenced events for a generation. He was a brilliant popularizer, a powerful pamphleteer, the most accomplished television performer of his time; yet his intellectual and philosophical system was so alien to the Australian pragmatic mainstream that he remained an Australian exotic. In the end he failed, perhaps the most complete failure of all the front-rank Australians of his time. The DLP, with its Catholic opportunists on the make, became as alien to him as the Labor Party led by the humanist Whitlam. In his eyes, the United States moved towards detente and decadence; the Church herself seemed in danger of succumbing to the prevailing hedonism. By 1977, his disillusion reached the stage when he could write:

> Nor am I reassured when I read in *Wharfie*, a communist waterfront publication, the observation on the statements issued in the name of the Catholic Commission for Justice and Peace: 'How times are changing, now that the Catholic Church is taking a more progressive and radical stand than the Waterside Workers' Federation.' How they have changed indeed.

To round off the irony, the political reconciliation, partial but crucial, between the Labor Party and the Catholic Church in Australia was due to one man more than any other — their *bête noir*, Edward Gough Whitlam.

On 7 March 1960, the Federal Parliamentary Labor Party (Caucus) elected Calwell to succeed Dr H. V. Evatt as Leader of the Opposition. To fill Calwell's place as Deputy Leader, the Parliamentary Party narrowly elected Whitlam, then 43, over the veteran Edward John Ward, a member of the Australian Parliament since 1931. The final vote was 38 to 34. Calwell was later to assert that Ward's defeat, or Whitlam's victory, had harmed Labor's electoral chances in the 1961 general election. Since Labor lost those elections by two seats, this amounted to saying that a Calwell-Ward team would have won the elections which the Calwell-Whitlam team just failed to win, and that Whitlam's election was the cause of Labor's defeat. Yet although Calwell himself voted for Ward, he welcomed the result. He knew too much about Ward's turbulence, independence and restlessness to believe that their partnership could ever have been smooth or successful. Whenever they had formed a temporary alliance, it had been destructive. Calwell also privately subscribed to the view urged by some of Ward's opponents against having four Catholics as the senior office-bearers — Calwell and Ward in the House of Representatives and Senators McKenna and Kennelly in the Senate. But the real objection made by those who opposed Ward was that he was a man of yesterday; he represented not just the past but a past increasingly unattractive to the emerging electorate.

Ward's constituency of East Sydney had expressed quintessentially the kind of Australia and the kind of Labor Party which Ward loved. It was almost an extension of himself in the old Sydney manner — touchy, turbulent, shrewdly generous, calculating, very gentle with women in their place, suspicious, fiercely loyal within the closed circle. Yet East Sydney, including Darlinghurst, Woolloomooloo and Paddington, was itself poised for a change in nature and style as fast as any urban area in Australia, to become not less typically Australian, but typical of an Australia which Ward did not want to know about or try to understand.

Ward was the greatest parliamentary pugilist Australia has produced, but by 1960 the fights had become reruns of all the old bouts. He was suspended from the House of Representatives a record forty-one times, the last time in May 1963, the last day Parliament sat before his death. With one of those delightful symmetries of irony which stud Australian politics, Ward's last defiant words to Parliament as the Serjeant-at-Arms escorted him off were: 'It was Pig-iron Bob's doing.' The old enemy remained. For Ward it was never Sir Robert Menzies, Companion of Honour, Knight of the Thistle, Lord Warden of the Cinque Ports; it was still just Pig-iron Bob, the brash failure whom he had described in Parliament a quarter of a century before as 'this posturing individual with the scowl of Mussolini, the bombast of Hitler and the physical proportions of Goering'.

Whatever Calwell's autumnal thoughts about Ward as his preferred deputy, the facts of the 1961 elections are on record: eight of the fifteen seats won by Labor were in Queensland. Whitlam bore the main burden of the Queensland campaign and began to create a personal presence in Queensland and a reputation there as a successful campaigner which was to be crucial in saving his career in 1966. The failure to win any new seats in Calwell's home state of Victoria, which recorded a swing less than half the national average of 4.6 per cent, was also a portent for the future of both Calwell and Whitlam.

It took a few days for even the best-informed politicians in the Labor Party to realize the full extent of the great gains Labor had made. But the shrewder men within the Party cut through the euphoria to the essence; that was that Whitlam was the first deputy or leader in living memory, not excluding Chifley or Curtin, to appear positively and demonstrably to have changed votes where votes were needed or could be changed. The 1961 election was the beginning of Whitlam's reputation as a vote-winner.

The election of 9 December 1961 was the peak of Calwell's political career. The rest was all downhill. The descent began rapidly and immediately. It began as it was to end — with a grotesque kind of war. It took ten days of counting before the final tally gave the Queensland seat of Moreton to the sitting Liberal member, Denis James Killen, by 130 votes (including 93 communist preferences) and gave the coalition parties 62 seats to Labor's 60. While political Australia had been preoccupied with this fascinating and altogether unexpected result, President Sukarno of Indonesia had stepped up his long, subtle campaign to oust the Dutch from the last and least remnant of their empire in the East Indies — West New Guinea. Three days after Australia's election, Sukarno appointed himself 'Commander-in-Chief of the liberation of West Irian'. Like all former European colonies, the Indonesians held fiercely to the idea of the integrity and validity of the boundaries set and claimed by the departing power. There was never any doubt in Indonesian minds that West Irian was part of the Dutch East Indies, part of the rightful inheritance and part of Indonesia. Successive Australian foreign ministers from Evatt under Labor to Sir Garfield Barwick under Menzies were uniformly ambivalent, unrealistic and hypocritical. All used the slogan of 'self-determination'. But this slogan was used to mask their real hope that the Dutch position in West New Guinea could be shored up indefinitely, perhaps even with encouragement from the United States, until some sort of incorporation with a nominally independent Papua-New Guinea could be arranged under Australian tutelage. The most prostituted and disastrous slogan of this century — 'self-determination' — the bastard orphan of Woodrow Wilson and Versailles, was used to promote this vicious delusion.

By Spring 1961, it was becoming increasingly clear that a combination of Sukarno's home troubles, Dutch weakness and American impatience with so untidy a situation would soon lead to incorporation, whatever forms and decencies might be observed. Canberra then moved to shift ground and accept the inevitable, though the political and economic preoccupations of 1961 meant that the usual process of preparing the opinion-makers for the change was skimpy and ineffective. It was not until January 1962 that senior officials of the Department of External Affairs visited Melbourne and Sydney for meetings with editorial writers and columnists to signal the change. A week earlier, shots had been fired in earnest; the Dutch fired on and reportedly sank two Indonesian torpedo-boats in New Guinea waters. In Melbourne, Calwell launched a broadside against Sukarno with rather less effect. He accused Sukarno of 'sabre-rattling reminiscent of Hitler at the time of Munich and just as menacing'. Calwell was almost alone in his outrage; he was able to strike no chord with the Australian public. The threat from the north, so potent a slogan in Australian politics, apparently lacked credibility in the mouth of a Labor leader.

The hasty official efforts to prepare opinion for the inevitable policy switch left one great gap, the magisterial *Sydney Morning Herald*. Never in that newspaper's long history had it been in a mood less likely to look kindly on a government's effort to explain away a major departure from established policy and principle. Alone among the Australian newspapers it had supported the Labor Party in the 1961 election, but its strength had been as the strength of ten. The impact of a switch to Labor by so staunchly conservative a paper was pervasive. Maxwell Newton, the editor of the *Financial Review*, stablemate of the *Sydney Morning Herald*, brilliant, aggressive and creative, wrote dozens of speeches for Calwell. Some of them were edited and set up in type before Calwell had seen them, much less spoken them. Newton was the author of the centre-piece of the Labor economic plan, the proposal for a Budget deficit of £100 million to revive the economy. There was never any deal between Calwell and the *Sydney Morning Herald*. There was no need for any deal. Long-standing dislike for Menzies, particularly on the part of Sir Warwick Fairfax, Chairman of Directors, and the catastrophic effect of the Government's economic policies upon the paper's advertising revenue, provided the complete and sufficient reason for the revolution of policy.

It is true that, in response to a suggestion from the Managing

Director, R. A. G. Henderson, Calwell undertook not to raise the question of nationalization during the lifetime of the new Parliament. But that was only a suggestion, not a condition of support.

The influence of the *Sydney Morning Herald* had been crucial, though not quite decisive. So near, yet still so far ... Was it possible that the final blow, so nearly given at the ballot-box, could after all be delivered in the Parliament itself? Three men at least remembered the lesson of October 1941, when two Independent members of Parliament had crossed the floor to make John Curtin Prime Minister of Australia. Menzies, Calwell and Henderson remembered it well. Could it be repeated? But was there a potential defector, and a potential issue to draw him out? A single defection would mean a deadlock and another election, and who could say where defection once started would end? The atmosphere of the aftermath of the 1961 election made such questions seem reasonable.

The man cast for Judas was Sir Wilfrid Kent Hughes, whose profession was loyalty. A veteran of two wars, a prisoner of war; in physique and bearing still the classic Anzac, he was deeply discontented with the Government's defence posture and obsessed with the orthodoxy of his generation that the whole island of New Guinea was the bastion of Australia's defence and should be under Australian control.

Menzies, the intended victim, and Kent Hughes, the intended traitor, failed to accept the interesting roles assigned to them. To have thought that Kent Hughes' dissatisfaction with aspects of his Government's defence and foreign policies could have cancelled out his hatred of Labor was, of course, absurd. From dissatisfaction to disaffection to defection was an unthinkable road for him.

On 7 February, Calwell issued and the *Sydney Morning Herald* by pre-arrangement ran in full a statement saying:

> If Indonesia seeks to use force to create a potential threat to Australia's security, then I say, with all due regard to the gravity of the situation, that threat must be faced.

A shudder went through the Labor Party. What on earth did it mean? On the face of it, it meant war. Calwell denied any such meaning. Menzies homed in. He saw intuitively how Calwell's implicit call for war against Indonesia had saved the Government from any challenge at all, however thin its majority. He moved immediately to pin the 'war-monger' label on the Labor leader. When Calwell denied the charge, Menzies was easily able to deride his opponent as a mere blusterer.

Thus are the fates of peoples disposed of. Sukarno had stepped up the pressure on West New Guinea to shore up his crumbling position at home; Calwell blundered into the West New Guinea controversy for a political advantage at home; Menzies used the issue to recover from his own near defeat at home. Within a year, Sukarno was to achieve incorporation of West New Guinea; over the next two years, Menzies steadily restored his supremacy; Calwell remained the Labor leader for four more years, but his decline from that day was deep, accelerating and irreversible.

Thirteen years later, in 1975, there was to be an echo of this strange affair. The Calwell catastrophe would deeply affect the attitude of Whitlam, as Prime Minister, to the problem posed by Indonesia's desire to incorporate East Timor. Whitlam, as Prime Minister, was determined never again to have a bar of the humbug humiliation and hypocrisy which had occurred over West Irian.

*

Calwell's conduct had two immediate consequences. His failure to consult any of his colleagues, including Whitlam, threw a party flushed and eager from its recent triumph into confusion and disunity. The great initiative was lost. The scent of victory was lost. Calwell's authority was damaged in its prime. A party two seats away from power went into the new Parliament embarrassed, bewildered, uncertain, defensive. With no apparent embarrassment, Menzies announced a series of economic measures which embodied the essence of the Calwell programme which he had denounced during the election campaign as 'wildly irresponsible and inflationary'. A feeble motion of no confidence, scattering Labor's shot over fifteen unrelated points, signalled Calwell's eclipse and Menzies' recovery.

The West New Guinea aftermath marked the beginning of a profound change in the relations between Calwell and Whitlam. More importantly, it changed Calwell's position in relation to the Party. In the years since the great Split of 1955, the Labor Party had maintained an uneasy coalition between its right and left wings. In the best and worst of times, the Labor Party is a coalition; but the agonies of the Split had sharpened the definitions members were expected to take and felt themselves expected to take. In 1960, Calwell's label was 'centre-right'. After 1962, after and because of West New Guinea, he began a long lurch leftwards. All sections of the Party were equally disturbed by his statements; no section found the implications acceptable or defensible. But open criticism was likely to come only from the 'left'. Calwell was especially uneasy about any possible reprisal from Ward. Ward was contemptuous of Calwell's campaign undertaking never to raise nationalization. The election result protected Calwell from serious attack on that score. West New Guinea would have provided new ammunition. For self-protection, Calwell sought to strengthen his position among members who identified themselves as left wing. The go-between to the 'left' was Leslie Haylen, whose supporters in the deputy leadership election of 1960 had unexpectedly switched their second vote to Whitlam rather than Ward, giving Whitlam an upset victory. Ward never forgave what he regarded as a betrayal. Haylen agreed to defend the Calwell line on West Irian.

Leslie Haylen, in his raffish and ragged way, was one of the most brilliant debaters in all the post-war Parliaments. There was ability, wit, charm, originality, and an odd but very Australian mixture of gentleness and cruelty; but laziness and lack of self-discipline nullified all. He was never really a leading parliamentarian; he was always, throughout his twenty years in Parliament, a leading character. He agreed with almost nothing Calwell had said on West New Guinea; his public support for the Calwell line was original and diverting. Its purpose was not to promote that particular cause in public, but to protect Calwell from internal criticism within the Party. And in that, Haylen did his job well. The price Calwell paid for protection from the 'left' was a deep loss of authority over Caucus.

Throughout the next year, the Labor Party found itself in deepening turmoil. The most difficult issues were the concept of a nuclear-free zone in the southern hemisphere and the establishment of a United States communication centre for nuclear submarines at the North-West Cape, Western Australia. Like Vietnam itself, once these questions became matters for actual decisions, there was never any real doubt about what the Labor Party would decide. The political damage was done by the manner in which the decisions were made. On the North-West Cape issue, which opened up the whole question of foreign bases on Australian territory, Calwell made the disastrous mistake of taking the decision-making away from the Parliamentary Party and inviting the Federal Executive of the Party to make the decision. This was the origin of the immensely damaging phrase coined by Menzies to describe the decision-making processes of the Australian Labor Party as 'the faceless men'. The irony was that the Parliamentary Party was perfectly competent under the rules of the Party to make all necessary decisions on its attitude to the legislation on the North-West Cape. The Parliamentary Party would have taken the same attitude ultimately taken by a Special Federal Conference in May 1963. The turmoil and damage caused by taking the decision away from the Parliamentary Party was lasting and needless. The decision was to accept the fact of the base, but to have a Labor Government renegotiate the terms with the United States Government to ensure an Australian share in the control, and to establish Australian legal sovereignty over Australian territory. Ten years later, a Labor Government was to carry out this policy or, more accurately, believed that it had carried out this policy.

From the time of the legislation to ban the Communist Party in 1950, Menzies had played on the theme that Labor parliamentarians were controlled by the non-elected 'outside body'. Calwell played right into his hands. Whitlam urged Calwell to keep the argument within the Parliamentary Party. Calwell, now unsure of his ability to control Caucus, sought a decision outside. This established a disastrous trend. By 1963, reference to the Federal Executive on any policy matter of significance had become routine. Finally, in October 1963, a meeting of the Federal Executive in Adelaide purported to instruct all Parliamentary Parties, above all, the Parliamentary Party in New South Wales, then forming Australia's longest-standing government, of the proper party attitude to State aid for non-government schools. This involved Australia's oldest, most bitter political and social issue. It set the tone for the relations between Calwell and Whitlam for the rest of their partnership. It destroyed the reputation of the New South Wales Labor Government. And it made up Menzies mind for him. He called an election for 30 November 1963.

Against this background, the Labor Party entered the campaign for the 1963 elections called by Menzies a year ahead of due time, but considering the political realities, hardly before time. Yet the Labor Party approached the elections with a confidence greater than at any time since 1954, despite the general economic recovery and despite its very public internal troubles. The opinion poll was promising. Whitlam's own misplaced confidence about the 1963 result made him thereafter reluctant to predict election results or trust too much to his own instincts on that score. In the event, Menzies won comfortably.

In every sense, the great loser was Arthur Calwell. He lost much more than the election. He lost half his heart. In a single weekend before polling day he was struck more savage blows than even the most hardened politician has any right to expect. The opinion poll showed an ominous downturn. The *Sydney Morning Herald* finally turned against him, with a bitter attack full of all the spite and venom of a deserting former ally. In Geraldton, Western Australia, he had to listen while a priest, a Monsignor of his Church, preached on the sinfulness of voting Labor. At 5 a.m. on 23 November in the Palace Hotel, Perth, he was told by the night porter that the President of the United States, John F. Kennedy, had been assassinated. Arthur Calwell and his campaign, begun with such hope, fell apart. In response to a garbled report that Menzies had invoked the name of Kennedy against the Labor Party, he wrote a statement, against all the pleas, even the tears, of his advisers: 'Menzies has smeared the blood of the Australian Labor Party over the coffin of the dead President.'

Calwell, heartbroken after 1963, was left only with a bitter determination to survive as Leader of the Labor Party. There was no immediate move for his resignation and nobody wished it. Everyone close to him except his wife and daughter, fiercely loyal, fiercely protective, believed that the proper course for Calwell in his own interests and in the interests of his Party was to prepare for his departure. Within the Labor Party, at all levels, there was a general expectation that this resignation would occur no later than the end of 1964. Menzies, by calling the early election of 1963 had broken the series of conjoint elections for the House of Representatives and the Senate.

Thus, a half-Senate election was likely between late 1964 and early 1965.

After the 1963 election, Whitlam wrote a report at the request of the New South Wales branch of the Party in which he said:

> In the intervening years (between 1961 and 1963) Mr Calwell did not speak or act as impressively as Prime Minister Menzies, the Labor members of Parliament did not show the same cohesion and solidarity as the Liberal ones, and the Labor organization and its affiliated bodies were not so self-effacing and discreet as the Liberal organization and its backers. The party's policies had been improved, but the manner of formulating and presenting them obscured their merits.

The partial publication of this report and the interpretation of disloyalty placed upon it ended anything which remained of the Calwell-Whitlam partnership. There were two simple and quite irreconcilable facts: Calwell was determined to have another chance at the Prime Ministership at the 1966 election, and Whitlam was convinced that such an attempt would be a total disaster for the Labor Party. Both happened. In a letter to his second son, Nicholas, then a student at Harvard University, Whitlam gave this account of the first open rift:

> Many individual members told me in December and January that I should oppose him because they were appalled at his statements that he would lead us at the next elections, when he will be 70. During the Denison by-election (to replace Athol Townley)

groups of members spoke thus to each other and to me. I told Arthur the following Monday, 17 February 1964, what I had told these groups what I had told individuals: I agreed that in his and the party's interests he should retire before the next elections. I didn't think it would be generous to him or advantageous to the party to oppose him now and I wouldn't do so. He didn't challenge me to get the numbers (as claimed in newspaper reports); nor did we or do we in the least talk to each other any less than before.

Calwell's published version does not contradict Whitlam's private account. Calwell adds: 'I rejected the suggestion with scorn and said that I would never be a caretaker leader.' Whitlam, however, underestimated Calwell's resentment and his determination to remain leader for the 1966 election. A relationship which had been constructive, reasonably successful, and, in view of the differences of background, interests, ideas, style and age, surprisingly warm then deteriorated rapidly. Two years later it was to reach its nadir with Calwell failing by the narrowest margin to have Whitlam expelled from the Labor Party. In that time, Australia had entered into its commitment with the United States in Vietnam. But it was a much more divisive issue which nearly brought down Whitlam. This dispute, too, was part of the Menzies inheritance.

2
'Is the Tumbril Ready?'

The oldest, deepest, most poisonous debate in Australian history has been about government aid to Church schools. The mystic incantation 'State aid' has broken governors, governments, parties, families and friendships throughout our history.

In 1852, Charles Perry, Anglican Bishop of Melbourne, wrote:

> The Roman Catholic Church and its Priests are aware of the impossibility of retaining their dominion over the consciences of an intelligent and independent people, such as the inhabitants of Victoria are likely to become. Great therefore in my opinion, is the guilt of assisting to prop up among us this system of Satanic delusion ...

In 1880, the Catholic Bishops wrote in a pastoral letter:

> This expenditure on godless education, this studding the colony with schools which the Church knows from experience will, in course of time, fill the country with indifferentists, not to speak of absolute infidels, this use of Catholic funds — of taxes paid out of Catholic pockets ... to sap the foundations of Christianity is an act so galling to every feeling of fair-play, that we do not see how any free man, with any spirit in him, can allow it to pass unchallenged ... it is a system of practical paganism.

In 1866, the New South Wales Public Schools Act provided for subsidies for church schools on certain conditions. It pleased nobody, neither the Protestant Ascendancy, the emerging secular humanism, nor the Catholic hierarchy. The Public Instruction Act of 1880 ended this first experiment in State aid for Church Schools and enshrined the principle of free, compulsory, secular education. The experience in the other colonies was similar.

The century-old failure of the Catholic Church in Australia to achieve her principal social aim is remarkable testimony to the political incompetence of her Bishops. Australian party politics are largely about conflict, co-operation and compromise between various pressure groups, some permanent, some temporary. The parties themselves are coalitions of tried and proved convenience between pressure interests, combining for commonly perceived and agreed purposes. The success or failure of the Australian system is measured by the degree to which it has been able to serve the significant groups. The unwritten agreement, the basic but unspoken condition of the national unity and identity, is that the needs and demands of each group will be accommodated within the system, if not met in full. The system embraces all groups which are numerous, or wealthy, or articulate, or well-organized. Those who lack these qualities are left outside the system. The aboriginals are the most notable example of a group outside the system. The system survives on the unwritten understanding that no powerful group shall be absolutely excluded, and no group shall get everything it wants.

The Catholic Church, in its long struggle for public finance for its schools system, seemed uniquely well-placed to succeed. It possessed all the essential attributes—numbers, wealth, argument, organization. Yet as a pressure group it failed in its most specific political objective. The achievement was in the end not to be made by the Church at all but through the political exigencies of the political parties. The State aid controversy was ended in favor of the Catholic Church after a century and a half by two men—Menzies, the self-proclaimed 'simple Presbyterian' and Whitlam, the exemplar of the humanist heresy.

The Church's failure was the more remarkable in view of its potential political strength within the Labor Party. Catholic preponderance within the Labor Party dated from World War I and the first great split over conscription. Protestant nonconformism had been a fundamental influence in the foundation years. That influence was gravely weakened by the split of 1916. Yet despite the disproportionate increase in members of the Labor Party who were also Catholics, the Church objective of State aid was in no way advanced. The Catholic ascendancy within the Labor Party was most marked in New South Wales and Queensland; education was then exclusively a State matter; yet Labor Ministries in those States, with a heavy Catholic membership, did nothing to promote the objective. Successive Catholic Premiers conceded their impotence to get State aid injected into the Party platform, policies or practice. The State aid controversy continued, bitter and endemic. Yet despite the preponderant Labor allegiance of Catholics, they were unable to secure their principal political aim. At its peak just before the split of 1955, Catholic representation in the Federal Caucus was 60 per cent. By contrast, Catholics were grossly under-represented in the conservative parties. They did not take part in the counsels of the conservative parties. Yet State aid never became a party question. There was bipartisanship on the issue by a refusal by all parties to acknowledge, much less accept, the Catholic case. The third great Labor split of 1955, for all its sectarian bitterness, did not immediately change this, although the breakaway party, the Democratic Labor Party, became the first significant Australian political grouping to include a plain and direct State aid plank in its platform. A century of resistance was to crumble almost overnight, and the Bishops were to have little share of the final, crucial battles. In the end, the economics of education bested both the Bishops and the bigots.

One of the basic reasons for the failure of the Catholic Bishops was the confused and contradictory way they put their case. For much of the formative colonial period they grounded their case on the godlessness of the secular schools. The pastoral letter of 1880 came perilously close to arguing against any public education from public monies at all. This was clearly an affront to the Protestant majority. Just as important, it was not what the ordinary lay Catholic thought or wanted. By the 1960s the growing cost of education was changing the whole nature of the question. It was becoming too costly and too important a matter to be left to the States. Menzies was the first to recognize this. Summing up his own career, he put education as his major domestic achievement. His great breakthrough was Federal Government acceptance of financial responsibility for universities, when he set up the Universities Commission in 1958.

Henceforth, education was no longer the preserve of the States. Menzies' rationale was the high cost of tertiary education. But once the principle of Federal involvement was accepted, its adoption for all levels of education, all increasingly costly and all beyond private means, was inevitable. Australia had developed three school systems — the State schools, the Catholic schools, and the fee-charging secondary schools, sponsored mainly by, though very vaguely associated with, the Protestant Churches and known as the Great Public Schools. Once this last category acknowledged that rising costs required government subsidies for their survival and began seeking State aid for wealthy Protestant schools, then State aid for Catholic schools became inevitable. But like so much that is inevitable, the last fight was the most bitter.

By 1963, resistance was crumbling at all points. Despite official adherence to the basic principle against State aid by governments of all parties, Church schools were in fact receiving assistance in a number of forms. Most of it was indirect through teacher training, book allowances, scholarships and bursaries. In the Australian Capital Territory, substantial help was being given in school-building programmes. But each new subsidy was represented as a special case. Whatever the actual practice, the basic principle was still deemed inviolable — no State aid.

In the Labor Party, the whole question was clouded by sectarian bitterness against the Church in the aftermath of the 1955 Split. The self-appointed and widely accepted keeper of the conscience of the Australian Labor Party on this matter was F. E. ('Joe') Chamberlain, its part-time Federal Secretary and also Secretary of its Western Australian branch. He had emerged from the Split as the chief tactician within the Labor organization and its most powerful figure. He was seen as both the sea-green incorruptible and the grey eminence of the Labor movement. His personal strength sprang from a singleness, even narrowness of mind, a terrier grip on the essential point of an argument, and a steely eye for the weaknesses of an opponent's case. He was literal-minded in interpreting texts and resolutions. This was important in a Party which had become obsessed with the nuances of the wording of policy declarations, a Party which had conceived a passion for textual analysis worthy of mediaeval scholars. Only once himself a candidate for public office — unsuccessfully — he professed a contempt for those parliamentarians who would have described themselves as pragmatists, and whom he would have described as unprincipled. He was a superb committeeman. He came to a committee thoroughly prepared, knowing exactly what he wanted and how to argue for it. In the 1960s, the Federal Executive of the ALP was its most powerful committee. Therefore, he was the Party's most powerful man.

The education policy of the Labor Party stated:

> The Australian Labor Party believes that it is the obligation of the State to provide a universal free compulsory secular system of education, open to all citizens . . . Citizens who do not wish to use the school facilities provided by the State, whether for conscientious or other reasons, shall have the absolute right to develop an independent system of schools of recognized standard provided they do so at their own cost.

Chamberlain had secured insertion of this interpretation in the Party's platform. Under the guise of conferring an absolute right, he had in fact imposed a total prohibition. His fight to keep his policy intact lasted until 1969. It was the cause of the destruction of a State Labor Government, of a near split in the Labor Party, the near expulsion of Whitlam, and was to become a major factor in the electoral débâcle of 1966.

In September 1963, the New South Wales Labor Government included in its annual Budget a proposal for means-tested allowances of £2, payable weekly to the parents of children in private secondary schools. Premier Heffron had taken the precaution of informing Calwell of the proposal some hours before the Budget was delivered and secured from Calwell a public declaration that the proposal accorded with the policy of the Australian Labor Party. The Federal Executive of the Party met in Adelaide a week later. Chamberlain placed on its agenda a motion denouncing the New South Wales Government. The New South Wales delegates, neither of them members of Parliament, refused to take any responsibility for the New South Wales Government's decision and only weakly defended it.

Calwell, attending the Executive by courtesy, failed to defend either the New South Wales Government or his own earlier statement of support. On 3 October, Chamberlain issued his last statement as Federal Secretary before handing over to a new appointee, Cyril Wyndham: 'The Federal Executive is confident that when the New South Wales Government is fully appreciative of federal policy it will not hesitate to recast its present plans. The amount allocated in the current budget can be used to further the State scholarship system whereby the student is directly assisted to further his education in accordance with policy.'

The New South Wales Government obeyed. It was a classic Chamberlain coup in committee — well-executed, highly-principled and absolutely disastrous. The Menzies tag of the 'faceless men' was fixed with a vengeance. Here was the 'outside body' not merely laying down Party policy but openly and crudely dictating to an elected government, a government of the senior State which the Labor Party had ruled for nearly a quarter of a century. Within days, Menzies announced a general election a year ahead of due time. He gave a specific State aid pledge to subsidize science buildings at private schools. He triumphed, particularly in New South Wales. During his campaign, he was surprised at the lack of any resistance to his State aid initiative. 'I confess I am amazed at the impotence of bigotry,' he said. The New South Wales Labor Government was doomed in its next election. It fell on 1 May 1965.

Menzies' science buildings proposal was in itself quite limited, but he had set an inevitable course for education in Australia: aid for all schools, nationally funded. By 1965, State aid legislation of diverse kinds in each of the seven Parliaments forced Labor members to the point of decision. In no case was the legislation opposed. Yet there was still Labor's policy. Chamberlain, determined to protect the policy, decided on a full frontal attack.

Although the people of the uniting colonies in their Constitution of 1901 had relied humbly on the blessing of Almighty God, they established nonetheless a secular State and determined in Section 116: 'The Commonwealth shall not make any law for establishing any religion or for imposing any religious observance or for prohibiting the free exercise of any religion and no religious test shall be required as a qualification for any office or public trust under the Commonwealth.' A similar provision in the First Amendment to the Constitution of the United States had recently been upheld by the United States Supreme Court to rule out bible-readings in public schools. Chamberlain put the proposition: might not the science blocks legislation be as grave an affront to the Australian Constitution as it was so obviously against the ALP Platform? Chamberlain was determined that the proposition must at least be tested.

A meeting of the Federal Executive of the Labor Party which began quite routinely on 9 February 1966 ended in crisis. Before most of the delegates had realized what they had done, they had precipitated disaster. Chamberlain alone knew exactly what he had done and why he had done it. At his instance, the Executive carried four resolutions denouncing all forms of State aid. Inspired by its disastrous triumph over the New South Wales Labor Government in October 1963, the Federal Executive had gone on to assume authority over Party affairs in matters of every detail. Yet, apart from Chamberlain, it was a body not notable for weight or stature in the Party. Half the members were paid Party officials. It was this group which took the Labor Party to the brink of disaster in February 1966.

Chamberlain came armed with his set of resolutions against State aid. The key one was a resolution instructing the Legal and Constitutional Committee of the Party to draw up an appeal to the High Court of Australia contesting the constitutional validity of the legislation providing government funds for science laboratories at Church schools. The resolutions were all carried.

Calwell had also come armed to the meeting. He could not vote but he could speak as Leader of the Parliamentary Party, and in an election year his voice might be deemed to count for something. He came armed with a letter from Archbishop James Carroll from New South Wales, who was one of the few friends of the Labor Party left in the Catholic hierarchy. Non-partisan himself, his heart cried at the damage done to his people and his Church by the increasingly partisan, anti-Labor attitude expressed by his brothers and sisters in the faith. Yet he saw that the Labor Party was not exactly making it easy for itself. So he wrote to Calwell in January 1966, setting out what he believed were the minimum demands for State aid — not his demands, but what he saw were the least the Labor Party could offer if he and people like him were in any way to be enabled to hold the line against mounting Catholic despair and hostility against the Australian Labor Party. Calwell was so impressed with the letter that he had it typed up in the form of a submission he intended to make on his own behalf to the Federal Executive. But when he saw Chamberlain's determination and how the numbers would probably go, he pocketed the submission and forgot about the letter. Suddenly, bigger, better game was afoot — not State aid, not education, not the next election, but Gough Whitlam.

Whitlam, also granted the privilege of attending as the Party's deputy leader, spoke strongly against the resolutions. In his view, the decision contradicted a decision of the 1965 Federal Conference that existing benefits should not be disturbed. For, clearly, a successful appeal would mean the ultimate, total disturbance of all existing benefits. Above all, he pointed to the humiliation involved for every Parliamentary Labor Party which had voted for existing benefits. Other agenda items dealing with the appointment of Party committees, from foreign affairs to northern development, indicated a massive move by Chamberlain to assert his own personal dominance, and the complete ascendancy of the Federal Executive over every other section of the Labor Party, including Federal Conference itself. It was not only what he regarded as the political stupidity of all these actions which incensed Whitlam. He was outraged by the spirit of vindictiveness, recklessness and prejudice he detected in them. He decided then and there upon his first great act of confrontation.

On the eve of the meeting, Calwell and Whitlam had jointly opened Labor's campaign for a by-election in the Central Queensland electorate of Dawson. The candidate, Dr Rex Patterson, formerly Director of the Northern Division in the Federal Department of National Development, had sacrificed high prospects in the Public Service to stand. He was a man with a cause. Prime Minister Holt described him as 'a crusader with a fanatical gleam in his eye'. The timing and nature of the conduct of the Executive filled Whitlam with a deep fury. He deliberately set himself on a collision course.

In a letter to the General Secretary, Cyril Wyndham, made public as intended, Whitlam gave his reasons for refusing to serve on the legal committee to examine the constitutionality of State aid:

—The High Court would not apply the United States Supreme Court decisions.
—The High Court would not invalidate capital aid which had been paid to mission schools in the Northern Territory and in New Guinea for decades, to schools in the Australian Capital Territory and in university colleges for a decade, and for science blocks throughout Australia since 1964.
—The State Attorney-Generals would not allow such a challenge.
—The Federal Parliamentary Party would not approve.

The guts of his letter set out the approach to education which he had been developing over the years, and which would provide the basis of the approach of the Labor Government after 1972. It is now the settled policy of all Australian Governments on education:

> I must reiterate my conviction that there will never be enough trained teachers and facilities and equipment in State or private schools in this country until the Commonwealth does the same and as much for teacher training and technical and secondary and even primary education as it does for universities and until it does so irrespective of whether the teachers and pupils are at State or private institutions ... No country with our resources should tolerate the present standard of our schools; no socialist party should tolerate the present inequality of opportunity for our school children. I shall not rest till the Federal Conference clears away the accumulated deadwood which prevent the Chifley Government's Education Act of 1945 from bearing fruit.

On Sunday 13 February, Calwell made a broadcast drafted by William Hartley, the Secretary of the Victorian branch of the Labor Party, defending in principle and in detail every aspect of the Federal Executive decisions. This young, eager Western

Australian had come to Melbourne as Chamberlain's protégé. Not for the last time, he was to involve a leader of the Labor Party in a great deal of trouble by his enthusiasm in championing him. His services have not always been without political cost to those accepting them. Whitlam's letter and Calwell's broadcast set the battle lines. In response to Calwell's broadcast, Whitlam broadened the crisis into a direct confrontation with the Federal Executive. On 15 February, he made his formal public challenge:

> The long-term future of the Australian Labor Party is now at stake. Continuance of present trends will reduce the greatest political party this country has known into a sectional rump. No party is immune from destruction. No party can afford to be controlled by people who want to use it as a vehicle for their own prejudice and vengeance. The issue is not between 'right' and 'left'. It is between those who want a broadly-based socialist and radical party and petty men who want to reduce it to their personal plaything. The present extremist controlling group has blatantly breached Federal policy as determined by the last Federal Conference. This extremist group has deliberately humiliated the Parliamentary Party. This extremist group breaches the party's policy; it humiliates the party's parliamentarians; it ignores the party's rank and file. It is neither representative nor responsible. It will and must be repudiated.

And to a reporter he said, 'Chamberlain's champions have had their last fling.'

The crisis developed against the background of the Dawson by-election. The seat, in the heart of the Queensland sugar industry, was traditionally and naturally a Country Party fief though with strong Labor pockets, particularly in the city of Mackay. However, a growing sense of neglect in the area, the perennial dissatisfaction of the North with Canberra and the splendid credentials of the Labor candidate made the contest interesting. For Harold Holt, facing his first electoral test as Prime Minister, the national crisis in the Labor Party seemed a bonus beyond belief. Labor's chances, slim at best, seemed to depend on insulating its local campaign from its national troubles. In view of the fact that Whitlam, the principal figure and immediate cause of those troubles, dominated the campaign by his presence throughout, this seemed an empty hope.

The journalist Peter Bowers wrote in the *Sydney Morning Herald* from Mackay:

> Mr Whitlam dashes to and fro trying to be everywhere at once — in Dawson fighting for Dr Patterson, in the south fighting for his political life. Government speakers, desperate to find any issue other than northern development, accuse Mr Whitlam of trying to win Dawson to bolster his bid for leadership. Mr Whitlam and Dr Patterson make a great team temperamentally. They are akin — impatient, peppery, at times boiling with frustration. They are feeding each other . . .

In between dashes to and from Mackay, Whitlam deliberately increased the level of crisis, particularly in a television interview given in Sydney, soon after his fighting mood had been given a sharp boost by a demonstration by party supporters at Brisbane airport. Two statements in particular outraged his opponents and dismayed his friends. On the conduct of the twelve-member Federal Executive, he said: 'I can only say we've just got rid of the thirty-six faceless men stigma to be faced with the twelve witless men.' Asked whether his leadership prospects had been harmed, he replied, 'Oh, I've been destined to be leader of the party for at least a year, as soon as there was a ballot for the position.' This last statement was the basis for the assertion which stuck to him thereafter that he was 'the self-proclaimed man of destiny'. The arrogance lay only in its accuracy.

At the urging of four of the State Executives, only New South Wales and Tasmania dissenting, an emergency meeting of the Federal Executive to discuss Whitlam's actions and statements was called for 2 March. Chamberlain and Calwell grasped the opportunity. Whitlam had given a clear case for action under Rule 7 (VII) of Party Rules stating that the Federal Executive shall 'have plenary powers to deal with and decide any matters which in the opinion of at least seven members of the Executive affects the general welfare of the Labor movement'. The chance to move decisively against Whitlam, even to the point of expulsion, was there; probably the numbers were there too.

Chamberlain decided to seek not merely censure but expulsion, to destroy Whitlam once and for all.

On 26 February, Rex Patterson triumphed in Dawson with a swing to Labor of more than 12 per cent. On 3 March, Whitlam was summoned to face the Executive in Canberra, charged with 'gross disloyalty'. Entering his appointed place of execution, Whitlam said to the pressmen: 'Is the tumbril ready?' Calwell waited in his Parliament House office; he and Chamberlain had counted the numbers—seven to five for Whitlam's expulsion. Calwell made the mistake of not keeping his knowledge to himself. That day, Patterson arrived in Canberra, not as any conquering hero but as a witness to an execution. A chance conversation with Allan Fraser, MHR, who was privy to the plot, confirmed what he sensed was afoot—the political destruction of the man who, alone of all the Federal members of Parliament, had assisted and encouraged him in his victory. He managed to inform the Queensland State Secretary, Tom Burns, what was about to happen. The Queensland Executive itself had already censured Whitlam but, especially in the afterglow of Dawson, would not have a bar of his expulsion. The Queensland delegates, themselves personally as bitter as any against Whitlam, were instructed to oppose expulsion. That reversal, however reluctant, saved Whitlam. The message which Calwell had awaited and prematurely disclosed never came. Whitlam had no illusion how close he had sailed to disaster. Years later as Prime Minister, he was to pay tribute to Burns, by then Leader of the Opposition in the Queensland Parliament, as 'the man who saved me'.

But the immediate cause of the crisis remained unresolved. Two very clear issues had been raised—State aid and the relations between the Parliamentary Party and the Federal Executive. The attempted humiliation of the Parliamentary Party went to the heart of the leadership question: Calwell had failed utterly to defend or protect the Parliamentary Party from the pretensions of the paid officers on the Federal Executive. As the crisis developed, Whitlam had made this central to his attack and his own defence. A challenge to Calwell was inevitable. In April, Calwell easily beat off a move to have the leadership positions declared vacant. He then announced what was scarcely a concession—that if the 1966 elections resulted in a Labor defeat, he would not again contest the leadership.

The ancient issue of State aid appropriately took rather longer to settle. Two Special Conferences in March and July 1966 resulted in a tentative settlement for the purposes of the 1966 election. Ironically, it was settled at Surfers Paradise, the citadel of Australian 'practical paganism'. The formula reached allowed members of Parliament to support any existing form of State aid, including capital works. Calwell, accredited as an ordinary delegate from Victoria, cast the decisive vote in favour. It was done reluctantly, hesitantly, yet it was his last really serviceable action for his Party and his Church. The Federal Secretary, Cyril Wyndham, counting the votes around the table had to ask, 'Is your hand up or down, Mr Calwell?' It was up, and the Labor Party had taken a big step towards complete acceptance of State aid.

The 1966 crisis is deeply instructive as an example of Whitlam's style, methods and character. All the elements were there. He later described his approach: 'You either crash through or crash'. This is an over-simplification. Between the two extremes lay a combination of tenacity and flexibility, a singleness of purpose reached by a variety of means, unmatched by any politician of his time. This was the pattern of his conduct in 1966 and the key to his future successes and failures. It should be emphasized that, as was so often to happen, Whitlam did not spark off the crisis, but he shaped it. Whitlam did not initiate the confrontation, but seized the opportunities and accepted the dangers once it became inevitable. There would have been no crisis except for the Chamberlain resolutions on State aid. If Whitlam had remained silent, if he had gone along, there would still have been a revolt of sorts within the Labor Party. In his defence at the March Executive called to expel him he said, 'The gravity of any situation determines the response.' Whitlam's response determined the shape and pace of the crisis. Essentially, his actions gave form and coherence to what would otherwise have been incoherent.

A diverse, turbulent party of the left will inevitably have many crises. The danger for the Labor Party lies not so much in the frequency of its crises, but that they may become uncontrollable. The vulnerability of the Labor Party lies in the paradox of its loose and diverse structure on one hand and the rigidity of the rules on the other. There was no fatal necessity about the Split of 1955. Nothing in the issues, ideologies or individuals involved made it inevitable. It was precisely because there was no clear commanding issue that events, not people, took control in 1955. In 1966, Whitlam ensured that the issues were clearly defined, and by doing that he helped control and contain the crisis. Whitlam's enduring contribution to the Labor Party's method is that he showed it was possible to have a tremendous row without an awful split.

The chief elements of his conduct in 1966 were these: first, he carefully defined the issue; second, he extended the specific issue, in this case, Labor Party policy on education, into a broader question — the role of the Parliamentary Party; third, he secured significant support from the rank-and-file members and supporters of the Labor Party; fourth, he deliberately provoked his opponents and made statements of great intemperance and great truth; fifth, he ignored those who counselled caution or retreat; sixth, he worked harder, physically and mentally, than his opponents; seventh, he was able to exploit an electoral success and a reputation as a vote-winner; eighth, he needed, and won, allies not necessarily in basic sympathy with him or views, but who accepted his value to the Party; ninth, he was prepared to stake all and lose all; tenth, he used to the limit the powers and prestige of his position; eleventh, he had good luck.

Whenever these factors have been significantly present, Whitlam has succeeded; whenever they have been significantly absent, he has failed. It is easy enough to say: 'Crash through or crash'. It is very hard to do. And nobody can do it every time.

In 1966, Whitlam had no illusions about the risks. He nicely calculated the risks and his strengths. In Australia, this has been called petulance or arrogance. Elsewhere, it is usually called courage. At the blackest moment of the crisis, when all the cards seemed stacked against him, when the numbers seemed cast for his expulsion, he amply displayed that characteristic which I, ten years later in a moment of deep emotion, was to single out as his finest characteristic, 'grace under pressure', to use Hemingway's definition of courage. The specific manifestation in Whitlam's case is a form of gallows humour, self-mocking on the brink of disaster, a distinctive Whitlam version of a distinctively Australian style. In 1966, in 1968, in 1975, in 1976 and 1977, in all his deep personal and political crises he was to display a very full measure of grace under pressure.

3
How Australia Got Into Vietnam

'Whatever you say Tom, I'm not going to let you bastards destroy the American alliance.' Thus, in August 1964, Arthur Calwell, with uncharacteristic crudeness of language, cut short an argument with Thomas Uren, MP, about the Gulf of Tonkin incident. Nobody in the Australian Government, the Australian Labor Party or the Australian press questioned the accuracy of the official American version of the events in that first week of August 1964 any more than did the leaders of the United States Congress. The only question was about the appropriate American response. The universally-accepted version was that North Vietnamese torpedo-boats had fired upon American destroyers on the high seas — an act of aggression comparable with Pearl Harbor. To suggest that the official American version might be based on a mistake or misunderstanding would have been a kind of heresy. Calwell's anger and Uren's anguish over what should be said about the Gulf of Tonkin incident pointed to the central dilemma of the Australian Labor Party on Vietnam: how to oppose the war without opposing the United States; how to condemn the war without condemning the United States.

The crucial importance of the American alliance was always common ground between the Australian political parties in the debate on Vietnam. As far as the political parties were concerned, the essence of the argument in favour of Australian involvement was that it would strengthen Australia's relations with the United States. The essence of the argument against Australian involvement was that the war would weaken the United States, and thus weaken the alliance. The Labor Party, as much as the Menzies Government and its advisers, saw Vietnam through the focus of the American alliance. Events in Vietnam itself were always secondary. There was normally a time-lag between opinion in the United States and Australia. In one respect only was Australia ever ahead of the United States on Vietnam: Australia lost interest first and forgot faster.

The attitude of Australians to Vietnam was not unlike their attitude 150 years before towards the convict system. As long as it clearly served the interests of the dominant group, the landowners, it was accepted. As long as the dominant power, England, financed it, it was accepted. Only when these two factors began to change did the questioning begin. Throughout, there was very little questioning of the meaning or morality of the institution itself. So, with many noble exceptions, the war in Vietnam became unpopular only when its failure became clear and its cost became burdensome.

*

The story of Australia's direct military intervention in Indo-China can be traced to October 1961. During that month, the Washington Embassy told Canberra that the Kennedy Administration was considering military intervention to save President Diem's government. The Australian Chargé d'Affaires in Saigon, B. W. Woodberry, told Canberra he had 'serious doubts about the efficacy of introducing Western ground troops'. In the whole military and diplomatic history of Australian involvement in the ensuing decade, this stands out as one of the few expressions of Australian official doubt about the enterprise. And that doubt was limited to its effectiveness, not its correctness.

Menzies then bought time with the American Administration because he had the 1961 election on his hands. He instructed our Ambassador in Washington, Sir Howard Beale, to tell Kennedy's Secretary of State, Dean Rusk, that the Australian Government hoped to announce a decision on assistance to South Vietnam after the election. Rusk in reply said that the United States was anxious for support 'to show it was not merely an American operation'. After his meeting with Rusk, Beale suggested to Canberra that the supply of Australian counter-insurgency personnel and small arms and ammunition would make 'a favorable impression' on the Kennedy Administration. Although these great matters were afoot, the 1961 campaign was the first election since 1946 which was not dominated by Communism and foreign threats. Vietnam was not then in the Australian vocabulary.

Beale, a former Liberal politician once touted as a rival to Menzies, became the foremost advocate of Australian military aid. Late in 1961 and in the early months of 1962, he repeatedly urged that 'even a handful of instructors would make Australia's mark with the United States Administration'. From the beginning, Australian officials recognized that Australian military intervention was wholly political—directed not to the military or even political situation in South Vietnam but to the political requirements of the Australian–American connection. Thirteen years later, the story of Australian involvement was to end as it had begun—as a question about domestic political advantage.

On 14 March 1962, the Defence Minister, Athol Townley, wrote to the Minister for External Affairs, Garfield Barwick, saying that '*if asked by the United States*', Australia could supply ten instructors in jungle training. On 31 March, President Diem, of South Vietnam, requested, in general terms, military assistance from 'the free world countries to help Vietnam from being overwhelmed'. At a meeting of the ANZUS Council in Canberra on 9 May, Menzies told Admiral H. D. Felt, Commander of the United States Fleet in the Pacific, that Australia was prepared to supply a few instructors. Felt replied that such assistance was welcome, although not necessary. At Menzies' request, Felt undertook to convey the Australian decision to Diem. On 24 May 1962, Townley announced that Australia was providing thirty military instructors 'at the invitation of the Government of the Republic of Vietnam'. Thus Australia started down the slippery slope, not only towards increasing involvement, but towards the calculated policy of public deception which marked every stage of that involvement. From the beginning, the ruling purpose of Australia's intervention was to ingratiate Australia with the American Administration.

This first commitment set the pattern for all future commitments. After Admiral Felt conveyed the Australian offer to Diem, Diem asked that the thirty instructors be sent to a northern area where he feared a Vietcong break-out. He likened such an Australian presence to 'the Anzac beach-head at Gallipoli'. The American military command rejected this idea and dispersed the Australian advisers over several provinces. Garfield Barwick complained to the Australian Defence Department that this dispersal reduced the political value of the commitment in the United States. For ten years, this was the overriding consideration.

Once the first commitment was made, the United States increased its demands almost monthly. In all cases, the demands came from the United States, never from South Vietnam. Usually the demands took the form of requests for equipment which Australia simply did not possess. For a decade, Australian defence spending had been pegged at a low level of the gross national product; the re-equipment programme got underway only with the grandiose decision to buy the F-111 bombers in 1963; and that decision was directed against Indonesia. Aware of its limitations and its lack of resources, preoccupied with Indonesia, the Defence Department was always far less enthusiastic about the Vietnam involvement than the Department of External Affairs. Each successive American demand was strongly supported, sometimes urged in anticipation, by External Affairs and resisted or only grudgingly accepted by Defence. Australia's war in Vietnam was very much the war of the Department of External Affairs. The Defence Department was, of course, in an irresistible position whenever the Americans requested something Australia did not have. The Americans seemed to have the impression that in 1964 Australia had a modern army. Anyone who reads the record must gain the overwhelming impression that, apart from the politicians, the guilty Australians were in External Affairs. As far as Australia's involvement was concerned, this was not a generals' war, but a diplomats' war.

In May 1964, the United States Embassy in Canberra delivered a note requesting another eighteen advisers, more jungle-training personnel, a helicopter unit, a reconnaisance unit, a fixed-wing transport unit and a surgical team. The Australian Embassy in Washington supported the request of its American counterpart in Canberra in these terms: 'South Vietnam is an area in which Australia can, without disproportionate expenditure, pick up a lot of credit with the United States.' The Embassy cable then set out this rationale: 'Our objective should be to achieve such an habitual closeness with the United States and a sense of mutual reliance that in our time of need — such as a crisis in our relations with Indonesia — the United States would have little option but to respond as we would want.' The current Minister for External Affairs, Paul Hasluck, cabled Washington: 'We should like the United States Government to know that we are anxious to reply promptly and sympathetically to their suggestions.' The proposals were then referred to the Inter-Departmental Joint Planning Committee, which in its report to the Government, written in the Department of External Affairs, set out the complete and classic rationale by which Australia's actions were to be justified for the next eight years: 'If South Vietnam falls, the West (sic) would be unlikely to hold Laos, Cambodia and Thailand. This would in turn make the future of Malaysia, Indonesia and The Philippines very uncertain.' Further, the defence and diplomatic planners said: 'Increased assistance would influence the obligation which the United States might feel to Australia in an emergency.' The Australian Government finally responded to the American request by offering another thirty NCOs for the army training team, an army dental team, an army workshop and six Caribou transport aircraft. It was at this point that Cabinet, for the first time, acknowledged privately that the commitment would inescapably involve Australian casualties in combat. The danger was not admitted publicly. The decision, made at a request transmitted through the United States Embassy in Canberra, supported by cables to and from Washington, was then notified to the Australian Embassy in Saigon with instructions to inform the South Vietnamese Government (the successor government, two or three times removed, to that of the assassinated Diem). On 9 June, a new Defence Minister, Senator Shane Paltridge, announced new military assistance 'in response to a request from the Government of South Vietnam'. Thus, from the beginning a pattern of official lies was established — that Australia acted in response to spontaneous requests from the Saigon Government.

Between 1961 and 1964, the Australian response to American demands for military aid to Saigon was governed by what Australia had available. But at the end of 1964, there was a revolution in the nature of Australian capability to meet American demands. For the first time, Australia was able to offer regular combat troops, ahead of an American request. More than that, the most recent American request excluded combat troops. The offers which led to the sending of an Australian battalion, firstly regular but ultimately conscripted, was an Australian initiative, altogether different in nature and form from America's request. On 11 December 1964, the Joint Planning Committee suggested to the Government that Australia could supply one infantry battalion, as well as ten more instructors, and two frigates. Before the Government could raise the matter with Washington, a letter arrived from President Johnson asking Australia for a further 200 advisers and an unspecified number of mine-sweepers, landing craft, salvage and repair ships and hospital ships — but no combat troops. Menzies replied to Johnson on 18 December that Australia was unable to supply the ships requested or a significant number of new advisers but was willing to send representatives to the United States to discuss 'the possible positioning of ground troops in Vietnam'. Thus the conclusion is inescapable: the crucial decision of Australia's generation-long involvement in Indo-China, the decision to commit regular forces on the ground whose numbers had to be conscripted, was made not because that was wanted or asked for, either by the United States or South Vietnam, but because that was the only thing Australia could supply if the political aims of Australia's intervention were to be met. The Menzies Government had anticipated a new American demand, but it met that demand in a way the Americans had not sought or expected. The American Administration had been willing to settle for more equipment, more instructors — basically symbols to show the Australian flag. Military or strategic aims, based on an assessment of the needs of South Vietnam, did not enter into consideration at all. Certainly, American pressure on Australia increased between November 1964 and April 1965, as Johnson personally involved himself more and more in the war. In the final analysis, 50,000 Australian soldiers went to Vietnam because the Menzies Government had nothing else to send, and decided it had to send something to get credit with Lyndon Baines Johnson.

On 25 October 1964, the Minister for the Army, Dr James Forbes, told the RSL National Congress in Hobart that while the Government did not have a 'closed mind on the subject' it had not introduced conscription because 'our military advisers have indicated in the clearest and most unmistakable terms that conscription is not the most effective way of creating the Army we need to meet the situation we face ... an Army composed entirely of long-term volunteers is a better one than one based on a mixture of volunteers and conscripts'. He assured the veterans, however, that an 'assessment is continuously undertaken by the Government'. On 10 November, the Government reintroduced conscription after a three-day defence review with the Military Board attending and advising. Menzies said, 'This is not just a politician's judgement.' The plan called for a call-up in 1965 of 4,200 men aged 20, chosen by ballot. Calwell described it as 'the lottery of death'. Calwell accepted Menzies' claim that 'the grave deterioration in the international situation since the last review in May' referred to Indonesia and her confrontation with newly-independent Malaysia. 'If Sukarno rattles the sabre, the Prime Minister will shake the lottery barrel at him.'

Until the end of 1964, Australia was much more concerned about the consequences of Indonesia's confrontation of the new Republic of Malaysia than about Vietnam. The sudden reversal of policy on conscription was justified by reference to confrontation. In the continuing controversy about the 1963 decision to buy the highly-sophisticated, highly-expensive F-111 bomber aircraft from the United States, the Indonesian situation was deemed to provide the clinching argument. The Menzies Government bought the F-111 in order to be able to bomb Jakarta.

The premature election of 1963 meant that a separate election for half the Senate was required in November 1964. Defence matters, particularly conscription, were central to the campaign, but Indonesia, not Indo-China, was the main talking-point. Australian diplomats and defence officials set out to convince the Menzies Government that confrontation and Indo-China were part of the same problem, and had the same solution. The problem was China; the solution was to place the might of the United States between China and Australia. If American intervention in Indonesia was out of the question, increased intervention in Indo-China was highly probable and would have the same result. If Australia took part, even a token part, in that increased intervention, there would almost certainly be an increased American interest in warning off Indonesia. The technicalities of the South East Asian Treaty Organization (SEATO) provided ample scope for entangling the United States in reciprocal obligations. Such were the arguments urged upon the Menzies Government in a stream of papers and cables written by Australian officials in late 1964. At no stage and at no level were the assumptions behind the theory, or the implications of carrying them through, subject to any rigid analysis. The meaning and nature of the war in Vietnam was never seriously considered. The argument which the officials put was immensely persuasive to a Government which needed little persuasion. American victory was automatically assumed, but what was meant by victory was never defined. The consequences of American failure were never raised, for failure was impossible. It no more occurred to Australian Ministers or officials to question whether America would win in Vietnam than it occurred to them to ask whether the F-111 bomber would be delivered on the date and at the price promised.

The process of conditioning the public to accept the new policy began in January 1965 with the visit of the Defence Minister, Senator Shane Paltridge, to South-East Asia. The *Sydney Morning Herald* obliged, on 20 January, with an editorial headed: 'Vietnam is our fight.' It said: 'It is to be hoped that the announcement of further aid which Senator Paltridge is expected to make when he visits Saigon will not just amount to the announcement of more "technical assistance". Limited as our military resources are, demanding as are the needs of the Malaysian crisis, South Vietnam is still our fight and it is past time that we showed a fuller practical recognition of this.' On the same day, the *Age* weighed in with: 'Self interest alone dictates a role for Australia to play in maintaining this unhappy country as a bulwark, no matter how shaky, against the downward sweep of China into South-East Asia. Confrontation may be closer to home, but the situation in South Vietnam poses a more immediate threat to Australia's strategic situation.' Sir Robert Menzies always acknowledged that the *Age* was his favourite paper; he failed to acknowledge its assistance in phrase-making. 'The downward sweep of China' was to become the settled justification for Australian intervention. As the *Age* said in its editorial of 20 January, referring to a current visit to Peking by Indonesia's Foreign Minister: 'Indonesia is free to pick its friends where it chooses, just as Australia has a duty to itself and to Malaysia to challenge confrontation. Dr Subandrio's presence in Peking is a reminder of what friendship and duty could mean if a collapsed South Vietnam gave China its first stepping stone towards Malaysia and Australia.' The officers of the Department of External Affairs had done their work well. They had established the grounds for the war they had so easily urged upon their political masters.

Menzies told Parliament on 4 May that the final decision to send the battalion was made on 7 April 1965. Full Cabinet was not consulted. On 20 April, the press revealed the decision. On 29 April, Menzies announced it to Parliament. The timing of the public announcement was determined by the premature press revelation. As a result of the press stories, a presumably grateful Government in Saigon learnt for the first time from the Australian Ambassador, Mr H. D. Anderson, that the Australian Government had decided to send a battalion of troops to its aid. Menzies decided that he must make his announcement on the night of 29 April to cut short the inevitable questioning and speculation. For the form of the thing, however, he needed an invitation from the Saigon Government now, briefly, headed by a Dr Quat. No request existed. It had to be procured.

Throughout the day of 29 April, cables of mounting alarm

from Canberra pressed upon the Saigon Embassy the need for the letter of invitation. Prime Minister Quat might have his doubts about the need but Prime Minister Menzies needed it. Menzies later explained to Parliament: 'I felt a great deal of embarrassment because the time has not quite arrived when I could feel that our relevant discussions had concluded. On Wednesday (28 April), a rumour was circulated that I would be making a statement on Thursday night (29 April) and that that statement would relate to the provision of an Australian battalion. At that time I literally did not know whether I could be ready by Thursday night.' In fact, he did not literally know whether he would be ready until 6 p.m., when a cable containing Dr Quat's 'invitation' finally reached and relieved Canberra. The tragedy and sacrifice of Australian intervention in Vietnam began with a characteristic element of farce, seldom absent, from our best or worst great occasions.

At 8 p.m. on 29 April, Menzies told the House of Representatives:

> The Australian Government is now in receipt of a request from the Government of South Vietnam for further military assistance.

The literal truth and essential dishonesty of that blessed 'now' was lost upon his audience. Menzies continued:

> We have decided – and this has been after close consultation with the Government of the United States – to provide an infantry battalion for service to South Vietnam. We decided in principle some weeks ago that we would be willing to do this if we had the necessary request from the Government of South Vietnam and the necessary collaboration from the United States . . . I am happy to tell the House that today I received from the President of the United States . . .

Johnson's letter began:

> Dear Mr Prime Minister,
> I am delighted at the decision of your Government to provide an infantry battalion for service in South Vietnam at the request of the Government of South Vietnam.

It was, though, no mere technical hitch that Johnson's letter was received in Canberra before Quat's so-called invitation. That was crucial to the whole operation. As late as 4 March, the United States State Department had instructed the United States Embassy in Saigon that the Government of South Vietnam was not to be informed or consulted on proposals to raise an international combat force. And even by 15 April, when only Australia, South Korea (as a mercenary force) and New Zealand (with deep and creditable misgivings) had agreed to fly their flags, the State Department cabled the American Embassy in Saigon: 'Do not tell Quat'.

The truth was implicit in the fact that Menzies could quote Johnson's letter, which *was* in his possession, but not Quat's letter, which was *not* in his possession. The Parliament and public did not learn of the contents of Quat's letter until six years later when the publication of the Pentagon Papers blew the lid off the whole affair. The letter, tabled by the then Prime Minister, William McMahon, stated:

'I wish to confirm my government's acceptance of this *offer* and to request the despatch of this force to Viet Nam.'

All the Australian records, published and unpublished, show that the decisions taken in Washington and Canberra were made without reference to the wishes of Saigon, much less at Saigon's urging. In the ordinary sense of the word, there was no request. Rather, the 'request' itself was requested.

Menzies 'embarrassment' led him into a gross parliamentary impropriety. The New South Wales State elections were due on Saturday 1 May. After nearly three decades of power, Labor was defeated, to return eleven years later to the day under Neville Wran. For personal and political reasons, Menzies wished to make his announcement after the New South Wales elections, preferably on Tuesday 4 May. Labor's defeat was important to him. While he had no doubts at all about the popularity of his decision, it was not the sort of issue to inject at the last minute into a State election campaign. The premature press stories pre-empted his choice of timing. On 29 April, Calwell and Whitlam were engaged for an election eve rally in Sydney. Alerted by the newspapers, Calwell sought throughout the day an answer to the simple question: was there to be a statement that night or not?

Further, if there was to be a statement, was it of such a nature to require the presence of the Leader of the Opposition or his Deputy? The harassed head of the Prime Minister's Department, Sir John Bunting, could only answer variously: perhaps yes; possibly no; probably not. At 5 p.m., it was still doubtful. Calwell and Whitlam went to Sydney. For Menzies, the impropriety was not unique. In April 1954, he had let Dr Evatt meet an engagement in Sydney, though he was that night to make the announcement of the defection of the Russian diplomat, Vladimir Petrov.

On the flight to Sydney, Calwell and Whitlam briefly, almost perfunctorily, discussed the Party's attitude to the forthcoming announcement. There was nothing much to discuss. There was no need for close discussion, because the Party's response was beyond question. The grotesque circumstances surrounding Menzies' announcement had no effect whatsoever on the Labor Party's attitude towards it. There was never any chance of bipartisanship. There was never the slightest doubt that the Labor Party would oppose the commitment of combat troops to Vietnam. No significant member of the Labor Party ever suggested that there could be any attitude other than opposition. The doubts and subsequent divisions within the Labor Party were about the role of the United States and the manner and timing of Australia's withdrawal.

Departing for Sydney, Calwell left behind a preliminary statement saying merely that parliamentary debate on the Prime Minister's statement must begin immediately, that is, on the following Tuesday afternoon when the House of Representatives next met, and that the Parliamentary Labor Party would meet on Tuesday morning to decide its attitude. The effect and the intended purpose of this statement was to preclude any meaningful discussion within the Party about its attitude. For there was no need for discussion. The only question was how Labor's opposition should be expressed. This was a matter only for the person who had to make the speech expressing that inevitable and automatic opposition. The physical task of preparing the speech was assigned to the author. Between Thursday and Monday 3 May, I spoke to two people about the speech: Dr J. F. Cairns, already the most concerned and best-informed member of the Parliamentary Labor Party about events in Vietnam, and John Menadue, then Whitlam's private secretary, and then and until 11 November 1975, whatever his position, an enduring source of sound and trustworthy advice. Nobody else was consulted about Calwell's speech. There was no need. On the morning of Tuesday 4 May, Calwell outlined to Caucus his general approach and without discussion his line was unanimously endorsed. After the meeting, two members of the Parliamentary Executive, Frank Stewart and Tony Luchetti, identified with the right wing of the Parliamentary Party, and recalling Korea, suggested that an explicit reference be made to Labor's willingness to participate in a United Nations peace-keeping force, should such a force ever be created. This was duly incorporated.

Calwell's speech of 4 May 1965 has a unique history. It was incorporated in its entirety into the platform of the Australian Labor Party by the Federal Conference in Sydney in August 1965. The speech was the source of Labor's basic continuing unity about Vietnam but also the source of continuing confusion about withdrawal. It succeeded in uniting the Labor Party against the commitment of Australian troops to Vietnam; it failed, by its silence on the subject, to get unity on the manner or timing of taking out the troops.

All the problems which were to face the United States in the future and all the problems which were to face the Australian Labor Party in the future were put in that speech. The two great problems for the Labor Party were the need to square its opposition to the commitment when that commitment was so clearly welcomed by the United States, and the problem of attacking a military decision without appearing unpatriotic. Calwell said:

> How long will it be before we are drawing upon our conscript youth to service these growing and endless requirements? Does the Government now say that conscripts will not be sent? If so, has it completely forgotten what it said about conscription last year?
>
> The basis of that decision was that the new conscripts would be completely integrated in the Regular Army. The voluntary

system was brought abruptly to an end. If the Government now says that conscripts will not be sent, this means that the 1st Battalion is never to be reinforced, replaced or replenished.

If this is not so, then the Government must have a new policy on the use of conscripts—a policy not yet announced. Or, if it has not changed its policy, the Government means that the 1st Battalion is not to be reinforced, replaced or replenished from the resources of the existing Regular Army. Which is it to be? There is now a commitment of 800. As the war drags on, who is to say that this will not go to 8,000, and that these will not be drawn from our voteless, conscripted twenty-year-olds?

Calwell's prediction came exactly true. He ended with an appeal to 'our fighting men who:

> are now by this decision, committed to the chances of war: our hearts and prayers are with you. Our minds and reason cannot support those who have made the decision to send you to this war, and we shall do our best to have that decision reversed. But we shall do our duty to the utmost in supporting you to do your duty. In terms of everything that an army in the field requires, we shall never deny you the aid and support that it is your right to expect in the service of your country. To the members of the Government, I say only this: if, by the process of misrepresentation of our motives, in which you are so expert, you try to further divide this nation for political purposes, yours will be a dreadful responsibility, and you will have taken a course which you will live to regret . . .
>
> I cannot promise you that easy popularity can be bought in times like these; nor are we looking for it . . . But if we are to have the courage of our convictions, then we must do our best to make the voice heard. I offer you the probability that you will be traduced, that your motives will be misrepresented, that your patriotism will be impugned, that your courage will be called into question. But I also offer you the sure and certain knowledge that we will be vindicated: that generations to come will record with gratitude that when a reckless Government wilfully endangered the security of this nation, the voice of the Australian Labor Party was heard, strong and clear, on the side of sanity and in the cause of humanity, and in the interests of Australia's security . . . We believe that America must not be humiliated and must not be forced to withdraw. But we are convinced that sooner or later the

dispute in Vietnam must be settled through the councils of the United Nations. If it is necessary to back with a peace force the authority of the United Nations, we would support Australian participation to the hilt. But we believe that the military involvement in the present form decided on by the Australian Government represents a threat to Australia's standing in Asia, to our power for good in Asia and above all to the security of this nation.'

At no time from 1965 on did Labor think that it was on an electoral winner on Vietnam. The political problem was how to present its opposition in such a way as to lessen its obvious electoral disadvantage. At least until 1968, the war in Vietnam was always popular with a majority of Australians. Menzies himself best defined Labor's dilemma with the courtly brutality which was the essence of his style and his character in his speech replying to Calwell on 4 May 1965:

> I recognized the somewhat pathetic note in the honorable member's speech when he turned to his own people and said metaphorically and literally: 'We will be unpopular but we will stick to it. You must remember that we are ready to suffer in an unpopular cause! All I can say is that I wish he were willing to suffer in a good cause of his own, because I have not the slightest doubt that on the merits not only we in Australia, but also all those governments and people with whom we are associated in this tremendously important exercise which is so significant for the security of our own country, are on the side of the great majority. If I may end on a horribly political note, it is a good thing occasionally to be in a big majority.

Menzies speech of 4 May, explaining his announcement of 29 April, was one of exceptional power. That speech contained the rationale for the origins of the 'request' from Dr Quat. He was as anxious as the Labor Party to invoke the United Nations. He said: 'It is because we believe that there has been a breach of international law (by North Vietnam) and a violation of the Charter of the United Nations that we have in relation to the present matter, notified the President of the Security Council of our decision.' He added to those words:

> This decision has been made at the request of the Government of the Republic of South Vietnam and it is in accordance with Australia's international obligations. That is the formal ground on which we stand.

In other words, the charade of 28–29 April 1965 was an elaborate exercise in building up a lawyer's case for a political decision which, among other things, was to lead directly to the death of 500 Australian soldiers.

Menzies, with fierce finesse, homed in on Labor's central political difficulty — the question of Australia's relations with the United States.

> Does the Leader of the Opposition really believe that the United States of America, of whose actions he has approved and re-approved, ought to be allowed to continue to carry this burden and that we, as one of the SEATO powers, with South Vietnam requesting our help, should say: 'Sorry, there is nothing we can do about it?' This is a very serious position, I venture to say, for the Leader of the Opposition to get into. It is certainly not a position that we want to get into. It is in the continuing interest of this country — to put it on no higher ground than that — to be regarded and to remain as a valued ally of the United States, which is, in this part of the world, our most powerful ally. I would hate to be the head of a government which had to say to the United States on an occasion like this: 'Sorry, we can do nothing about it. We will help you with debate in the United Nations. We will offer some fine words and some good sentiments. But, as for practical action "No. That is for you." American soldiers from the Middle West can go and fight and die in South Vietnam, but that is not for us.' I think that is a disastrous proposition for any opposition to put forward.

In all the hundreds of columns of Hansard which Menzies occupies, there is no paragraph which better shows his skill and mastery in debate and better explains his power over the Australian electorate. More than a decade later, when every premise of that paragraph has been exploded, and when two of its central points — the 'request' and the invocation of SEATO — have been exposed as disingenuous, the diabolic power of the thing shows through: the statement of the obvious joined with the statement of high principle; self-interest joined with highest obligations; above all, the jugular technique of going straight for the most sensitive part of his opponent's case.

The grand simplications by Menzies gained further power from the fact that the Labor Party in its efforts to straddle public opinion had already half conceded their truth. In February 1965, the Parliamentary Foreign Affairs Committee met in Calwell's Sydney office, under the chairmanship of Allan Fraser, MP. The Committee, including Cairns, accepted a declaration drafted by Kim Beazley, MP, which basically supported the American position. It stated: 'In its statement to the Security Council on 7 February, reporting air strikes against military installations in the south of North Vietnam, America insisted that its object in South Vietnam, while resisting aggression, is to achieve a peaceful settlement maintained by the presence of international peace-keeping machinery and that it would not allow the situation to be changed by terror and violence. *This statement of American purposes is unexceptionable.*'

Calwell welcomed this statement with unqualified delight, dictating it himself to Party secretary Wyndham in Canberra to ensure maximum distribution under party auspices. Naturally, Menzies used this pronouncement with great effect in his speech of 4 May. He seized upon the basic difficulty the Labor Party had in reaching an unambiguous attitude about Vietnam. Labor's ambivalence persisted until 1969.

Until the Tet offensive of January 1968, American official statements about American purposes were accepted in good faith. The Party platform itself gave primacy to the American alliance. This was the reference point of all policy statements made on Vietnam up to 1972. There could be no outright denunciation of the United States. Labor sought to solve its dilemma by advancing the concept of two Americas, and the 'real' interests of the United States as opposed to short-term Administration interests. In 1965, this was extraordinarily difficult to sell, but after the Tet offensive in 1968, it became increasingly credible.

In one sense, the commitment of Australian forces in strength solved Labor's policy problem, because there was no difficulty at all in reaching a firm and clear position on that issue. Once

Australian forces were there, a new difficulty arose — how and when to get them out. But withdrawal was never the stuff of which splits are made. Once the decision to commit combat troops was made, it was the end of any chance of a serious split, or significant defections, over Vietnam.

Calwell, Whitlam and Cairns each approached the problem differently. Calwell probably had the greatest emotional turmoil of anybody. The threat from the North, the fear of China, the persecution of Catholics, racial contempt for Asians, the dream of the American century, were all things to which he had a gut attachment. His release from doubt became complete only when Harold Holt sent conscripts to Vietnam in 1966. From that time on, Calwell threw himself into the fight with a passion sufficient to make up for any past shortcomings. Cairns drew inspiration, encouragement and example from the small but growing movement of protest within the United States. On this great issue, he saw further and clearer than anybody in the Labor Party. Like all his colleagues, he was inhibited from outright denunciation of the United States. By taking charge of the popular protest, he ensured that it did not become purely a channel for anti-Americanism. Under his leadership, the Moratorium movement developed into the most sophisticated and disciplined instrument of popular protest ever fashioned in Australia.

Whitlam is on record as the first Australian politician to warn against Australian collaboration in American intervention in Indo-China; not in 1964 but in 1954. The occasion is revealing about the times and the people involved. The Geneva Conference on Indo-China was about to convene. The French Army was besieged but still undefeated at Diem Bien Phu. The Australian elections were due on 8 May 1954, with the opinion poll strongly indicating a Labor victory. On 8 April, a letter signed by Professor C. P. Fitzgerald, one of the world's leading sinologists, Professor Manning Clark, Australia's greatest historian, and the Anglican Bishop of Goulburn and Canberra, Dr Burgmann, appeared in the *Age*, and the *Canberra Times*. Their letter pointed out, moderately enough, that the Vietminh forces were nationalist as much as communist. On the adjournment debate that evening, William Charles Wentworth, the Parliament's chief 'red-baiter', denounced the letter as 'partly designed to paralyse Australian policy with regard to Indo-China'. Calwell attacked Wentworth for 'the sort of McCarthyism that is creeping into this country'. He said: 'Honourable members have their own views about the Vietminh. They desire to see peace brought to Vietnam. They want to see a community established there with the aid of the French Republic that will be of lasting value to the people of Vietnam and save them from incorporation behind the bamboo curtain, the iron curtain or any other part of the Communist world.'

Whitlam followed Calwell saying:

> I am confident that . . . the vast majority of the Australian people believe that before Australia becomes involved in Indo-China the whole matter should be thrashed out not between the Americans and the French on one hand and the Vietminh and Chinese People's Republic on the other, but by the United Nations . . . There is a threat to the peace of the world in Indo-China . . . if the peace of the world is to be maintained it should be maintained through the United Nations. If the United Nations says that we should intervene with police action, then in my view Australian troops should go to Indo-China. But if the United Nations does not issue any such verdict we should keep out of the matter.

The Minister for External Affairs, R. G. Casey, made a violent riposte, singling out Whitlam's remarks:

> The article finished on a note which would lead uninformed people to the belief that in the opinion of the four distinguished writers, the Vietminh movement was nationalist and not communist. That is quite simple and unequivocal and I say that it is wholly and absolutely untrue . . . I condemn them for what they have done, and I condemn the honourable member for Werriwa for the speech in which he supported them tonight . . . It is easy for honourable members opposite to condemn the United States of America, as they frequently do, in one form of words or another . . . The honourable member for Werriwa is a fairly typical well-informed, educated example of the exponent of that form of snide international politics. I hope he is proud of himself for what he has done . . . The United States of America is on our side. It is on the

side of democracy, decency and right, and the forces of darkness opposed to it are very apparent and very powerful. The world may have a show-down at any time between our form of life and the forces of darkness . . . I think all decent people — and I direct this remark largely to the honourable member for Werriwa — should decide in their minds where they stand, where the interests of their country, of their children, if they have any, and of all democratic countries lies, instead of playing at party politics snatched out of the air of a very touchy international situation.

Casey later wrote in his diary on 18 April 1954:

> A prominent Frenchman said to me a year ago that there was no military solution to Indo-China's problem but that the only solution was a political one. This visit to Saigon has renewed this statement in my mind and I tend to believe it is largely true.
>
> I have been impressed by the inter-action of military and political factors in Indo-China. There seems little doubt that the majority of Vietnamese tend to be pro-Viet Minh when they have any political views one way or the other.

As a result of his speech in 1954, Whitlam was given an early insight into the emotions behind the issues on which the Labor Party would split in 1955. John Michael Mullens, one of the seven Federal parliamentarians to be expelled and to form the Anti-Communist Labor Party (later the Democratic Labor Party) in 1955, confronted Whitlam in the Caucus room where they had adjoining lockers. He said, 'That's all very logical, Gough, but I can't forget that there are two million of my fellow-Catholics there who are going to be massacred.' The assumption that a blood-bath, particularly of Catholics, would automatically follow a communist victory in Vietnam persisted until 1975, and became the last line of defence for foreign intervention in Vietnam.

Whitlam was also the first Australian parliamentarian to perceive the significance of Congressional opinion about Vietnam. He was the first to try to bring home to Australians that neither United States policy-making nor United States opinion was monolithic: Congress was also important. In 1966, the point made little headway. The United States was the President of the United States and United States policy was what the President said it was. Visiting Washington in 1966, Whitlam learnt that at least half the Senators were deeply uneasy about Vietnam and perhaps one third were opposed outright. Congressional opposition was very slow to crystallize. This was very largely because opposition came principally from Democrats and the Democrats were in power. The war in Vietnam was a war of the Democratic Party. The 'best and the brightest' were mainly Democrats, serving Democratic presidents, using Democratic rhetoric. It was not until December 1967 that Senator Eugene McCarthy declared himself as a candidate for the Presidency and gave national leadership to the anti-war Democrats. It was not until March 1968, when Senator Robert Kennedy made his presidential bid, beginning in ambiguity and ending in tragedy, that Whitlam's claims about the depth of American anti-war feeling were really heeded or accepted in Australia. Until 1968, the big difference between the anti-war movement in Australia and in the United States was this: in Australia, throughout the war, a major party officially and openly provided the main channel and the natural vehicle for opposition; in the United States, until Richard Nixon became President, the responsibility for the war and opposition to the war lay with the same party. American Democrats faced a conflict of loyalties in a way Australian Labor Party supporters did not.

In only one case did conflict of loyalties lead to political penalty—of a sort. In August 1966, Captain Sam Benson, member for Batman, then a marginally safe Labor Melbourne seat, a merchant seaman of marked gentleness and decency, automatically expelled himself for his refusal to resign from the distinctly hawkish Defend Australia Committee, which had been proscribed by the Federal Executive. Neither side was disposed to compromise in a situation where there was plenty of room for compromise. But Benson was the only Victorian member to express outright support for Whitlam in his abortive 'spill' attempt of April. At the same time, Benson could not be blind to the fact that, as things were shaping, he would have at least as good a chance of retaining his seat as an Independent as the official Labor candidate. So in fact it turned out. He was expelled from the Labor Party and retained his seat in 1966.

In the long-term, the most important group of opponents to the war were Liberal supporters. In 1966, some of them formed the Liberal Reform Group, stating in their manifesto:

> A great many Liberals do not support the Vietnamese war nor the Government's use of selective conscription to raise the troops to fight it. On the other hand few of them regard with enthusiasm the alternative of voting for the Labor Party.

This group was the forerunner of the Australia Party. More importantly it was the bridge over which a new generation of middle-class Australians were to cross from the conservative to the radical side of Australian politics.

On 21 June 1966, Peter Raymond Kocan, aged 19, attended a meeting at Mosman Town Hall, Sydney, at which Calwell was the principal speaker. It was not specifically a Vietnam protest rally, although Calwell took conscription for Vietnam as his topic. After the meeting Calwell saw this young man running towards his car. 'I thought he might just be another well-wisher who was making a belated effort to tell me how much he supported the Labor Party's attitude to Vietnam,' Calwell recorded. 'So I started to wind down the car window to shake hands with him. But before I could get the window open, there was a loud bang and I suddenly realized a shot had been fired at me. The thick glass window shattered into particles and I felt a stinging sensation in the lower part of my face.' This was the first assassination attempt on an Australian political leader. Kocan told police that he had decided to shoot someone important because he wanted to prove that he was different from 'all the other nobodies'. While it is clear that the assassination attempt was not directly politically motivated, it forever changed attitudes towards the security of Australian political leaders. Previously, Prime Ministers had received perfunctory security and Leaders of Opposition none at all, even during election campaigns. After 21 June 1966, nothing would ever be quite the same again.

The real tightening of Australian security consciousness came with the visit of President Johnson in October 1966. In Canberra, Sydney, Melbourne, Brisbane and Townsville, the security resources of two nations were pressed to the limit for his protection. In Melbourne, two young men managed to break through the screen to throw a can of paint over the Presidential limousine, flown out for the tour. In Sydney, Premier Askin endeared himself to the President and immortalized himself by instructing the driver to 'run over the bastards' who lay in the path of the car. The visit was a political triumph; for both Johnson and Harold Holt it represented the peak of public endorsement for their Vietnam policies. There was very little evidence that the Australians were embarrassed by Holt's declaration, given at the White House in July, 'All the way with LBJ'.

The Johnson visit opened the campaign for the House of Representatives elections on 26 November. The deliberate failure of the Labor Party to set out a fixed schedule for withdrawal of Australian troops had its inevitable consequence. There was no commonly-agreed formula. Any difference in wording about Labor's intention could easily be seized upon as a split. Calwell's policy speech referred only to conscripts. A statement by Whitlam a week before polling day implying that regular troops might be retained indefinitely was seized upon by the Government and press as further evidence of the split between Calwell and Whitlam. The real trouble lay in the Party's lack of precision on this crucial point. Calwell could not point to any policy declaration contravened by Whitlam's statement, Whitlam could point to any number of statements by Calwell which supported his interpretation of policy.

This incident merely confirmed the appearance of disarray which the Labor Party had given throughout 1966. Vietnam was not the decisive issue in November 1965. Calwell correctly identified the principal cause of the débâcle as 'the disunity in our ranks on personal and policy questions carried on during the whole lifetime of the 25th Parliament'. Vietnam was not a basic cause of that disunity. If anything, it was a source of unity. The leadership itself was the cause of disunity and disarray. The 1966 election was not a plebiscite on Vietnam. It was a decisive rejection of the leadership—not just of Calwell, for the people had decided that in 1963—but of the whole way the Party had developed in the previous decade, particularly in the relationship between the parliamentary leadership and the leadership of the Party organization.

Perhaps because he had come so late to the leadership, Calwell's concept of his task was essentially passive. He did not seek to try to impose his stamp either on policies or events. After 1963, he was much more interested in the honour and prestige of the title 'Leader of the Opposition' than in the use of its powers. He clung to the title because it conferred a nominal claim to an even higher honorific — Prime Minister of Australia. In his brighter days, he had often stopped admiringly before the bust of Alfred Deakin in King's Hall, Parliament House, to point out to young journalists that Deakin had refused to accept the Privy Councillorship normally conferred on Australian Prime Ministers; thus Deakin alone of Australian Prime Ministers could not call himself 'Right Honourable'. Yet in 1970, Calwell accepted a Privy Councillorship on the recommendation of a Liberal Prime Minister. Arthur Calwell spent most of his political career waiting — waiting to become a member of Parliament, waiting to become Leader of the Labor Party and waiting to become Prime Minister. Nevertheless, the plebiscite on the leadership of the Labor Party had been already taken. And, whether they knew it or not, the Australians in confirming Harold Holt as their continuing Prime Minister had also vested Gough Whitlam with the leadership of the Australian Labor Party.

4
The Leader

Edward Gough Whitlam was born in Kew, Melbourne, on 11 July 1916. His father, Harry Frederick Ernest Whitlam, was a middle-ranking officer in the Commonwealth Crown Solicitor's Office and had also served in the Victorian State Public Service. When Whitlam was two, the family moved to Sydney. When he was eleven, his father, by then Assistant Crown Solicitor to Sir Robert Garran, transferred to the new national capital, Canberra, and the family moved there in 1928. In Sydney, Whitlam attended a private preparatory school and Knox Grammar School. In Canberra, he attended Telopea Park High School, a government school, and Canberra Grammar School. From 1935, he lived at St Paul's College in the University of Sydney. In December 1941, he enlisted in the Royal Australian Air Force and four months later, in April 1942, he married Margaret Dovey, daughter of a leading Sydney barrister, later Justice Wilfred Dovey. In 1945, still in uniform, he joined the Australian Labor Party in Sydney. In 1946, he completed his law degree and was called to the Sydney bar. In 1950, he sought and won Labor pre-selection for the State seat of Sutherland in Sydney's outer suburban south, but failed to win the seat. In 1952, he was endorsed as the Labor candidate for the Federal seat of Werriwa to succeed H.P. Lazzarini, who proposed to retire at the 1954 Federal election. Lazzarini died in 1952 and Whitlam won the seat at the by-election on 29 November, significantly increasing Lazzarini's last majority. Whitlam took his seat in Federal Parliament on 17 February 1953.

Whitlam's background is typical of an Australian Labor leader. To put it another way, there is no typical background for a Labor leader. There has been no special set of qualifications. There would appear to be only one serious disqualification—high office in the trade union movement. None of the eleven Labor leaders sought power through the hierarchy of trade unionism. None has been a paid party official. The most powerful trade union leaders have either declined to enter Parliament or have declined in it. It has been unusual for Labor leaders to have been blue-collar workers. Chifley, a former engine-driver, was quite atypical. Perhaps for that reason he is regarded as the typical Labor leader. The only generalization that can be made is that each leader has been more or less representative of a significant class of Australians of his generation. As in most things, Hughes was an exception. In the company of Australian Labor leaders, he is the only exotic. The essentially middle-class nature of the Labor leadership has reflected the essentially middle-class nature of the Labor parliamentary parties, Federal and State.

However, it can be argued that of the outstanding Labor leaders—Watson, Fisher, Hughes, Curtin, Chifley, Evatt and Calwell—Whitlam is the most normal in the sense that there is the smallest gap between his background and his attainments. In this special sense, his career is the least remarkable of them all. Given his background, education, talents and reasonable good fortune, he could never have been less than a High Court judge, a permanent head of a Commonwealth department, or most likely of all, a Vice-Chancellor of a university. There was never any time when it would have been inherently improbable to say, 'This fellow will be Prime Minister.' What Whitlam could never have been was a captain of industry. He could never have succeeded, or dreamt of succeeding, in business as incorporated in Australia. The same background and cast of mind which closed off the option of a business career made it certain that if he followed a political career it would be inevitably on the Labor side.

In a B.B.C. interview with Lord Chalfont, a former Minister in the Wilson Labor Government, in September 1973, he gave an account of his political development:

My background in English terms would be that my parents

would have been people voting not Conservative or Labor but Liberal. In the Australian context they would vote Labor as the party of change and public responsibility—things being done by elected persons rather than by self-perpetuating directorates.

Chalfont: But can you put your finger on the moment or general time in your life when these ideas began to form?

Whitlam: During the war there was a period of very great flux in Australia. During the 1930s Australian politics were quite hopeless and federally, extraordinarily drab, futile. When the war came and particularly when there was a Labor Government elected in New South Wales under McKell and Labor came to office in Canberra under Curtin there was a real upsurge—exaltation. I suppose I would have been blooded during the referendum of 1944 to enlarge the authority and the jurisdiction of the Federal Parliament. I was appalled by the spurious arguments put against the referendum by the ancestors of the Liberal Party, the conservative elements in Australia. I thought, 'Well, this can't be allowed to continue.' I propagandized in my R.A.A.F. squadron in favour of the referendum. I was really very upset and discouraged when it was beaten. From then on, I decided to do something about it.

Chalfont: When you joined the ALP did you ever think that one day you would lead it?

Whitlam: Never. I don't think I thought I'd be in Parliament. I just wanted to be a supporter. It didn't occur to me that I'd end up in Parliament. These things snowball immensely. I wanted to support what it was trying to do. I was so disillusioned or alienated by what the non-Labor interests were doing, I thought, shallow and cynical, so I decided to help to give things a move along.

Whitlam's opponents have used his family background as a stick with which to beat him. Detractors within the Labor Party have used the tag 'silver-tail'; detractors outside the Labor Party have described him as a 'class traitor'. The real surprise would have been for Whitlam to have been attracted into the Menzies' 'free-enterprise' Liberals. Whitlam's crucial decision was to become politically active; once that decision was made, the choice for Labor was automatic.

Whitlam's family background and his father's career had three crucial influences on his thinking: on the role and nature of the Federal Government, the role and nature of the public service, and the problems of urban life in a new suburb. Speaking in November 1973, delivering the Robert Garran Memorial Lecture which his father inaugurated in 1959, Whitlam said:

> It is not without significance in my own career in my attitude to the Public Service, to the role of the Public Service, the duties and responsibilities of the Public Service and to the role of Government, that I lived my boyhood here in Canberra as the son of a great public servant, among whose colleagues were great public servants, and that I am Australia's first Prime Minister with that particular background.

He took certain propositions as self-evident and among these were: that Canberra was the political and administrative centre of Australia; that the national Parliament was the only really important parliament in Australia; that the role of government was constructive, positive and benevolent; that action by governments, through Parliament and the public service, was the normal and natural approach for the solution of Australian problems; that the public service was a creative, active, partner in the government of the nation; that senior public servants were by definition loyal, capable, creative and estimable. Other Australian politicians, conservative as well as Labor, have come to accept these propositions more or less. But Whitlam absorbed them as part of the early furnishing of his mind. A contemporary observer of Canberra, the journalist Warren Denning, wrote in 1937 in his brief classic, *Caucus Crisis—the Rise and Fall of the Scullin Government*: 'One feels that Canberra helped to ruin the Labor Government as it helped ruin the Bruce Government and as it will tend to ruin governments of the future, unless they become alive to this danger. Had the Scullin Ministry been compelled to live and work, from day to day, among the people whose destinies it was managing; had it seen the daily course of unemployment with its miseries; had it kept closer to the heart of the people whence it sprang, its story might have been a different one. In Canberra, it was all too fatally easy to forget the brutal realities of the world outside; too fatally easy to draw plans on paper, to make fine speeches to a crowded Press Gallery, in the knowledge that your words, your heroics, were being flashed to all the corners of the continent; too fatally easy to feel that these paper plans, these glowing words, stood for things done, for sufferings alleviated, for tragedies repaired, for hopes stimulated and fears allayed. In this city of trees and roses, politics tend to become all too easily, just a game.'

Forty years later, nobody involved in the Whitlam experience can read those words without the shock of recognition.

To be a boy growing up in Canberra was to live in the only place in Australia in the thirties where the community experience was in any way comparable to that to be shared by families in the post-war growth suburbs of Sydney and Melbourne in the fifties and sixties. The economic circumstances of Canberra in the thirties were quite different from those of the rest of Australia but its social circumstances were quite comparable to those of the Sydney western suburbs in the fifties and sixties. Whitlam's concern with the problems of creating a decent community life in new suburbs alienated from the metropolis owed much to his Canberra boyhood and adolescence.

Two other aspects of his character were drawn from his family background, and specifically from his father's influence — the pursuit of personal excellence and his contempt for race pride or prejudice. Whitlam is not an ambitious man in the sense of having any vaulting drive for power, but he is undoubtedly attracted to the idea of distinction and excellence in any matter to which he seriously addresses himself. Throughout his political career, this has sometimes been self-defeating, as the pursuit of excellence in a matter of detail can lead to wasted energy. Unlike Menzies, Whitlam has never accepted that deliberate laziness can often be the most productive course. His contempt for Menzies' gross simplifications can push him into pedantry. His standing instruction and his standard complaint to people preparing speech material for him is: 'More matter and less art.'

Inaugurating the Robert Garran Lecture fourteen years before his son spoke in the same series, Fred Whitlam said:

The task before Australia is honourable and its efficient discharge would make for a dynamic peace; to it, all the resources, skills and energy that Australia can command deserve to be committed. The honourable task, however, could become majestic and infinitely inspiring, and the peace could become creative, deep and rich and enduring, if there be added what I have termed Excellence. Excellence in all its fullness.

Without attempting any psychological analysis of the relationship between father and son, we may safely conclude that in the Whitlam household, words and concepts like peace, honour, efficiency, skills, creativity and excellence were used with some meaning and without embarrassment.

The household headed by Fred Whitlam was full of good cooking, good books and good public servants; it was notably lacking in drink, tobacco and profanity. Ever since, Whitlam has been at home with good food, good books and almost any public servant and uncomfortable with drinking, smoking and swearing. He adjusted quite well to the fact that he would have to be closely associated with men and women who would smoke heavily and work in a system in which heavy drinking was the needful solvent for most activity. He proved far less able to handle profanity, which he discovered as a means of self-expression relatively late in life. His reaction to a blue joke is not unlike that of a fourteen-year-old boy at Canberra Grammar School circa 1930, a kind of guilty chortling to hide his basic embarrassment and puzzlement. He is totally incapable of remembering or telling one.

A remarkably perceptive valedictory appeared in 1942 in the *Pauline*, the College journal of which Whitlam had been editor:

> E. G. WHITLAM, B.A. (1935–Lent, 1942) — Divinely tall, and more divinely dark and handsome, it is unfortunate that Gough could not claim direct descent from Olympus. Red Hill was not quite the same thing, but he made the most of his origin from Canberra's most exclusive quarter. Rather too perfect sartorially, Gough would admit that perhaps Robert Taylor did resemble him slightly. His manner and attitude to life was reminiscent of Pooh-Bah. A continuous and fairly successful effort to conceal his real contempt for his fellows was Gough's constant practice. He

was above all a classical scholar and ordered his life according to this conception of values. From his deep reading and close study of modern arbiters of fashion and manners he developed a sense of what to do on any possible formal occasion. We will all miss his timely sallies in general meetings and debates. Few men have done more for the College during their stay here. His activities were legion and included the positions of Chapel Warden, Honorary Secretary and Senior Student, and we will all remember the occasions when he would put on 'di V' and lend his presence to the revue. In the sporting line he was a stylist, as might be expected; he filled five seats very prettily in the College Eight, and also rowed in two Challenge Crews. Every edition of *The Pauline* bore unspeaking witness to his painstaking care, for it is as Editor that most of us will remember Gough. Consistent hard work and attention to detail put his issues among the best on record. We should mention specifically his work in systematizing the College Records. The foundations he laid were such that the continuance of his methods will ensure the keeping of a detailed record of the College for all time. No account of the man would be complete if we omitted to mention Gough's ability to sum up his fellow men. His opinions concerning notabilities of the day and current world events were always of interest, as not only did he bring to the study of them a broad classical background, but he was singularly free from dogma and partisanship. It is all the more to be regretted that possessing this insight he had not a greater gift in his relationships with other men. When he completes his service with the Air Force we know that his spotless appearance, his acute brain, and his mastery of the English language, plus the strongest possible personal links with the profession, will assure him of success in his chosen vocation.

His war-time service as an Air Force navigator was not for Whitlam, as it was for so many of his generation, the dominant experience of his life, though it was broadening and maturing, particularly in his understanding of his fellow Australians. In his mental and political development, the failure of the 1944 referendum was undoubtedly the most important of his war-related experiences. He said in his Curtin Memorial Lecture in 1961:

> My interest in constitutional matters stems from the time when John Curtin was Prime Minister. The Commonwealth Parliament's powers were then at their most ample and it was constitutionally if not always politically more open to a Labor Government to carry out its policies than it is in peace time. John Curtin, however, saw that he was presiding over a passing phase. He was not content with the paradox that the Labor Party was free to enact its policies in times of war alone. Accordingly, in 1944, he sponsored a referendum to give the Federal Parliament post-war powers. His motives for holding the referendum were based on patriotism and experience. He argued the case with his full logic and eloquence. The opposition to the referendum was spurious and selfish. The arguments were false. My hopes were dashed by the outcome and from that moment I determined to do all I could to modernize the Australian Constitution.

The outstanding characteristic of Whitlam's thinking is its consistent growth. Consistency is by no means the highest virtue in politics, but Whitlam's capacity to learn and to grow has saved him from narrowness. In his development, the key word is 'relevance'. His consistency lies in a deliberate and continuing effort to ensure that his position on any matter should be relevant to his previous positions. His growth comes from an insistence that the statement of a position should be relevant to actual circumstances and practical needs.

The essence of his method lies in a continuing search for a formula, the form of words which will say exactly what he means and which embodies a plan for practical action. The process of perfecting the formula has sometimes taken years. In some cases, on some issues, the search for the formula is still on. The process began publicly with his maiden speech on 19 March 1953. Many of the issues he raised were those which would preoccupy him for the next twenty years: Commonwealth-State relations, Constitutional reform, electoral equality, housing, schools, hospitals, and above all, the problems of the new suburbs, of which his own electorate of Werriwa was the prime example. That constituency, in its narrower sense, was the microcosm of the wider constituency to which he would appeal in 1969, 1972 and 1974 and upon whose support his success would always depend.

While Whitlam was learning all the time he was also teaching. In his interview with Lord Chalfont he said: 'A public meeting is part of a continuous educative process that politicians have to engage in.' This is an unusual, not to say idiosyncratic, view of public meetings. It is nonetheless a seriously held view. Of Gladstone, Walter Bagehot wrote:

> He has the didactic impulse. He has the courage of his ideas. He will convince the audience. He knows an argument which will be effective, he has an enthusiasm which he feels will rouse the apathetic, a demonstration which he thinks must convert the incredulous, an illustration which he hopes will drive his meaning even into the heads of the stolid. At any rate, he will try. He is sure, if they only knew what he knows, they would feel as he feels, and believe as he believes.

This is an exact description of Whitlam's method; but it can, at times, come close to pedantry. 'He kept on refining while others thought of dining' or, at least, while some of them might be keeping an anxious eye on the clock as it neared closing time. But the seriousness and intellectual respect with which Whitlam approached his audiences gained him a growing reputation within the Labor movement in the years following the Split of 1955. Within the rank-and-file membership he appealed in particular to two groups: those who were wearied with the barren, backward-looking recriminations of the leadership and those who were concerned that Labor policies as they stood offered no relevant or realistic alternative to Liberalism. With those members and supporters, he shared a deep frustration.

> I was concerned (in 1957) by the way in which the Labor Party's failure to move on, to look ahead, to attempt to find new ways towards reform was shortchanging the Australian people and shortchanging the Party itself. In 1957, the Platform, policies and structure of the Australian Labor Party had remained basically unchanged since the days of the Chifley Government . . . The Party became obsessed with the idea that rather than being about

renewal for the future, its purpose was to return to a more comfortable past — not renovation but mere restoration. As a result, both the achievements of the past and the hopes for the future receded equally. The Party stagnated and the Platform was stultified.

In developing his own thinking, in developing policies and later in getting them accepted as Party policy, Whitlam's aim was to find a definition of democratic socialism which could be translated into practical action, and more specifically, to find a practical alternative to nationalization.

> The advocacy of policies and methods which we know the Constitution does not permit is a fraud on our supporters and a boon to our opponents. It will still secure cheers but it will still not produce results.

Earlier, for example in his Chifley Lecture of 1957, he was pessimistic about the prospects for public enterprise until the Constitution could be changed.

> The way of the reformer is hard in Australia. Our Parliaments work within a constitutional framework which enshrines Liberal policy and bans Labor policy. Labor has to persuade the electorate to take two steps before it can implement its reforms: first, to elect a Labor Government, then to alter the Constitution.

Ironically, it was Menzies who helped show him a way through by his act of directing Commonwealth aid to universities. this had been done in 1958 through Section 96, empowering the Australian Parliaments to 'grant financial assistance to any State on such terms and conditions as the Parliament thinks fit'. Whitlam was the first to realize the potency of Section 96 as a means of achieving Labor objectives. In 1961, he proclaimed:

> In our obsession with Section 92 which is held up as the bulwark of private enterprise, we forget Section 96, which is the charter of public enterprise.

If Section 96 had been forgotten in the past, he was determined that it would not be again.

It has to be emphasized that Whitlam, less than any Labor

leader except Chifley, was to use the Constitution as an alibi for failure to try to implement Labor policies. He was as contemptuous as E. J. Ward had been, for different reasons, of Calwell's undertaking in 1961 not to 'raise the question of nationalization'. To Ward, that was betrayal; to Whitlam, it was stupid and shallow, a confession of intellectual bankruptcy. Clearly the nationalization plank as it stood in the Party platform was anachronistic and irrelevant, placing the nationalization of the banking system on the same footing as nationalization of sugar refineries, and it was counter-productive where it called for the nationalization of health as an excuse for not working out a health programme. It was self-defeating and defeatist to say that that was the end of the matter. Nor was the preamble to the Party platform especially helpful. It was written by Kim Beazley in 1957 and remains an unexceptional expression of excellent sentiments excellently expressed. In the context of 1957 it served its intended purpose to prove that the Australian Labor Party was anti-communist. There is little in it which would today cause surprise or difficulty if the words 'Liberal' and 'Labor' were interchanged.

Whitlam's answer to Labor's dilemma was given expression in 1961 in these terms:

> The Australian Labor Party's attitude to more equitable and efficient regulation of the economy is often thought or alleged to centre chiefly or solely on nationalization. The Party's objective is 'the democratic socialization of industry, production, distribution and exchange to the extent necessary to eliminate exploitation and other anti-social aspects in those fields'. This is a limited, negative and apologetic definition which makes little allowance for the creative scope of socialist measures. Socialists should not be content with nationalizing where necessary; they should be intent on competing where possible and initiating where desirable.
>
> A more fruitful and complete use can be made of Australia's human and material resources through the initiation of public enterprise than the regulation of private enterprise. The Australian government is as constitutionally free as any other national government to initiate public enterprise internally or internationally. Public enterprise is not only the best but probably the only means of now staving off or counteracting private

monopoly in Australia and providing continued competition where there is still competition.

Whitlam's proposals were not put up to tempt the Labor Party away from the virtuous path of nationalization. Since the failure of bank nationalization, enthusiasm for the nationalization plank was decidedly low. Given that Section 92 prohibited direct nationalization, Whitlam saw two courses: to change the Constitution or to develop policies of social reform within the existing Constitution to achieve at least some of the purposes of nationalization. The first step towards either or both of these courses was to form a Labor Government. Up to 1958, the first course occupied most of his attention. After Labor's failure in the 1958 election, he set the second course as the first priority. He recognized that this would require a substantial rewriting of the Party platform. The reason for this should be made very clear. The chief fault of the platform was not that it was electorally unattractive. Its fault lay in its irrelevance. It had become irrelevant to winning, or losing, elections. More importantly, it had become irrelevant as a plan of action for any Labor Government. In framing their election policy speeches, Evatt in 1954, 1955 and 1958 and Calwell in 1961 and 1963 largely ignored the printed platform. They were forced to do this because the printed platform was quite unhelpful as a guide to action. It had become a mass of unrelated proposals, reflecting the interests, concerns or whims of Labor conferences going back to the 1920s and beyond. In matters like education, health and foreign policy, it was far clearer on what should not be done than on what should be done.

Australian Marxists despise the ALP for not being what it has never claimed to be. As far back as 1913 Lenin wrote: 'Those liberals in Europe and in Russia who try to teach the people the needlessness of class-war by the example of the (Australian Labor Party) only deceive themselves and others ... the Australian Labor Party does not even claim to be a socialist party. As a matter of fact it is a liberal-bourgeois party and the so-called Liberals in Australia are really conservatives.' Accepting that analysis, Lenin would have placed Whitlam firmly in the tradition of Watson and Fisher. As Whitlam came to gain a wider national audience, he increasingly attracted criticism from the young, able and articulate Marxist economists. Their basic criticism was, and is, that Whitlam was not a socialist but a social democrat, that his programme was not socialism but liberal reformism. Within the Marxist framework, that is quite true. But in this, again, Whitlam is firmly in the Labor leadership tradition and in the mainstream of eighty years of social democracy in Australia. His departure is that he began in the 1960s to express the tradition in contemporary terms, in a more precise and comprehensive way than had been attempted at least since W. M. Hughes wrote *The Case for Labor* in the first decade. *The Case for Labor* was written as a defence against the 'socialist tiger' campaign launched in 1906 by Sir George Reid. Hughes' purpose was to show that Labor was not revolutionary but reformist and, above all, that Labor could be safely entrusted with government. It is perfectly accurate for Marxists to deny Labor's right to proclaim itself as a socialist party in any Marxist sense. It is totally inaccurate to accuse Whitlam of having tried to dilute the socialist element in Labor's policies. There was precious little to dilute. Marxist hostility to Whitlam as a leader has the same cause as right-wing hostility to him: by making Labor reformism more attractive to the voters, he threatened to make the Labor Party more successful.

The word most commonly applied to Whitlam is pragmatic. In fact, his thinking is not so much pragmatic as programmatic. He does not take up a day-to-day practical approach to each political question. He tries to relate each particular issue to a general, long-term frame, relevant to the likely future outcome of that particular issue. The essence of the programme he began to develop in the early sixties was the expansion of the public sector in order to promote equality.

He stated his underlying philosophy in this way:

> ... in modern communities, even the wealthiest family cannot provide its members with the best education, with the best medical treatment, the best environment, unaided by the community. Increasingly, the basic services and opportunities which determine the real standard of life of a family or an individual can only be provided by the community and only to the

extent to which the community is willing to provide them. Either the community provides them or they will not be provided at all. In the Australian context, this means that the community, through the national government, must finance them or they will not be financed at all.

This concept of equality—what I call positive equality—does not have as its goal equality of personal income. Its goal is greater equality of the services which the community provides. This approach not merely accepts the pluralistic nature of our system, with the private sector continuing to play the greater part in providing employment and growth; it positively requires private affluence to prevent public squalor. The approach is based on this concept: increasingly a citizen's real standard of living, the health of himself and his family, his children's opportunity for education and self-improvement, his access to employment opportunities, his ability to enjoy the nation's resources for recreation and cultural activity, his ability to participate in the decisions and actions of the community, are determined not so much by his income but by the availability and accessibility of the services which the community alone can provide and ensure.

The quality of life depends less on the things which individuals obtain for themselves and can purchase for themselves from their personal incomes and depends more on the things which the community provides for all its members from the combined resources of the community.

In looking to the national government and national power for the implementation of Labor's programme, Whitlam spoke not only from instinct and conviction but also from necessity. Throughout the creative years of Whitlam's programme when he was Leader of the Opposition, the State Labor Parties were politically impotent. After the defeat of the old Labor Government of New South Wales in 1965, the two great States seemed political dead ends for Labor for at least a generation. However weak Labor may have seemed nationally, however bleak its national prospects, Labor in the States looked weaker and bleaker. Every State Labor Government was defeated during this period, and in Victoria, New South Wales and Queensland, the State Labor Party never looked like gaining office. The ascendancy of Federal power was not only a doctrine; it was a political necessity. This should be remembered before the

Whitlam programme is damned as obsessively centralist.

There is one important reason why Whitlam's thinking turned on programmes of social reform rather than economic theory. From 1953, the end of the war in Korea, to 1970, when the real economic cost of the war in Vietnam began to bite deep, the Western economy shared a long period of sustained growth. It was easy to accept that the problem of full employment and inflation had been solved. The post-war recovery of Western Europe, the 'miracle' of Japan, and above all the economic and technological might of the United States created unlimited optimism. In Britain, West Germany and Australia, social democrats embraced the idea of indicative planning within the capitalist system. Whitlam summed up the prevailing belief: 'Socialists no longer have to ration scarcity but plan for abundance.'

The most common criticism of Whitlam's abilities has been that he lacks economic expertise. When this charge was first made early in his leadership, he was inclined to dismiss such critics by referring to the examples of Curtin and Menzies, both successful Prime Ministers without any special claims to economic skills. He could not accept that economic expertise was a necessary qualification for political leadership. Indeed, in Australia, it has not even been regarded as a necessary qualification for being Treasurer; there has never been any obvious relationship between the success of an Australian Treasurer and prior training in economics. In a parliamentary democracy within a capitalist system, economic decisions reflect political priorities and political choices. In the fifties and sixties, it was quite possible for politicians to believe that the days of the primacy of economics were past. These years were the hey-day of the idea of 'fine tuning', the belief that the only choice to be made was between a little more inflation or a little less employment. The philosopher's stone had been found at last: growth. For social democrats in Britain and Western Europe and liberals in the United States, growth was the new touchstone, solving the dilemma between the need for profits and the demands of public welfare within the 'mixed economy'. As the English writer, Henry Fairlie, wrote:

> The liberal and politically-minded economists had found one answer on which they could all agree: Growth. It became a prominent word about 1958; it was a clearer word by 1960; and it remained clear until the conservationists about 1968 began to say that it was not only a dirty word, but the cause of most of the dirt which was suffocating the planet . . . for most liberals, Growth was enough, a fine word for a platform; a gutteral to start it off, a roll to carry it forward, a long vowel to hold it, and a dental to cut it off.

The internal dilemmas of the social democrat parties in the late seventies are very much the result of the failure of the golden promise of growth as an idea and as a reality. But in the sixties, its appeal was overwhelming. In a sense, faith in growth was to blame for America's Vietnam calamity. Had he been forced to choose, Lyndon Johnson would certainly have chosen the Great Society ahead of the Vietnam escapade. He thought he could have both. All the expert advice coming to him told him he could have both. Growth would take care of everything. In Australia, the concept was never questioned. Everything in the Australian experience in the sixties encouraged the delusion that the ordinary, old-fashioned problems of economic management were solved. This was why Whitlam and all his contemporaries turned their attention to the development of programmes for social engineering rather than economic planning. Whitlam, however, always resented the assertion that he was not interested in economic matters. He would ask: 'Are Federal-State financial relations not economic matters; is the distribution of wealth not the economic question of our time or any time? What am I talking about on any of these things—education, health, cities, transport, if I am not talking about economic matters.' The truth is that he delved more deeply and, by his persistent questioning of Ministers, secured more information, in these years, on all economic matters than professors and journalists paid for that specific task all year round.

*

Whitlam's faith in Parliament as an instrument for change springs from his belief in Parliament as a living and effective institution. To dismiss Parliament as a rubber-stamp for the

Executive Government is to mistake its nature and purpose. Exactly 100 years before Whitlam became Leader of the Opposition in the Australian House of Representatives, Walter Bagehot set out the functions of the House of Commons:

> The main function of the House of Commons is one which we know quite well, though our common constitutional speech does not recognize it. The House of Commons is an electoral chamber, it is the assembly which chooses our president . . . The elective is now the most important function of the House of Commons . . .
>
> The second function of the House of Commons is what I may call an expressive function. It is its office to express the mind of the English people on all matters which come before it . . .
>
> The third function of Parliament is what I may call . . . the teaching function . . .
>
> Fourthly, the House of Commons has what may be called an informing function . . . it makes us hear what otherwise we should not . . .
>
> Lastly, there is the function of legislation . . .

Bagehot wrote at the height of the imagined Golden Age of the House of Commons (significantly, in this passage he uses Parliament and House of Commons interchangeably) during the years of the splendour of Gladstone and Disraeli. Nor did he claim that, even in 1867, the House of Commons carried out its functions efficiently, except by electing the Prime Minister-President. This definition of functions is still as relevant for the Australian House of Representatives as it was for the House of Commons a century ago. The growth of the power of the Executive, the bureaucratic explosion and the revolution in communications have left the functions of Parliament unchanged.

More than any other Australian parliamentarian, Whitlam has committed himself to and exploited Parliament's teaching and informing functions. For Menzies, Parliament was a stage. For Whitlam, it is the best platform in Australia. Television, radio, the press, party meetings, conferences, seminars, public meetings were all important, but Parliament remained the supreme forum. Of Whitlam's contemporaries, Menzies had a greater presence over the House; Ward was more feared;

McEwen commanded greater authority; Allan Fraser could be a better debater; Kim Beazley could be more eloquent; but in his recognition of Parliament's teaching and informing role, and in his sustained use of Parliament for the exposition of a cause, Whitlam can well lay claim to the title of the supreme parliamentarian of his time.

Whitlam's other claim to be a great parliamentarian rests on his advocacy of the two-party system. The Parliamentary system depends on the strength of the party system. The cause of the two-party system is the cause of responsible parliamentary government. The non-partisan parliamentarian is a contradiction in terms. By their rejection of Independents, the Australian people have instinctively recognized this. The paradox of the parliamentary party system is that while its essence is the adversary relationship between the two opposing parties (and there is always a certain artificiality in this) the system is a unifying force. In a nation like Australia, with few unifying bonds of shared memories, culture, institutions or traditions, the parties are the major force for unity. In Australia, partisanship is practical patriotism.

Above all, Whitlam saw Parliament as an instrument for the creation of equality. He is rare among Australian Labor politicians in taking the doctrine of equality seriously. He is unique among Australian politicians in taking Parliament seriously as the chief means of achieving equality. As a man, as a party man or as a politician, Whitlam cannot be understood except on the basis of his commitment to the concept of equality. Anyone who cares to follow the development of his thinking over twenty years will find that equality is the consistent theme. Yet, within the consistency, there remains an unresolved contradiction. For the equality which Whitlam envisages involves the existence of an élite, either the elected élite of Parliament or the appointed élite of the bureaucracy. In this lies the unresolved contradiction of Whitlam's political programme.

An early exposition of Whitlam's concept of Parliament as an instrument for equality was given at the 21st Summer School of the Australian Institute of Political Science in Canberra on 31 January 1955. It is more instructive because it was a reply to a defiant statement of conservatism by Professor F. A. Bland, formerly professor of Public Administration at Sydney University, then member for Warringah, bluest of blue-ribbon conservative seats on Sydney's North Shore and the father of public servant, Sir Henry Bland. Bland's paper expressed perfectly the proprietorial view of Parliament which was at the heart of the conservative refusal to accept the legitimacy of an elected Labor Government twenty years later. He said: 'The working of our scheme of Parliamentary Government has entirely changed with the advent of socialist parties offering electors an alternative philosophy and government. Until that occurred, parties could afford to bicker over policies because they were all agreed as to the nature and purpose of the State. A change of Government was merely a change of emphasis . . . Now there is no common ground between socialism and a government believing in the sanctity of private property. And so a change of Government is a change of direction, which must make for political instability.' Professor Bland went on to proclaim for conservatives a superior brand of loyalty in terms echoed twenty years later by Malcolm Fraser. Bland said: 'The developments in party government are a natural result of identifying politics with economics, the extension of the franchise, the size of electorates and the cost of conducting an electoral campaign. The identification of politics with economics has been especially disastrous. It has allowed the socialist parties in Australia to exact from their members an allegiance to the socialist movement which overrides all other loyalties, whether to the electorate, to the Government, to the Parliament, or even to the sanctity of the ballot box.'

In reply, Whitlam identified the conservative nature of the Federal system: 'The aspect of parliamentary government to which Professor Bland directed most attention was the Australian Federal system which he regards as a bulwark of liberty. I take issue with him at once. The United Kingdom and New Zealand, both unitary states, know no less freedom and no fewer liberties than Australia.' In what might serve as the epitaph on his own relations with the States as Prime Minister two decades later, he said: 'Federalism does not make for freedom so much as friction, or for liberty so much as for litigation ... The buck-passing which Professor Bland rightly condemned in bureaucrats and politicians is intensified under a Federal system, where the several governments can all find plausible excuses for evading or transferring responsibility.' Opposing Bland's extraordinary separation of politics and economics, Whitlam said: 'Parliament has been our great liberating force ... There is no freedom without equality. To redistribute and equalize liberty has been one of the principal functions of Parliament. Parliament alone can give equality of opportunity and thereby increase liberty for all. If we are to have economic equality of opportunity, which is the next stage in the advance of liberty, we must have effective parliamentary government and, accordingly, dispense with fetters on Parliament rather than contrive them. For every person whose liberty has been prejudiced by governmental action, there are many whose liberty has been enhanced.'

Whitlam then asserted that Parliament itself was deeply flawed as an instrument for equality as long as it remained itself based on inequality: 'Parliamentary government has the dual function of giving effect to the will of the majority and safeguarding the rights of the minorities. In Australia, parliamentary government is not adapted to give expression to the will of the majority in any of our seven parliaments or to safeguard the views of more than a very few minorities, and those entrenched ones, in those seven parliaments ... as long as we have seven parliaments instead of one, we should at least have the safeguard in our Constitution that electorates shall be of equal population.' Nineteen years later, Whitlam's effort to write the principle of equal representation into the Constitution failed.

Towards the end of 1975, in the tumultuous last weeks of his Prime Ministership, a sceptical member of the Labor Party's advertising agency asked Whitlam to give a concrete example of what he meant by equality. In a revealing response, Whitlam said: 'I want every kid to have a desk, with a lamp, and his own room to study.' Not so robust, perhaps, as the wish of Henri Quatre that every Frenchman should have a chicken in the pot. Yet, one can see what Whitlam is getting at. It is through the creation of the physical conditions for equality that the ideal of equality may be approached. The 'kid with a desk lamp', and the outrageous proposition that sewerage was a proper question for the national Parliament are part of Whitlam's concept of equality through parliamentary means.

After the catastrophe of 1975, it was natural that many in the Labor Party exhausted by the turbulence of recent years should yearn for a leader who would give the party peace and quiet. Clyde Cameron would call for even greater curbs on the powers and prerogatives of a Labor leader. The leadership of Chifley took on a golden afterglow, a period of calm, camaraderie and consensus. Such a picture calls for a considerable revision of the facts and much forgetfulness or ignorance of what actually happened. A government which included Evatt, Calwell, Ward and Dedman was not exactly a band of brothers; and while Chifley's style of leadership differed greatly from Whitlam's, it should be remembered that Chifley's decision to go ahead with bank nationalization was taken and announced without any Cabinet consultation at all. It was an assertion of personal authority far greater than anything Whitlam has ever attempted and its consequences reached further than any purely personal action of Whitlam's. It is always attractive for a political leader to dream of creating a lasting 'consensus' within the party or the nation. It is an impossible dream, because it denies the inevitability of conflict of interests and ideas. For a Labor leader it is doubly impossible, because a consensus, something on which it may be supposed that 70 per cent are agreed, must be essentially conservative.

In Whitlam's rise to the Labor leadership, the crucial step was winning the deputy leadership in 1960. The deputy leadership itself confers no right to succession. But it does confer prominence. If a deputy has the ability and will to use that prominence, he automatically becomes the leading candidate for the leadership. Being the leading candidate does not guarantee the succession, but it helps. Traditionally, the Parliamentary Labor Party has gone for the obviously outstanding candidate. Not since Curtin beat Frank Forde by one vote in 1935, has there been a real upset, a choice other than the expected one. On the day, Chifley, Evatt and Calwell were the easy favourites. And in that as in so much else, Whitlam was to follow the Labor tradition. Watson, Fisher, Hughes, Scullin, Chifley, Evatt and Calwell could each have stated that they were certain to be elected as leader whenever the appropriate elections were held. The difference between them and Whitlam was that, living in an age when the question was crudely pressed, he was honest enough to state clearly his expectations. Only in that sense did Whitlam believe or proclaim himself to be a man of destiny.

5
'The Party, the Policy, the People'

At a dinner party in Adelaide a few days before polling day 1966, Arthur Calwell and Clyde Cameron discussed the shape of the next Labor Cabinet, the Cabinet which Calwell at least had convinced himself would be elected on the coming Saturday. This kind of self-delusion by a politician on the edge of disaster is a necessary part of his psychological survival kit. Whatever mild disagreements there may have been between Calwell and Cameron in the allocation of portfolios, agreement was complete on the crucial matter: Whitlam would not remain deputy and would not be Minister for External Affairs. Miss Joan O'Donnell, Calwell's private secretary since his days as Minister for Immigration and one of that remarkable race of Australian women, now almost extinct, who served a single Minister or leader quietly, efficiently, faithfully, throughout a working lifetime, glumly surveyed this scene. 'You don't seem to be saying much, Joan,' said Calwell. 'No, because it's all damned nonsense.' Cameron, of course, knew as much; but he was pleased to encourage the Leader he knew to be doomed.

The results confirmed it. It is important to emphasize that the actual result repeated the findings of successive opinion polls for months previously. Throughout 1966, the polls showed support for Labor at 40 per cent. So it turned out. There is no reason to believe that the campaign itself improved or worsened Labor's position. The election had been lost in the way it was lost and to the extent it was lost in February 1966.

Calwell, who had declared himself a non-candidate for the leadership, delayed convening Caucus until 8 February 1967. That day's meeting in Canberra was delayed two hours by the disastrous bushfires of 6–8 February around Hobart. The meeting was further delayed from 2 p.m. to 4 p.m. to allow the arrival of Tasmanian members smoke-bound at Hobart airport. This short delay had a long-lasting consequence.

Whitlam won the leadership predictably and easily. Against Cairns (15 first preferences), Crean (12), Daly (6), Beazley (3), Whitlam had 32 first preferences. After the elimination of Daly and Beazley, Whitlam was declared elected with 39 against 29 for Cairns and Crean. In the ballot for deputy leadership, Whitlam's closest confidant, Lance Barnard, won against Cairns, 35 to 33. This narrow and rather surprising victory led to a certain complacency in the Whitlam-Barnard camp. Because of the late start in the voting, Caucus adjourned for dinner before proceeding to ballot for the positions of leader and deputy leader in the Senate. While the right relaxed and the centre celebrated, the left wing organized. When voting resumed, Senator Lionel Murphy, Q.C., defeated the incumbent Senate leader, Don Willesee, who had held the position since the previous February. Senator Sam Cohen, nominally at least associated with the left wing, was elected deputy leader. The sudden rise of Senator Murphy was to have enduring consequences in the later fortunes of Whitlam, of the Labor Party and ultimately of the Labor Government. It would not have occurred on 8 February 1967 if the balloting for the four principal offices had not been delayed and then interrupted because of the Tasmanian bushfires.

In politics, there are few more effective ways to damn a man with faint praise than to brand him with the label of 'the loyal lieutenant'. To Whitlam, Lance Barnard was indeed a loyal lieutenant, but he was much more than that. Typically, the lieutenant in politics is a dependant; his loyalty is to his master and patron to whom he owes his position. Such was the lieutenancy and loyalty of Harold Holt to Menzies. At no time and in no way did Barnard owe his position to Whitlam. The burden of debt lay rather the other way. It was Barnard who first suggested and first supported Whitlam for the deputy leadership in 1960. Then, in September 1965, he and his fellow Tasmanian, Ron Davies (Braddon), tried to whip up a Caucus revolt against Calwell. Whitlam was not directly involved but he was clearly the intended beneficiary. Throughout the horrible months of 1966, Barnard held the line for Whitlam. He won the deputyship not in the wake of Whitlam's victory but through his own hard-won and well-organized support. He differed from the typical loyal lieutenant in possessing a genuine power base of his own. In fact, he had two bases—in the Caucus itself and in the Tasmanian Labor Party machine. His family exercised great influence in Northern Tasmania. His father had been a Minister in the Chifley Government. In his personal support and regional associations, his ability to call upon old loyalties and established relationships at all levels of the Party, particularly in Tasmania, he was more the lord-lieutenant.

The Whitlam-Barnard partnership was unique in Australian politics. As Curtin's deputy, Frank Forde had served with equal loyalty but hardly with equal effectiveness. Barnard brought to the partnership an instinctive sense of what the Party, particularly the Caucus, would wear. He had closer social contact with more members and knew more about their moods and their gripes than Whitlam. The true measure of his loyalty is that he never attempted to convert his own support and strength in the Caucus into a claim for the leadership, though he would have claimed and won the succession if Whitlam had for any reason been removed. He had made the conscious decision as early as in 1959 that Whitlam was the man to lead the Labor Party to victory More than any person except Whitlam himself he made his own prophecy come true. The relationship says much about both men; it was equally creditable to both.

In his first statement as Leader of the Labor Party, Whitlam set down the basic principles from which he never thereafter swerved. He was saying exactly the same thing on 8 February 1967 that he was saying on 11 November 1975:

> It is a high privilege to be elected to lead the Labor Party at so decisive a time in its history and the history of the nation. Of course, it is, and has been for many decades, a cliché of political and editorial rhetoric that 'the next ten years' are to be the most important and decisive in our history. There are nevertheless some unusually urgent reasons why it now possesses more than the usual amount of truth. For the Labor Party, what is clearly at stake is its

future role within the Australian Parliamentary System. That a Labor Party — whether defined as a party in close association with some trade unions or as a democratic socialist party putting forward candidates for election to parliament — will long continue to survive cannot be a matter of doubt. But our actions in the next few years must determine whether it continues to survive as a truly effective parliamentary force, capable of governing and actually governing. Clearly the Labor Party cannot exist as a credible alternative government unless we are, and are seen to be, willing and capable of gaining and holding the power of government in 1969. We cannot afford an indefinite continuation of the defeats of the past eight elections. The people who work for us and vote for us year in and out expect, and are entitled to, something better. I believe the Australian people want and are entitled to something better than this. The parliamentary system, fundamentally a two-party system, cannot long survive without the existence of genuine prospects for a change of government. The alternative is not indeed a perpetuation of Liberal/Country Party Government; it is the collapse of the two-party system as we know it. So I deeply believe, and would believe even if this were the only reason, that the future success of the ALP and the future welfare of Australian democracy are deeply involved together.

The very completeness of Labor's election defeat gave a sense of purpose to Whitlam's supporters. The more romantic of those around him frequently asserted that the next three years would decide whether or not the Labor Party would continue to exist. The author was the chief exponent of this line. It may be that our sense of drama outran our understanding of the Labor Party. Nevertheless, the sense of drama, the idea of being involved in a great and decisive phase of Australian history, added to the zest with which we all approached the work. Certainly, everybody involved in the work needed all the encouragement they could get. The soft spots in the massive conservative victory were not at all apparent. That such a victory had been won without Menzies only confirmed the certainty that the Australian electorate had made a further shift to the right. Whitlam himself drew encouragement from Menzies' example after the 1943 landslide to Labor.

John Menadue, returning as Whitlam's private secretary after unsuccessfully contesting the rural seat of Hume (N.S.W.), sketched a timetable: 'This year, the party; next year, the policy; 1969—the people'. Of course, such formulae are not meant to be applied rigidly. It was, however, a handy way of giving a degree of coherence and order to the incoherent and disorderly practice of politics. It also accurately expressed the practical priorities. The three tasks of party reorganization, policy development and public acceptance went together and had to be carried on together. But in any ordering of priorities, the state of the party came first.

Menzies' jibe at the 'thirty-six faceless men' of the Federal Conference of the ALP was devastating. It was coined by the journalist Alan Reid after the Special Federal Conference which agonized over the proposal for an American naval communications base at North-West Cape for two disastrous days at the Kingston Hotel, Canberra, in April 1963. Photographs, also astutely organized by Reid, showing Calwell and Whitlam waiting outside the Kingston Hotel during the Conference emphasized the exclusion of the Parliamentary leadership from direct participation in the proceedings. The humiliation lay much more in the appearance than in the substance. Both Calwell and Whitlam had been active, and largely successful, in framing the Conference resolution which conditionally accepted the establishment of the base. But the rules clearly stated the Federal Conference was the supreme governing authority. There for all to see was the proof that the Parliamentary leaders were literally outside the decision-making process. As a result of this humiliation, it was agreed that henceforth the Parliamentary leader and the deputy leader could attend the two-yearly Conference and meetings of the Federal Executive and could speak, but not vote.

In theory, the exclusion of the Parliamentary leadership emphasized democratic control by the party membership, the rank and file. Large and representative, State Conference would elect a representative delegation of six from each State; each stage of the process was governed by the majority rule. Nothing could be more democratic; nothing could be more representative. That was the theory. In practice, the very group supposed to control the policy-makers and the politicians — the rank-and-file membership — was precisely the group totally excluded. The Conference had become controlled absolutely by persons who earned their income from the Labor movement. The Conference which met in Sydney in August 1965 may be taken as an example. Of the thirty-six delegates, fifteen were full-time union officials, six were full-time party officers, nine were Federal Parliamentarians and four were members of State Parliaments, one was the Lord Mayor of Brisbane, Clem Jones, one was a self-employed chemist, J. M. Berinson, later MHR for Swan (W.A.) and a Minister in the Whitlam Government. Most of the parliamentarians had been paid union officials or paid party officers or both. The delegates as a whole were altogether unrepresentative of the Party as a whole. Those delegates who happened to be members of Parliament were unrepresentative of their parliamentary peers, taken as a whole. The dominance of paid officials was even more pronounced in the Federal Executive. This economic monolithism, far more than any ideological considerations, determined the character of the Labor Party's decision-making processes.

The need for reform had been long accepted in principle. Ironically, the politically disastrous Federal Executive meeting in Adelaide in October 1963 had taken the first important and constructive step towards reform. For the first time in its history, the Labor Party appointed a full-time national secretary, with headquarters in the national capital. The Executive set up a National Organization and Planning Committee. In August 1964, the Executive instructed the national secretary to carry out a full review of the 'Federal structure' of the Party, thus recognizing that the party which made special claim to be national did not itself have a really national structure. By February 1966, the Labor Party, for the first time in its history, had a national leader and a national secretary both deeply committed to party reform.

The first full-time national secretary was Cyril Wyndham. In 1958, Wyndham, a young officer at Transport House, London, headquarters of the British Labour Party, had been invited by Evatt to come to Australia as his press secretary. Wyndham was the only member of Evatt's staff kept on by Calwell when he succeeded Evatt in 1960. Late in 1960, Calwell helped arrange for Wyndham to be appointed secretary of the Victorian branch. His promotion to the national secretaryship in 1963 recognized that the Party needed an administrator rather than an operator. To the task of creating something out of nothing, he brought administrative skills of high order. Apart from anything else, he could actually take a record in shorthand and, unlike almost everybody else connected with the Labor Party, thought that it was rather important to present orderly records of proceedings.

Wyndham's work of reorganization before 1967 ensured that Whitlam had a ready-made report to work on from the beginning. The Whitlam-Wyndham plan had six objectives:

1. To get direct representation of the parliamentary leadership at Federal Conference and Federal Executive.
2. To improve the policy-making processes.
3. To strengthen the national structure and organization.
4. To reduce the power of the paid officials.
5. To change the public perception of the Labor Party as union-dominated.
6. To get direct representation for the rank-and-file membership.

The first four of these objectives were to be attained, wholly or partly. The last has still not been achieved. Yet it was through the appeal of the idea of rank-and-file representation that Whitlam was able to achieve so much on the other four. Irony is indeed one of the themes of this story.

Whitlam's campaign was a deliberate mixture of persuasion and confrontation. There was little hope that the Federal Conference would reform itself without strong outside pressure. In May, the Federal Executive put the Wyndham proposals into cold storage. This was a clear signal to Whitlam that reform would not come from within. His only hope was to bring pressure from outside. There were three possible sources of pressure — the Parliamentary Party, the State parties and the rank and file. Only the last of these offered much hope. The States' branches in particular were likely to resist reform. By definition, their machines had the deepest vested interest in the fragmented federal structure. A strong national organization, transcending State interests, would threaten their dominance, both within their own States and in the national party. The task was to stir up the membership to generate pressure for reform. Whitlam used all his prestige as a new leader in that task. He took every opportunity — branch meetings, union meetings and party conferences — to hammer the theme of reform.

The centre of resistance to reform was the Victorian Central Executive. That the proposals came from Whitlam would have been sufficient reason for the VCE to oppose them. The VCE correctly saw Whitlam as a threat to its power in Victoria and its influence on the national party. Whitlam blamed the VCE for the loss of the 1961 elections when, in a nation-wide landslide, Victoria failed to deliver a single additional seat. Victoria was still dominated by the memories and the consequences of the Split of 1955. The VCE depicted itself as under siege against the 'groupers', a cover-all for any opponent, the Democratic Labor Party, right-wing reaction, the National Civic Council and above all, Santamaria. Whitlam was seen as the Trojan Horse.

The ruling group within the Victorian Central Executive kept its tight control by posing as the last defence against the DLP. Yet Whitlam, in his first statement as the new Parliamentary Leader, had already closed the door against the DLP. At his first formal press conference on 9 February, he was asked if he was in favour of reconciliation with the DLP. He replied, 'There is no short cut to getting a Labor government by reconciliation, whatever that may mean, with the DLP. The only way that we can get a Labor government, federally, and this also applies in some of the states, is by getting votes from the Liberals. In short, we have to persuade a quarter of the people who voted for the Liberals at the last Federal elections to vote for the Labor Party at the next elections. If you add up all the Labor votes and all the DLP votes you still don't get a majority of the Australian electors. We operate under a two-party system. People have a choice between Labor and Liberal. The DLP no longer determines the result of elections.' To the question, 'Does this mean you will not consider under any circumstances coalition with another party?' Whitlam replied, 'There is only one other party in the House where governments are made, the House of Representatives. That is the Country Party, and I don't believe the Labor Party would wish a coalition with the Country Party and I don't expect the Country Party would wish it with the Labor Party. In a two-party system we govern in our own right or not at all.'

Whitlam's preliminary activities were all planned to lead up to a big push during the Queen's birthday weekend in June, which the political calendar sets aside for State conferences in Melbourne, Adelaide and Sydney. For New South Wales and South Australia, the note would be persuasion; for Victoria, confrontation.

The atmosphere was set by an incident on the eve of the Conferences. Returning from a tour of South Vietnam, Barnard stated that Labor policy on troop withdrawal should be reviewed. Although he had undoubtedly been impressed by the outward sign of American military success, Barnard was saying very little as the Labor Party had no clear policy on troop withdrawal. It was precisely the lack of clear policy one way or the other which led to the confusion of emphasis in statements by Calwell and Whitlam during the 1966 campaign. Chamberlain, reacting immediately to the newspaper reports, implied that Barnard had collected his information around 'the cocktail circuit and bars of Saigon', in Barnard's case a suggestion of the highest improbability.

Whitlam's rebuke was written in the form of a letter to the Western Australian Executive, of which Chamberlain was secretary, to preserve the elaborate fiction that party contention should be kept within the Party. Whitlam, on a week-long tour of North-West Australia, wrote his philippic in the pre-dawn hours of 6 June at a Geraldton hotel. With evident relish, he drew up his indictment against his old tormentor:

> It is not Mr Chamberlain's prerogative to interpret or enunciate the Federal policy of the ALP. If there is a Federal Labor government, he will neither state nor dictate its policy. Until he apologizes to Mr Barnard, I shall have no further communication with him.
>
> On 17 March, Mr Chamberlain attacked me through the press for advocating that the ALP should have a National Conference

> consisting of delegates chosen by the party members in Federal electorates and by affiliated unions instead of a Federal Conference consisting of delegates from State branches. I ignored him. On 21 March, he had to apologize to his State Executive. His conduct last week establishes how essential it is that in matters concerning the Federal Parliament, the ALP should be based on the rank and file and not on State bureaucracies. If other State secretaries were as autocratic and idiosyncratic as Mr Chamberlain, the party's membership would be as reduced and frustrated in other States as it is in Western Australia . . .
>
> Mr Chamberlain chose to attack Mr Barnard on the basis of a reported interview in Saigon . . . It is intolerable that Mr Chamberlain, without inquiry or consultation, should put me in the dilemma of disowning a party functionary or my own deputy . . .
>
> Mr Chamberlain says that Mr Barnard would not find evidence of crop destruction in Saigon saloons and at cocktail parties. Such disgraceful and disloyal insinuations must be repudiated by the Western Australian Branch if the Federal leaders are effectively to co-operate in the Senate campaign in this State.

The attack on Chamberlain set the tone for the Melbourne confrontation. Whitlam's express intention was to shock and on that level, it was a wild success. As he strode to the lectern, on floor level with the delegates, beneath the gaze of Calwell sitting in the front row of the gallery, the applause was polite enough if perfunctory. The mood swiftly changed from polite interest to uneasy expectation, to derision, to anger and fury, as Whitlam first scoffed at the pretensions of the VCE to be the guardians of the true faith. Then, defining his view of the aims of the Party, he went on to attack directly the controllers of the Party in Victoria.

> We construct a philosophy of failure, which finds in defeat a form of justification and a proof of the purity of our principles. Certainly, the impotent are pure. This party was not conceived in failure, brought forth by failure or consecrated to failure. Let us have none of this nonsense that defeat is in some way more moral than victory.

To foot-shufflings, laughter and jeers, Whitlam proceeded to enunciate his central principle: the primacy of parliamentary power.

This Conference, comprising a greater percentage of Union delegates than any democratic socialist party on earth, must reject any idea that constant defeat and permanent opposition do not matter, that the Labor movement can be strong and effective industrially, while being weak and ineffectual politically. On the narrowest industrial views, this is patently false. There is not a union affiliated with the ALP which has not called for the removal of the penal clauses of the Arbitration Act; many of them have suffered severe and unfair penalties because of the operation of the clauses. Yet there is not the slightest chance of their being removed until Labor wins power in the National Parliament . . .

On the wider field, the economic and social policies which can enhance or destroy industrial gains are the responsibility of governments, and in particular, of the National Government.

The National Government alone determines and applies economic policies which can make wage advances worthwhile, or can render them meaningless by inflation. Its economic policies determine whether unions will be in a strong bargaining position under full employment, or weakened by the threat of unemployment. Its views at wage hearings will carry powerfully with the Arbitration Commission. Its attitude towards its employees will profoundly affect the conditions under which all other employees work.

Wages and conditions are no longer the chief determinants of real living standards. The health and housing of ourselves and our families, our children's education, represent the most important social capital this nation has to offer. These are the things which determine the quality of life. They are matters of particular concern to us as democratic socialists, because the things which governments provide are those that determine most whether the community enjoys a measure of true equality.

The quality of life and equality in life is what socialism is all about. In Australia, the means of raising the quality and securing equality lie in the hands of the National Government. Are we as socialists to permit that power to rest, or rather sleep, indefinitely in the hands of parties whose philosophy is not quality and equality, but philistinism and favouritism? Is this what being 'true to our principles' really means?

In the weeks after I was elected Leader on February 8, every trap and inducement was laid for me to concentrate on the two questions of when I would get out of Vietnam and when I would go into Victoria. I was intent on de-escalating these controversies until it would be possible for the Party to have rational discussions and to make constructive decisions on these crucial external and internal issues in the proper place and atmosphere. In the last week, attempts have been made inside this State and outside it to give the impression that the Party's Vietnam policy would be preserved only if the Party's present Executive in Victoria were preserved and that the Party's Victorian Executive must be preserved to preserve the Party's Vietnam policy.

It is true that some parties can exist only as pressure groups. The Communists support this view because they do not want to win Parliamentary representation or power; the DLP supports it because it cannot win Parliamentary representation or power. Neither our traditions nor our purpose permit us to adopt this role for ourselves. We are in business to serve and preserve democracy, Parliamentary democracy.

I did not seek and do not want the leadership of Australia's largest pressure group. I propose to follow the traditions of those of our leaders who have seen the role of our Party as striving to achieve, and achieving, the National Government of Australia.

The meeting dissolved in an explosion of boos and jeers. When Whitlam closed with a reference to the forthcoming by-election for the seat of Corio, over a huge roar came a shout: 'You've lost it, you bastard!' In the gallery, Calwell's broadening smile switched to frank and uncomplicated laughter. After the meeting he accosted the author with rekindled affability: 'You won't be working for your new boss for long now.'

Yet Whitlam had achieved exactly what he set out to achieve. From that night on, nothing could ever be the same again in the Labor Party. He had established the terms of reference. Nobody involved in Labor Party affairs could ever again ignore those terms of reference. It was perfectly possible that they would get rid of him. But it was impossible for them to ignore what he was saying. By placing the confrontation in Melbourne, against the VCE, Whitlam had made a statement about his leadership from which neither he nor the VCE could ever escape: either he must fall or the VCE must fall. That statement was to be the central fact of Labor Party internal affairs for the next four years.

Arriving next day at Adelaide airport, Whitlam was welcomed

by the Labor Premier of South Australia, Don Dunstan. Only recently installed as Premier, Dunstan was worried about Whitlam's intentions. Although the South Australian Party had a good electoral record, its organization was conservative. Its leadership was exclusive and oligarchical and concerned with preserving the existing structure. Its merit and the key to its success, was that the leadership made a conscious effort to give a reasonable share of influence, or at least the appearance of it, to all the various factions. Relations between the Parliamentary Party and the organization were close and co-operative. The maintenance of the existing set-up within South Australia, whatever might happen elsewhere, was essential to Dunstan, still feeling his way as a new Premier in an inherited premiership. In terms of attitudes, approach and ideas, he was close to Whitlam. He shared Whitlam's view of the damage done by the Victorian Central Executive. But he could not afford any situation where he might be forced to choose between Whitlam and his own local organization. His presence at the Adelaide airport was more than a fraternal courtesy; he was there to warn Whitlam against any repeat of the Melbourne performance. He need not have worried. Having deliberately affronted Victoria, Whitlam needed all the allies he could get. He had no wish to alienate South Australia. The prepared speech text which Whitlam showed Dunstan was a measured explanation of the case for national reorganization. Dunstan considered even some of this too strong. On the wind-swept tarmac, Whitlam stood by meekly as Dunstan, using Whitlam's car for a desk, put his pen through a dozen paragraphs and tore out whole pages. Impartial in his censorship, he threw out a passage which read:

> All my adult life, I have held the view which Don Dunstan has expressed that ultimately the Federal system as we know it will give way to a national government and to large metropolitan or regional governments. This is the logic of history. Is our Party, in its organization, to set its face against such trends and resist all moves to make it more effective, more modern and more national?

What did survive was largely Whitlam's first attempt to describe to a party assembly his proposals for health reform. Perhaps

Dunstan was prescient as well as cautious. Certainly, Whitlam was to have far greater success in changing Australia's health insurance system than in reorganizing the Labor Party structure along national lines or altering Australia's federal system.

This torrid Queen's birthday weekend was rounded off more congenially in Whitlam's home State of New South Wales. Whitlam then departed for his first overseas journey as Leader of the Opposition, giving everyone time for reflection on what had happened.

Aptly and ironically, the first electoral test for Whitlam's leadership came in Victoria when the transfer of Hubert Opperman, a Minister in the Holt Government, to Australia's High Commission at Malta left his seat of Corio vacant. Corio had a symbolic meaning for the Labor Party. Its capture by John Dedman at another by-election in 1940 was crucial in the events leading to the fall of Menzies' first ministry and later the formation of the Curtin Labor Government. Dedman became one of the most controversial and most constructive members of the Chifley Government. As Minister for Post-War Reconstruction, he was responsible for the White Paper on Full Employment which established the prime economic goal of successive Australian Governments for the next quarter of a century. His defeat by Opperman in 1949 seemed a bitter rejection of Labor's proudest achievements. Corio is based on the great industrial centre of Geelong. Labor's failure to win back Corio in subsequent elections underlined its unacceptability even in an industrial constituency.

In May, Whitlam pressed Robert Hawke, the brilliant young industrial advocate for the Australian Council of Trade Unions, to contest the coming by-election. Hawke, who had been the candidate in 1963, declined, preferring to wait to succeed Albert Monk as President of the ACTU. The pre-selection went to Gordon Scholes, a young engine-driver, union official and Geelong city councillor who had been Labor's candidate in 1966. This was quite acceptable to Whitlam.

*

The win in Corio was one of the most important in Whitlam's

career. To have won a by-election within nine months of the November débâcle anywhere in Australia would have been a considerable achievement. To have won it in Victoria was a double boost. The confrontation with the VCE seemed to have paid off handsomely. An objective analysis of the result might find any number of factors contributing to it, but in the delight of the moment, there was no disposition anywhere to go beyond any cause than the obvious: Whitlam was what his supporters had always claimed him to be — a winner.

With Corio won, Whitlam went to the Federal Conference immensely strengthened. His tactics in stirring up argument about reorganization meant that the issue could not be swept under the carpet. The success in Corio vastly enhanced his chances of securing a favourable outcome.

A further unexpected advantage, something of a coup, had been achieved in Western Australia. Returning from his overseas journey via Perth, Whitlam persuaded the Western Australian Conference to accept a resolution favourable to national reorganization. This was a sharp defeat for Chamberlain on his home ground. Conference applauded, with as much relief as approval, when Whitlam and Chamberlain posed for a photograph of formal reconciliation. Communication had been restored.

Whitlam thus went to the Adelaide Conference on 31 July the victor of Corio and with the explicit support of three States, New South Wales, Tasmania and Western Australia, for party reform. Whitlam's aim was to secure appointment of a special commission comprising one representative from each State Executive, with himself, Barnard and Wyndham to report on all reform proposals. At this stage, Whitlam envisaged a process taking about a year — the special commission, debate by State Conference, then a Special National Conference. With three States already committed, eighteen votes out of thirty-six, the key lay in the Queensland delegation. Pre-conference lobbying, led by Clyde Cameron, made it certain that Queensland would stand firm against the Whitlam-Wyndham proposals. If the issue was taken directly to Conference, deadlock was certain.

The decisions of the Adelaide Conference were made behind

closed doors, often literally, in the smoke-filled rooms of the Hotel Australia. The result of days of Byzantine manoeuvres was a compromise. Whitlam's proposal for a Special Commission was quietly shelved. In return, the Chamberlain-Cameron group gave ground on the key proposal to strengthen the parliamentary representation on the Federal Executive and at the Federal Conference. Henceforth, the four Federal Parliamentary officers —the leader and deputy leader in the House of Representatives, and the leader and deputy leader in the Senate—would be delegates to Federal Executive by right. This was to prove a decisive shift in the balance of power towards the parliamentary group and away from the paid officials. The architects of the compromise, principally Clyde Cameron, did not completely foresee its consequences. Cameron's main objective was to preserve the existing balance towards the 'left' on the Federal Executive, the Party's key power body. He calculated that Murphy and Cohen would cancel out the votes of Whitlam and Barnard on crucial issues. This equation did in fact hold until Cohen's premature death during the 1969 election campaign and his replacement by Don Willesee as deputy leader in the Senate. Nevertheless, the Adelaide Conference was Whitlam's first big breakthrough, for he had won his basic point—full representation by right of the Parliamentary leadership, federal and state.

'The party ... the policy ... the people.' The three aims were being pushed ahead together. In the early months of 1967, John Menadue undertook the formation of a comprehensive range of policy advisory groups. The contacts he made were wide, though drawn most heavily from the universities in Sydney, Melbourne and Adelaide. In a memorandum to Whitlam on 25 May, Menadue warned:

> The groups must be kept activated and a free and continuous flow of ideas in both directions must be ensured. Otherwise they will wither away. This is a very real danger at the moment.

The warning was ignored and the prophecy fulfilled. While many of the recruits continued to collaborate individually and informally, only the group on health worked with anything like the effectiveness sought. After Menadue resigned as private secretary in July, most of the groups he had created disintegrated. Thereafter, the development of new policy depended overwhelmingly on Whitlam himself.

Four areas of policy illustrate in different ways the Whitlam approach. In each of these four — cities, health, education and foreign policy — the Whitlam stamp is indelible. But each required a different approach. Urban affairs, the idea of national involvement in the growth and shape of cities, new growth centres, local government finance, a national sewerage programme, urban public transport, were all matters on which Labor policy was vague or silent. The task was not to alter policy but to create one. In the case of education, the task was the exact opposite. The need, as Whitlam saw it for both electoral and educational reasons, was to remove the prohibitions and inhibitions against national involvement imposed by the State rights argument or the State aid arguments. Thus, the main effort had to be undertaken within the Party. In health, the task was to flesh out existing policy and to work out practical ways of restoring the free hospital system established by the Chifley Government and dismantled by the Menzies Government. In foreign policy, all three approaches had to be combined, creating some new policies, changing some existing ones, giving new expression to established principles.

The rewriting of Labor's platform and policies represents an unequalled personal achievement by Whitlam. Yet, in government, the approach was to hold certain disadvantages. Firstly, though the policies went through all the normal constitutional processes of the Party, the high element of Whitlam's personal preferences and preoccupations remained. They were not fully absorbed by the Party as a whole, often not fully understood and rarely deeply felt. Thus the fight to implement his policies too often became a fight for Whitlam alone. This was the real origin of the so-called communications gap between the Labor Government and the Labor Party. Secondly, some important areas in which Whitlam was less interested, involved or informed tended to be neglected. Thirdly, the priorities of the government, in terms of energy and money, tended to be geared to the programmes which Whitlam himself had worked hardest for, and with which he was most closely identified. The seeds of some of Labor's difficulties in office lay in the success of the years in Opposition.

*

The origins of Labor's health programme deserve some study in detail as an example of policy development under Whitlam.

The provision of free treatment at public hospitals was always the core of Whitlam's health programme. In a sense, Whitlam and the Labor Party became side-tracked on the health insurance issue because the middle class have always been much more concerned about insuring themselves for private treatment than in obtaining free treatment.

As far back as 1957, Whitlam had complained in his Chifley Memorial Lecture, 'The Constitution versus Labor', that: 'The present constitutional position is quite unsatisfactory in that the Commonwealth has to pay more and more for the running of hospitals and still has no say in running them, patients are unable to afford medical and hospital treatment and the medical profession participates in any scheme only on its own terms.' His dissatisfaction prompted a major rethinking of the ALP approach to health and passed into policy at the Party's Federal Conference in May 1959. Summarizing this approach, Whitlam argued in his 1961 Curtin Memorial Lecture that: 'While the constitutional position precludes the socialization of doctors, it permits the socialization of hospitals . . . The best way to achieve a proper National Health Service is to establish a National Hospital System. Quite apart from the economic advantages there is the great social advantage of providing both patients and doctors with an alternative to the present system. Patients would be free to consult the salaried staffs; doctors would be free to join them. Doctors and patients and communities are alike unable to provide an alternative. Only the Commonwealth can give them a choice. There is no sphere in which government initiative can be such a liberating force.' Whitlam, Menadue, Cass and Mrs Ruth Inall, of the Australian National University, were to work together closely in developing arguments in favour of the policy and amassing the statistics and international comparisons with which to sustain them. In 1968 and 1969, the burden of the work fell overwhelmingly on Race Mathews, who succeeded Menadue as private secretary and was elected to the House of Representatives in 1972.

In debate on the National Health Bill of 1961, Whitlam pointed out that the motivation of Liberal health insurance was 'not the desire to reduce the expenses of patients, but the desire to guarantee the fees of doctors'. In a seminal speech delivered for the annual meeting of the Rochester and District War Memorial Hospital on 28 August 1964, he said: 'We are spending one third as much as they are in England and Wales on a less comprehensive service for a population which is not even one quarter of theirs.' 'But', Whitlam continued, 'it is more important to nationalize hospitals than to nationalize the medical profession.'

While Whitlam and his Labor associates were developing their critique of Liberal health insurance on the basis of international comparisons and replies by Ministers to questions on notice, a similar undertaking was being launched based on exhaustive economic analysis by two research workers in the Institute of Applied Economic and Social Research at Melbourne University. In 1965, the Institute had drawn John Deeble from his post as deputy manager of Melbourne's Peter MacCallum Clinic, and a year later it had taken Dick Scotton from the Commercial Banking Company of Sydney where he was employed as an economist.

In 1967, both men were invited by Moss Cass to his home for a meeting with Whitlam and Menadue at which the basis for an alliance was established. As the meeting was breaking up, Whitlam asked Deeble and Scotton if they had developed in addition to their criticisms of Liberal insurance, ideas about the arrangements by which its place should be taken. He was told that they had ideas, but so far there was no system. In January 1968, a Joint Committee of the Victorian Council of Social Service and the Victorian branch of the Australian Association of Social Workers brought down a report which argued strongly for making health insurance obligatory. By June, Deeble and

Scotton had developed proposals for an obligatory insurance system, and in July these proposals provided the core of the definitive address on 'The Alternative National Health Programme' delivered by Whitlam at Sydney's Royal Prince Alfred Hospital and subsequently published by invitation in the Medical Journal of Australia.

In 'The Alternative National Health Programme', Whitlam focused public attention for the first time on the finding by Deeble and Scotton that the health funds were inefficient, their competition was wasteful, that some of their managements spent contributors' money on executive aircraft and political campaigning, that $85 million (in June 1966) of contributors' funds was tied up in reserves, that 17 per cent of Australians had no medical cover at all and 15 per cent had no hospital cover and that because of tax rebates, the poorest paid most and the wealthiest paid least.

He then outlined what was to become the basic plan for Medibank:

> The Commonwealth Government could replace the existing system of voluntary health insurance with a system of universal insurance. This scheme would be administered by a Commonwealth Health Insurance Commission. The Commission would draw its revenue from a National Health Insurance Fund. The Fund would be financed by a health insurance contribution, assessed and collected as a 1.25 per cent surcharge on income tax; a matching Commonwealth Government contribution, in part funded by withdrawing income tax concessions now granted for voluntary contributions to insurance funds; and by a levy on compulsory insurers equal to their present liability for medical and hospital care under third party insurance and workers' compensation insurance.
>
> Representatives of the Commonwealth Government, the Health Insurance Commission and the Australian Medical Association would negotiate a schedule of benefits designed to cover 85 per cent of current standard fees. Doctors in private practice would then have the option of charging their customary fee to patients, who would then be entitled to recoup 85 per cent of that fee from the Health Insurance Fund; or to bill the Fund regularly for the scheduled 85 per cent in full settlement of their fees.

The Commonwealth Health Insurance Commission would provide hospital benefits at the level required to fully finance public treatment, without a means test, in public hospitals. Whether particular patients are accommodated in single-bed wards or multi-bed wards should be determined purely on medical grounds. The full public-bed benefit should, however, be available for those who choose private or intermediate bed care in either public or private hospitals, with voluntary insurance funds providing cover for the optional additional cost.

The foundations for what became Medibank were well laid in the first twelve months of Whitlam's leadership. The rest of the story — the further refinement of the programme, the campaign for public understanding of the scheme, the response of the Gorton Government, the resistance by the health funds and the Australian Medical Association, the long and skilful negotiations by Bill Hayden as Minister for Social Security in the Whitlam Government to convert the plan into reality — lies in the future. But essentially the foundations for what became Medibank were well laid in the first twelve months of Whitlam's leadership.

Throughout his long campaign against the Liberal system of private health insurance, Whitlam drew a parallel between his own situation as a high-income earner and that of his Commonwealth drivers, George Bevitt and Robert Millar: 'The tax rebate is worth twice as much to me as it is to my driver on a third of my income, so I pay much less for my health insurance than my driver.' For his audiences, it was an effective illustration of the inequity of the existing system.

By the spring of 1967, the first two elements of the rough sketch of a plan of action had been given substance. The reform of the Party and the renewal of the policy had begun. The proof of Whitlam's success, if any, would lie with the third and most important element — the people.

6
The Decline of Holt

Menzies' long mastery disguised a major characteristic of the conservative party in Australia — disloyalty to its leader. Paradoxically, Menzies himself had been in 1941 the most famous victim of them all. Every conservative leader from Lyons to Snedden suffered from this chronic disloyalty. Three leaders — Menzies, Gorton and Snedden — were disposed of. Death almost certainly saved Lyons and possibly Holt from removal by other means. The Labor Party is traditionally much more tolerant and patient. For all the Caucus revolts, intrigues and turbulence, no Federal Labor leader has ever been voted out. Perhaps the reason why Labor is so tolerant towards failure is that it is used to it and expects it as its normal lot. The conservatives' definition of what is normal and what is successful and their low threshold of tolerance is shown by the fact that the three leaders they have deposed since 1941 either retained government or won seats in the election before their dismissal. Nonetheless, the smoothness of Holt's succession and his massive victory in 1966 seemed to assure him of a long leadership and loyal support. Yet within nine months of his triumph he was a leader under siege; three months after that he was dead.

The size of his victory carried its penalty. The new Liberal members knew little of the old days of defeat and struggle before 1949. They knew little about Holt and he knew little about them. The disarray of the Opposition was so great, the causes of its defeat so obvious that the new Liberals held little personal gratitude towards Holt. No member of the Government believed that Holt was the architect of its victory. Certainly there was great goodwill towards him, but that was because he was such a decent, likeable fellow. No one regarded him as the indispensable man.

Notable, though not typical, among the new Liberal members was Edward St John, a Sydney lawyer. An independent liberal, son of a country vicar, he proudly claimed descent from Oliver St John, Hampden's defence lawyer in the Ship Money case in 1637. In seeking pre-selection for the blue-ribbon seat of Warringah, he had beaten back a vicious campaign of rumour and denigration organized by a right-wing racist group who hated him for his outspoken views against race discrimination in Australia and apartheid in Southern Africa. He owed nothing to Holt or the official Liberal Party. He was to play a decisive role in the fall of two Liberal Prime Ministers — Holt and Gorton. The price he was to pay was his own political destruction.

It is seldom that the beginning of a politician's decline can be placed with any precision. In Holt's case it can be put down to the minute — 8.07 p.m. on 16 May 1967 — the moment at which Holt interjected on St John's maiden speech.

On the night of 10 February 1964, the flagship of the Australian Fleet, the aircraft carrier H.M.A.S. *Melbourne*, collided with and sank the destroyer, H.M.A.S. *Voyager*, off the coast of New South Wales. Eighty-one men aboard the *Voyager*, including Captain Duncan Stevens, were drowned. The first Royal Commission was unable to determine the cause of the tragedy but made some criticism of Captain R. J. Robertson, Captain of the *Melbourne*. For two years, several Liberal members, notably John Jess (Latrobe, Vic.), believing an injustice had been done to Robertson, sought a reopening of the Inquiry. Early in 1967, a former Naval Officer who had served with Captain Stevens, Peter Cabban, made allegations suggesting that Stevens had a serious drinking problem. Jess moved for the appointment of a Select Committee to inquire into Cabban's allegations. St John, impressed by Cabban's courage and sincerity, decided to make his maiden speech in the debate on the motion. It was a brilliant performance, a coldly penetrating analysis of an apparent cover-up by the Government and the Navy, combined with a passionate plea that justice be at last done. The power of his argument brought Holt hurrying into the

House, angered, red-faced. With visibly mounting impatience, he listened as St John dealt with Cabban's allegations: 'Is it irrelevant that the Captain of the destroyer when in port was perpetually drunk? Or have I lost the meaning of the word "irrelevant"?' Holt could restrain himself no longer: 'And what is the meaning of the word "evidence"?' St John paused briefly and then said quietly: 'I didn't expect to be interrupted by the Prime Minister.' Nor did anyone else expect it. It was the first serious breach of the convention that maiden speeches are heard in silence since McEwen had interrupted Whitlam's maiden speech in 1953. Holt knew he had blundered and subsided into an embarrassment shared by his colleagues around and behind him. The effect of St John's speech, enhanced greatly by Holt's gaffe, was to raise pressure for a new inquiry to the point of irresistibility. Three days later, Holt announced the appointment of a second Royal Commission to investigate Cabban's statement.

The political significance of the *Voyager* was that, on a deeply emotional issue, big fissures opened for the first time in the hitherto impregnable front of the coalition. Its personal significance for Holt was that it showed, in a fleeting but memorable incident, how the pressures of his job were beginning to eat away at a man distinguished above all by his innate courtesy and dignity.

The first electoral appeal involving Holt and Whitlam found them not in contest but in collaboration. In May, the Government submitted to the people two proposals to amend the Constitution. One asked the people to give the Commonwealth responsibility for aborigines; the other sought to break the 'nexus' between the Senate and the House of Representatives. Section 24 states that as nearly as practicable, the number of members of the House of Representatives shall be twice the number of senators. This is called the nexus. Its practical effect is to limit the size of the House of Representatives. Any significant increase in the House of Representatives must be accompanied by a massive increase in the Senate. Thus in 1948, when the Labor Government increased the number of members from 75 to 120, it had to double the number of senators from 36 to 60. The

Government proposal to break the nexus was fully supported by Labor. The DLP, with a deep interest in maintaining the relative strength of the Senate, the only Chamber in which it had been able to secure representation since 1955, campaigned against the proposal on the crude but effective slogan 'No more politicians'. The referendum on aborigines was carried massively. The referendum on the nexus was defeated decisively, despite the support of both Government and Opposition. Nobody blamed Holt, but the fact that he, as Prime Minister, had sponsored the proposal and chosen the timing of the referendum raised some mild questioning of his leadership.

The Corio by-election was the first real electoral test between Holt and Whitlam. In Corio, Whitlam used the technique which was to be successful throughout the years in Opposition — concentration on specific local issues but relating those issues to broad national policy. He first developed this technique in the Dawson by-election. Whitlam was to be fortunate that the key by-elections before 1972 happened to occur in electorates based on provincial centres — Dawson, Corio, Capricornia and Bendigo. They were ideally suited for his technique and style. The method calls for a great deal of preparation and a lot of hard work. It also takes a lot of time, calling on the pillars of local society, such as the local newspaper editor, the RSL and the two Bishops. It is much easier to run this sort of campaign from the Opposition. RSL clubs take it for granted that a Prime Minister will visit them; Bishops are grateful when even a leader of the Opposition does. Whitlam and his staff — Peter Cullen, Barbara Stuart, Carol Summerhayes, the author and his Melbourne driver, Bruce Taylor, established themselves in a Geelong motel and lived there for a fortnight. It was in this campaign, in the discomfort of Victorian mid-winter, that Whitlam and his new staff established a pattern of relationships towards each other and to the work which was to mould them into a formidable team.

Holt's campaign was perfunctory. The Liberal candidate was an import from Melbourne. Holt avoided parochial issues and campaigned on Vietnam and the announcement by the British Labour Government of its intention to phase out Britain's military presence east of Suez. In the 1940 Corio by-election,

Labor's opponents had used the slogan 'Hitler's eyes are on Corio'. Twenty-seven years later, the citizens of Geelong were similarly unimpressed by the suggestion that Ho Chi Minh's eyes were on Corio in 1967. The only significant Liberal to see the trend was Senator John Grey Gorton, Minister for Education and Science. Returning to Melbourne after a day's campaigning around Geelong, he told his driver, 'We're wrong if we think this is in the bag.'

The result on 22 July was an increase of Labor's vote by 9.5 per cent and an absolute majority for Scholes with 50.26 per cent of the vote.

The political damage done to Holt and his Government by the defeat in Corio was small. He was protected by the accepted convention, not always true, that by-elections go heavily against governments. Within the Liberal Party there was no disposition towards recrimination or self-examination. Much more damaging, politically and personally, was the result of the next by-election, Capricornia, based on Rockhampton and Gladstone in Central Queensland. It was made necessary by the death of the Labor member, George Grey, a moderate Labor man, exceptionally hard-working and a popular local member. He had retained the seat only narrowly in the general swamping of 1966. Between 1949 and 1961, the seat had been held by a Liberal member. The Liberal candidate for the by-election, a well-known and well-liked figure closely identified with the popular local issue of regional development, was regarded by southern politicians and the press as an ideal 'horse for the course'. The Labor candidate had been chosen by the local branches against the express wishes of Whitlam, who had personally and unsuccessfully intervened on behalf of another contestant. The endorsed candidate, Dr Douglas Everingham, was an avowed atheist who had recently written an article for a communist journal exploring the common ground shared by communists and humanists. He was also an habitual writer of 'letters to the editor'. This is considered eccentric. The political 'experts', including myself, regarded all these things as grave disadvantages for a Labor candidate in that area in that year. For all these reasons, Harold Holt decided to invest a great amount of political and personal capital in the Capricornia by-election.

What the southern experts, both Liberal and Labor, did not know was that Everingham, a general practitioner who had treated hundreds of patients in the Rockhampton area free of charge, was widely respected for his integrity and even loved for his humanity. The fact that he was the brother-in-law of the Liberal candidate may even have proved his respectability. When the Federal Treasurer, William McMahon, influenced by poor advice and worse champagne, spoke at a Liberal rally in Rockhampton of the dangers of atheistic communism, the Anglican and Catholic Bishops issued statements defending Everingham as their friend and 'a better Christian than many claiming the name'. Everingham obediently accepted the role assigned to him by the campaign organization to keep as quiet as possible. But there were limits to his acquiescence. In the old Australian fashion, he had given his house a name. It was 'Ingersoll', after the 19th century American humanist, polemicist and politician, Robert Ingersoll. Officials suggested that the name-plaque should come down, lest some alert pressman should spot it and revive the argument about Everingham's atheism. Everingham was adamant. 'I've done everything you've asked so far, but some things are sacred.'

The result on 30 September was that Everingham more than held Grey's vote, with an absolute majority of 52.96 per cent.

There had never, in fact, been any chance of Holt securing a Government victory in a by-election in an established Labor seat. The conditions for victory never existed. But he had unwisely committed himself, not just in public to his supporters, but privately to himself. He really believed he could win. The failure to win Capricornia hurt him psychologically and harmed him politically far more than the actual loss of Corio had done. Harold Holt lost something of himself in Capricornia.

He was to lose more over what became known as the VIP affair. The 34th Squadron of the Royal Australian Air Force is identified by the pompous acronym VIP – 'Very Important Person'. This alone would make its existence and use a matter of suspicion, derision and envy. The Squadron is used by the Governor-General and the Prime Minister and, subject to the

Prime Minister's approval, senior Ministers and foreign politicians. In certain circumstances — election campaigns, or visits to difficult areas such as the North-West or Papua-New Guinea an aircraft is made available to the Leader of the Opposition.

What became known as 'the VIP Affair' began, with exquisite irony, with a flight made by Calwell. In September 1965, he had received Menzies' permission to travel to Perth for a meeting of the Federal Executive. At that time it was customary that the person in whose name the flight was booked could invite whoever he wished to travel with him. Calwell invited the Federal Secretary, Wyndham, and the Victorian Secretary, Hartley.

The veteran Sydney Labor member of Parliament, Fred Daly (Grayndler), was at this time at odds with Calwell. Seeking to embarrass Calwell over his Perth trip, he placed on the notice paper a question seeking information from Holt about the use and cost of VIP flights during the preceding twelve months. This sort of thing is the small change of politics. There was no real reason why Holt should have feared embarrassment. In the way of politics, Daly and Calwell had a reconciliation before the question was answered. Calwell casually remarked to Holt, 'Fred's not going to press you for an answer on the VIP.' So with his guard down, Holt accepted from his Department an answer using the following formula:

> Passengers' names are recorded only so that aircraft may be safely and properly loaded. After a flight is completed, the list of names is of no value and is not retained for long. For similar reasons, no records are kept of the places to which aircraft in the VIP flight have taken VIP passengers. The answers to these questions are therefore not available.

Each sentence was untrue. The answer, buried in a mass of questions answered after the Parliament rose for the winter holiday in May 1966, was forgotten. But the untruthful formula was enshrined in Hansard. There it lay for a year, a small time-bomb quietly ticking away.

In 1967, questions about the use of the 34th Squadron were again raised, notably by Senator McManus, Deputy Leader of the DLP, and Senator Turnbull, a Tasmanian Independent. The

Minister for Air, Peter Howson (Casey, Vic.), repeated the formula devised a year before by the Prime Minister's Department. He locked the Government into a lie.

Word got around that the Air Department was, naturally enough, unhappy at being associated with the deception. The Opposition Leader in the Senate, Murphy, threatened to call the Secretary of the Department, Archibald McFarlane, to the bar of the Senate. The game was up. The Government Leader in the Senate, John Grey Gorton, realized from his own war-time experience as a fighter pilot that R.A.A.F. procedure required the keeping of such records, the existence of which was now being denied. Without consulting Holt he demanded that the records be produced and forthwith tabled the lot. Howson, returning to Australia after leading a parliamentary delegation in Africa, offered his resignation, which Holt refused to accept. Any censure of Howson must equally have involved Holt himself. Gorton, whose apparently bold candour had brought him for the first time into national prominence, was to prove the real beneficiary of the affair. Whitlam, himself an ex-airforceman, was rather rueful that he had not realized earlier that the damaging answers could not be true. The tragedy for Holt was that the lie was altogether needless. The records proved the substantial part of the Government's case that there had been no abuse of the VIP flights. The whole affair was petty, seedy and naive. That was why it was so damaging.

*

Yet all his difficulties together — the *Voyager* disaster, the rising cost of the F-111, the VIP affair, the loss of the referendum and the by-elections — do not account fully for the gloom and depression into which Holt sank. He and his government were going through a bad patch; but what Prime Minister had not? There had been no fatal blow, nothing beyond recovery. But there was upon Harold Holt in his last months a terrible pressure — Vietnam. Indeed it was something more formidable than the war itself; it was LBJ. Simply, he was being bullied by Lyndon Johnson. Holt had taken over Menzies' commitment with far greater enthusiasm than its author. He had made the decision to send conscripts. He had increased the troop level three times, in March 1966, December 1966 and October 1967, until it reached 8,000. And LBJ was still asking for more. The greater the difficulties, the greater his demands. In early October, the Treasurer, William McMahon, visited Washington to be subjected to a Johnsonian tirade about the 'need to stand up and be counted', the same argument R. G. Casey had used against Whitlam in 1954. The Australian Ambassador in Washington cabled home that he could never remember such strong pressure being brought by the Americans. Submitting, Holt told Johnson that the new addition would put Australia 'at the full stretch of our present and planned military capacity'. This decision was taken in the face of doubts by Ministers and Service Chiefs of the Defence Committee. Holt carried the day. The pressure from Washington still continued. 'All the way with LBJ' was becoming an expensive slogan. Where would it lead? When would it end? In early December, Cabinet accepted the Chiefs of Staff insistence that further commitments were impossible. Holt knew he had reached the end of the line. He also knew that Johnson would keep on making impossible demands. The Australian Embassy in Washington failed utterly to warn the Australian Government of a growing American disenchantment with Vietnam, especially in Congress. Embassy contacts were largely limited to the State Department. Whitlam gave great offence by describing the Embassy as 'the post office' for the State Department. Visiting Washington in June 1967, he was impressed by the depth of Congressional opposition to the war. The leader of the Democratic majority, Senator Mansfield, told him that 'at least half the Congress was opposed to the President'. The chairman of the powerful Senate Foreign Relations Committee, Senator Fulbright, believing Johnson had betrayed him over the Gulf of Tonkin resolution in 1964, declared that he would henceforth oppose any foreign aid appropriation request made by the Administration, because Johnson could not be trusted.

In a sense, the bombing of North Vietnam did for Whitlam in Vietnam what the sending of conscripts to South Vietnam did for Calwell. The specific evil brought the general evil into sharper focus. The 1967 Labor Conference in Adelaide had declared for a cessation of the bombing, conversion of the military effort into a holding operation pending negotiations, and recognition of the National Liberation Front as a party to any negotiation. McMahon, fortified by his session with Johnson, denounced this policy as 'treason'. There is no evidence that this language was thought to be excessive. It was par for the course in 1967.

In Australia's political calendar, there is, or used to be before 1975, no event so calculated to fill political practitioners with the sense of the futility of their existence as an election for half the Senate. Such an election alone cannot change the Government (although under the post-1975 dispensation it is now a crucial part of the process of changing governments). Because of the proportional system of voting, four of the five vacancies in each State are predetermined. The principal interest in the result lies in the total percentage vote cast for the main parties as a pointer to the likely result of the next House of Representatives election. Yet the politicians and the parties must go through the ritual as if their world depended upon it. By calling a House of Representatives election a year ahead of due time in 1963 Menzies had put the elections for the two houses out of joint. That half of the Senate elected in the conjoint election of 1961, the last 'normal' election Australia has had, was now due to retire. Despite its landslide victory of 1966 and its massive House of Representatives majority, the Government was in a technical minority in a Senate reflecting as it did the elections of 1961 and the half-Senate elections of 1964. Its working majority in the Senate depended on the support of the DLP. This situation led to a dispute in Caucus, which would echo down the years to the constitutional crisis of 1975.

There were several factors to make the 1967 Senate election more interesting than usual. For Whitlam, it was his first national campaign test as leader. For the Labor Party, it was an urgent measure of what recovery, if any, had taken place since the 1966 disaster. For Holt, it was his first test against Whitlam. And because of the postal charges dispute and Gorton's conduct in the VIP affair, public attention had been focused on the Senate more than usual.

In conjunction with the 1967 Budget, the Government had introduced measures for increases in postal charges. The DLP senators announced their intention to oppose the increases. Two government senators made a similar threat. Holt made it clear that a defeat would be treated as refusal of supply and grounds for an election. The consequence would have been an election for the House of Representatives but not for the Senate. Murphy, as Labor leader in the Senate, had already done considerable work in developing a network of active committees in the Senate and in general lifting this body from the torpor of generations. It may have been laudable to try to raise the prestige of the Senate; its real result was to raise its pretensions. Murphy argued vigorously in the Senate and the Caucus for the equal rights of the Senate. Whitlam was joined by Calwell in a successful resistance. It was the last act of the old partnership and the last time Calwell spoke with effect in the Caucus which he had entered in 1940. Assuming the mantle of prophecy, he foretold the disasters a Labor Government would face if the Senate were conceded power over money bills. He warned that if the Labor Party helped raise the pretensions of the Senate, the day would come when a hostile Senate would turn and rend a Labor Government. In the best sense, it was his last hurrah.

*

The election results were politically inconclusive but psychologically damaging to Holt and the Government. The Labor Party won 45 per cent of the national vote, exactly what it had been for the Senate in the general elections of 1961 and the separate Senate election of 1964. But it represented a 5 per cent increase over Labor's House of Representatives vote in 1966. The Government parties share fell from 51 per cent in 1966 to 43 per cent. The DLP vote rose from 7 per cent to 10 per cent. There was something in the figures for everyone. But the psychological advantage lay with Labor. Labor was no longer a spent force.

In the last week of the campaign, trivialities were swept aside by the need for a genuine act of statecraft. On 18 November, Britain devalued sterling by 14 per cent. With excellent despatch, the Government decided not to follow Britain down.

The ease and efficiency of this decision owed much to the absence overseas of the Leader of the Country Party, Sir John McEwen. It was a notable victory for the Treasurer, William McMahon. Throughout 1967, McEwen had sought McMahon's political destruction. In his efforts to expand the Country Party's declining rural base, McEwen sought to forge an alliance between farmers and manufacturers on the basis of 'all-round protection' of high tariffs and big subsidies. He saw McMahon as the principal opponent of this plan, in both its political and economic aspects. He pursued McMahon with single-minded vindictiveness. Returning home after the Senate elections, he issued on 11 December a long statement canvassing the decision not to devalue: 'It is sad and serious that the decision strikes in a most selective manner at our wealth-producing industries, both primary and secondary.' Holt repudiated McEwen immediately and forcefully. It was a worthy action of a Prime Minister protecting the currency and asserting the principle of Cabinet responsibility; it was a courageous action of a coalition leader asserting the collective policy against a sectional interest and a party faction. In a year of drift and decline, it was Holt's best day. Five days later, he disappeared in boiling seas off Portsea, Victoria.

7
The Coming of John Grey Gorton

John McEwen decided who would not be Prime Minister after Holt. Whitlam indirectly determined who would be Prime Minister. Because of McEwen, it could not be McMahon; because of Whitlam, it was Gorton.

McEwen, sworn in as interim Prime Minister, failed in a transparent manoeuvre to obtain the leadership himself by involving both parties of the coalition in the election process. When McMahon, as acting leader of the Liberal Party, established with other potential candidates that the Liberals and the Liberals alone would choose the new Prime Minister, McEwen let it be known that he would not serve under McMahon. At a press conference on 20 December, two days after Holt's disappearance and two days before the memorial service in Melbourne, McEwen said, 'I have told Mr McMahon that neither I nor my Country Party colleagues would be prepared to serve under him as Prime Minister ... I had a tormenting problem for myself, knowing that this was the attitude of myself and my Country Party colleagues in deciding whether I should disclose before the Liberal Party's election at the cost of being accused of seeking to influence it; or alternatively not disclose my attitude, and, it being what it is, if Mr McMahon were elected leader of the Liberal Party, then undoubtedly produce a very serious national crisis by only at that time indicating that we could not work under his prime ministership.'

With McMahon removed, four candidates emerged: Senator John Gorton; Paul Hasluck, Minister for External Affairs; Leslie Bury, Minister for Labour and National Service; and Billy Mackie Snedden, Minister for Immigration. Snedden damaged himself badly at the outset by a fatuous statement claiming to be 'on the wavelength of the era'. The principal claim on behalf of Bury, a lanky, languid Englishman, was that he represented a Sydney electorate. The real contest was between Gorton and Hasluck.

Paul Meernaa Caedwalla Hasluck, journalist, poet, teacher, historian, diplomat, seemed superbly equipped for the highest office. Since 1951, he had served as Minister for Territories, Defence and External Affairs. Son of Salvation Army parents, he assumed a not unattractive patrician air. The press said he was aloof. This was hardly a disqualification for the office he sought, but in the atmosphere of intrigue, bitterness and bewilderment after Holt's death, it was a disadvantage. But even his closest supporters could not have known that by 1968 he believed himself to be, politically, a burnt-out case. He was to write in 1975:

> Territories killed me politically and I knew all the time that it was killing me, but what else could one do but stick at a job that no one else wanted . . . Twelve years a Minister for Territories killed in me all personal political ambition and deadened my political interest.

Gorton, by contrast, keenly wanted the job and openly worked hard to get it. He had three disadvantages. He was a senator; he had no national reputation; and Menzies preferred Hasluck. The managers of the Gorton campaign were the Liberal Chief Whip, Dudley Erwin, and the Minister for the Army, Malcolm Fraser, and Senator Malcolm Scott (W.A.). But Gorton's chief weapon was television. Alone of all the candidates he realized that, in the extraordinary circumstances existing in the nation, the numbers in the Party room could be influenced by pressure from the people. And that meant using television.

Alan Reid, whose book *The Power Struggle* will stand as the definitive account of the period, writes:

> Gough Whitlam . . . had emerged as the Liberals' electoral ogre because of his yet minor but growing impressive successes against Holt in parliamentary performances, in by-elections and at the Senate elections. Gorton, whose parliamentary existence in the

Senate had been remote from Whitlam's, had to create the impression that he had the ability to outmatch Whitlam if there should be a direct confrontation as there inevitably would be once Gorton was raised to the prime ministership and entered the House of Representatives, Whitlam's political habitat ... This view of Gorton as a TV performer reflected the view of a significant number of Liberal parliamentarians who voted in the elections. They had no hesitation in saying that if TV gave Gorton such an advantage over his Liberal rivals he should be able to use TV to outshine Whitlam also.

The numbers game being vulgar, the Liberals do not publish their Party room votes. Reid believes that after the elimination of Bury and Snedden the second ballot gave Gorton 43 and Hasluck 38. The idea that Gorton was the man to match Whitlam was decisive.

Hasluck offered stability and experience. The post-1966 Liberals, secure in their massive majority but thrown off-balance by Holt's death, chose novelty and experimentation. If they had chosen Hasluck, the great gains by Labor in 1969 would have been impossible; Labor's victory in 1972 would have been difficult. The Liberals lost the 1972 elections on 9 January, 1968. By indirectly influencing their decision in favour of Gorton, Whitlam unknowingly had guaranteed his six-year timetable.

If the Liberals chose Gorton partly because they thought he could equal Whitlam, it was equality by contrast, a balance of opposites. Gorton came to the prime ministership less prepared for its rigours, intellectually and psychologically, than any of his eighteen predecessors; Whitlam's career to 1968 had already been a long exercise in self-training. Gorton had no deep attachment to his party and ultimately left it; for all his fights and confrontations, Whitlam's relationship with his party was fundamental. Gorton had a cavalier attitude to Parliament; Whitlam was a Parliamentarian through and through. Gorton was contemptuous of most of the public service; Whitlam inherited a respect for it. Gorton scorned detail; Whitlam had a passion for it. Gorton worked on hunches and instincts; Whitlam advanced from a carefully prepared position, even when seeming to act most rashly. Gorton struggled in the serpentine coils of a language which frequently overwhelmed him; Whitlam loved the language for its precision. Gorton, as a senator, represented no special constituency. Whitlam was emerging as the spokesman for a vast post-war suburban constituency. Gorton was a generalist; Whitlam worked from the particular to the general principle. Gorton was given to slow long anger and lasting grudges, a man of the vendetta; Whitlam, to short sharp outbursts, not given to grudges, forgetting little but forgiving much. Gorton had a lonely, unsettled boyhood, shuttled between boarding schools, his father's orchard property and Melbourne hotel suites and the working-class cottage of his natural mother at Port Melbourne, a curious mix of luxury and austerity; Whitlam's background was one of uncomplicated affection, solid comfort, middle-class stability and straightforward respectability. Gorton was a heavy drinker; Whitlam in 1968 had still to learn, as one newspaperman put it, to 'hold his beer-glass convincingly'. Gorton's crumpled face, the result of a war-time injury, contrasted, appealingly to many, with Whitlam's more conventional good looks. Gorton was confident and aggressive with women; if ever Whitlam was aggressive with women, it masked an inner shyness. Gorton was by nature suspicious and was sparing in his confidences; Whitlam was essentially trusting and gave his confidence too easily and indiscriminately. Yet both were to be brought down by men they trusted. Even in their attitude to their shared war-time experience they differed. For Whitlam, it was marginal in moulding his character and attitudes; for Gorton, it was the central experience and he was essentially a man of the officers' mess, even in 1968. The one quality they shared was political courage of the highest order. The interaction of these two contrasting men dominated Australian politics for the next three years and permanently changed the content and style of the Australian political debate.

From Osaka, Japan, Whitlam cabled the new Prime Minister:

> Your colleagues have given me a formidable opponent. I look forward with zest to our contest in the cause of our parties and our co-operation in the cause of our country.

For Whitlam, the chief difficulty was that he knew practically nothing about Gorton. He had been able to establish a clear ascendancy over Holt; but he knew and understood Holt. As Deputy Leader he had collaborated closely with Holt as Leader of the House. He had never heard Gorton speak. Their contact in what passes for social life in Canberra had been minimal. Whitlam was as ignorant of Gorton's policies or ideas as most of the rest of Australia. He knew that Gorton had South African connections and had defended the Smith regime in Rhodesia; on an otherwise blank sheet of Whitlam's knowledge, there was that one black mark. Whitlam approached the forthcoming contest not so much with zest as with curiosity.

In the eyes of all observers, Gorton was very much top dog throughout 1968. Far from winning ascendancy, Whitlam seemed uneasy, erratic, imperilled. All this was surface; beneath, Gorton's conduct, Whitlam's action, and mighty events abroad, were nibbling away at the Liberal hegemony.

Gorton's great defect as a Liberal leader was that he had no reverence for the myths and little understanding of the facts upon which the Liberal Party's power were based. When he was most correct, he did most damage to himself. He saw the weakness and inefficiency of the public service, and particularly the flabbiness of his own Prime Minister's Department. He did not realize how deeply conservative governments depended on its co-operation. By posing a threat to entrenched interests he created enemies within the citadel of power itself. He saw the need for a Prime Minister to have a small personal staff of loyalty and dedication: but by doing so, he caused envy and gossip, especially when he appointed an intelligent, energetic woman, Ainslie Gotto. He believed that a Prime Minister had a right to a private life; he did not realize how much the conservative Establishment demanded an outward show of respectability. He had an instinctive feeling that Australia had come of age and that State divisions were artificial. He did not see how much conservative power rested in the Federal system and how much the great business interests in Melbourne and Sydney depended upon the Federal structure to preserve and promote their interests. He did not like being bullied by the DLP, and thought that it was on the way out, sooner or later, as a political force. He did not realize how much the Liberal dominance of the past decade depended on the DLP. He thought the Australian nationalism would respond to the idea that Australian companies should have a greater share in the exploitation and development of Australia's mineral resources. He did not realize that capital is neither patriotic nor national. Within a year, Gorton had offended the public service, the Liberal Party organization, the DLP, the mining corporations, the State Premiers and the United States State Department. All these were key elements of the Liberal hegemony. The only thing Gorton had to put in the balance against that was his continuing popularity with the people which, in the emotional aftermath of Holt's disappearance, was real and deep. Should that begin to erode, Gorton was doomed.

Gorton's basic and irretrievable mistake was not to take advantage of his popular support before it eroded and before he made powerful enemies. He made the mistake twice in 1968 by not calling an election in May, as he could have, on the grounds that as a new leader he needed a mandate, and by not calling an election in November, as he could have, on the grounds that his new policies required confirmation. He really believed that his honeymoon with the public would endure. He did not see how quickly the basis of his popular support would erode or how much he depended upon old established interests which he was offending. The best of leaders, the most powerful of Prime Ministers will get into difficult patches; then they need support. Gorton's problem was that he weakened the traditional support of a Liberal leader and failed to get new areas of support. Gorton thought that being Prime Minister was easy. He conceded the difficulty of getting there; but once achieved, the rest was easy enough. All that was needed was common sense; if one knew the right thing to do, then it would be done, and the people would see that it was being done. He based everything on the way he had become Prime Minister. He thought that popular support could be made a continuing thing and that through it he could control the Liberal Party in Parliament and outside Parliament. He never understood the real nature of the Liberal Party. He began by believing that forthright honesty, the frank admission of doubt or difficulty, the approach he had used so effectively in his leadership campaign, would serve his purposes as Prime Minister. Yet this approach made his supporters uneasy from the outset. It was in the field of international affairs, so basic to Liberal prestige, that the first weakness appeared.

In an ABC interview on 21 January, he said: 'Australia could not move in and fill the gap that Britain had left east of Suez. The five power talks are something that we would attend, but not something which we would initiate.' It was no more than a statement of the obvious but to a generation trained in the rhetoric of the 'downward thrust of China between the Pacific and Indian Oceans' it was strange and disturbing.

As with so much in Australia's story in this decade, Gorton's destiny was deeply influenced by Vietnam. He was another Vietnam casualty. On 31 January 1968, the Vietcong launched their great Tet offensive. The Tet offensive devastated large areas of South Vietnam and the White House. It destroyed Hue, Vietnam's ancient capital. It also destroyed Lyndon Johnson, President of the United States. The Tet offensive was a military failure which won the war. It achieved the first aim of military effort—to weaken the morale of the enemy.

In the second week of January 1968, Whitlam spent ten days in Vietnam. By an accident of timing, his visit coincided with the twice-yearly tour by the Australian Ambassador of every place where there could be said to be any sort of Australian military presence. This took him from the Mekong Delta in the south to Hue in the north. For ten days, he was subject to unrelenting propaganda by the Americans. Four months later, Senator George Romney was to destroy any chances he may have had of winning the Republican nomination for the United States presidency by admitting that he had been 'brainwashed' in Vietnam. Only those who were subjected to it can understand what he really meant and how powerful and skilful and unremitting was the pressure and conditioning. In a dozen provinces, Whitlam was told the exact number of people protected, pigs distributed and fish put in ponds under new 'resettlement' programmes; but could it really be true, Whitlam queried, that the number of fish was exactly 21,173 rather than say, 20,000 or 30,000?

The precision seemed phoney. The Australian Ambassador, L. H. Border, a dedicated defender of the Government to which he was accredited, said that such questions were unfair. But the doubts remained. Back in Saigon, Whitlam's visit, like that of the troop of visiting ministers and congressmen and parliamentarians before him, culminated in the great set-piece, interviews with the American Ambassador, the patrician Ellsworth Bunker, and the American Military Commander, the superbly clean-cut General Westmoreland. Westmoreland asserted: 'There is no place in South Vietnam which could not be overrun within the next twenty-four hours; there is no place in Vietnam where the Vietcong could last more than twenty-four hours.' Two weeks later, the Vietcong took Hue and held it for six days before American forces saved Hue by destroying half of it. They invaded and held for more than twenty-four hours the American Embassy in Saigon itself.

At the height of the Tet offensive, Gorton held a press conference on 12 February in Canberra. He was asked, 'In view of these latest developments is there any suggestion that Australia will increase its commitment?' Gorton replied, 'Australia won't increase its commitment.' 'Is that a final statement?' Gorton said, 'As far as I am concerned, it is.' In fact, Gorton was only saying publicly in February what Holt had said privately to President Johnson in October. The Australian Chiefs of Staff had made it clear to Holt that Australia's military capacity was overstretched by the despatch of the third battalion in October; Holt had written to Johnson making it clear that the third battalion was the limit of Australia's commitment and capacity. Yet by stating no more than the truth, Gorton miscalculated badly. By presenting settled Government policy as a personal one, he seemed to be changing policy without consulting his colleagues. By pre-empting any further request from the United States he seemed a less wholehearted ally than Holt or Menzies, an ally who refused help at the very time when help was needed most. It was the first discordant note in the carefully orchestrated rhetoric of the past six years. In particular, he offended the DLP, which had welcomed his election in the belief that he was an uncomplicated hawk on Vietnam. It was this incident which sowed the seeds of the distrust and suspicion between Gorton and the DLP, which was to become a major factor in his decline. It caused the first stirring of doubt on the part of his Cabinet colleagues, particularly his friend and backer Malcolm Fraser, now Minister for Defence. That all this should have happened because he stated no more than the facts and the necessity of the case is a measure of the extent to which the Liberal Government had become a prisoner of its own rhetoric. The lack of true debate about Vietnam meant that it had become dangerous even for a Liberal Prime Minister to state a simple fact.

Returning from Vietnam, Whitlam stated in a television interview that in 1966 both Calwell and Holt had 'debauched the debate' on Vietnam:

> No self-respecting country could have pursued the policy which Mr Holt or Mr Calwell put in the House of Representatives elections in November 1966. Mr Gorton is doing his best to disown his predecessor's presentation of this. I, of course, have done the same. I have had to do the same. But let us deal with things as they are now. The important thing is, we ought to be trying to bring about an end to hostilities in South Vietnam. We will not achieve this by continuing an attack on North Vietnam. The policy which the Australian Government has aided and abetted America in pursuing for the last three years is in ruins.

Whitlam's remark about Calwell inevitably triggered off another spate of criticism against Whitlam in the Victorian branch. In itself a small incident, one of Whitlam's by now notorious television asides, it was to be part of a chain leading to another explosion which imperilled Whitlam's leadership. It was part of the poison that flowed from the Vietnam tragedy and which left no one who touched it unscarred.

Gorton's first test as the newly-elected member for Higgins in the House of Representatives came over an incident in Vietnam. His performance was not impressive. In March, newspaper allegations appeared that a woman suspected of being a member of the Vietcong had been tortured by an Australian army interrogator. The new Minister for the Army, Phillip Lynch, on the advice of Army officials asserted there was not a 'scintilla of evidence' to support the allegations. Evidence from reporters and photographers on the spot at Nui Dat, the Australian base in

South Vietnam, quickly forced Lynch to offer an inquiry. In the subsequent debate, Gorton adopted a tone patronizing to the Parliament and perfunctory in its handling of the issues involved. The woman, he said, might have ended up the interview 'a little wet'. It was less than a Prime Ministerial performance.

In the eyes of his supporters, he redeemed himself with a competent effort in the last of the debates about the *Voyager* disaster. The report of the second Royal Commission had dealt with allegations about the drinking habits of the *Voyager*'s captain but threw little new light on the actual cause of the disaster. Its long-term political significance was that it further alienated Edward St John from the mainstream of the Liberal Party. In what might have been a warning to Gorton he said, 'I must confess that I have felt at times when reading the report that the Commissioners were a little too tolerant of human weakness, a trifle too benign in their attitude . . .' The feeling in the Liberal Party was overwhelmingly one of relief to have done at last with the whole business. They were grateful to Gorton for his part in helping to bring it to an end.

Whitlam's own speech, described by the *Sydney Morning Herald* as 'highly damaging and in many respects unanswerable', had been marred in the eyes of a House whose mood was generally one of forgiveness and reconciliation, by a reference to the Navy official Secretary as 'this creature'. This was deemed not to be playing the game. In criticizing anyone involved, however mildly, Whitlam and St John opposed the general mood summed up by Gorton's new Minister for the Navy, C. R. Kelly, closing the debate: 'Let us now rule the book off and let the Navy get on with its job.'

But the *Sydney Morning Herald* editorial on the debate was nearer the mark:

> Truly the *Voyager* has proved unsinkable. Four years after it went down it remains a scar on our conscience. Two Royal Commissions have been held, one contradicting the other, and neither inspiring our full confidence. Three parliamentary debates have run their course . . . Two Navy ministers have been replaced. Two Chiefs of the Naval Staff have moved on. And throughout this long period of recrimination, reputations have been tarnished, Navy

morale has suffered, public money has been wasted, sad memories have been kept alive. It is the longest and sorriest story in our naval—and perhaps in our parliamentary—history.

When next day Parliament adjourned for Easter, Gorton's colleagues generally felt satisfied with their choice. Their judgement about his ability to match Whitlam seemed justified. A fortnight later Whitlam was facing a mortal challenge to his leadership.

8
Resignation

Late on Friday 19 April 1968, Labor Federal Parliamentarians were variously stunned, appalled, or delighted, to learn that Whitlam had resigned the leadership. In a telegram to all members he said:

> Am firmly convinced that I cannot face the Parliament or the public with confidence unless the Caucus shows its confidence in me. You will have read that three times at Federal Executive yesterday, Lance Barnard, Harry Webb and I were outvoted on a show of hands by Lionel Murphy, Sam Cohen, Martin Nicholls and Jim Keeffe. Such damaging division entirely nullifies McEwen-McMahon and comparable splits among our opponents. Am therefore calling meeting of Caucus for midday Tuesday April 30 when I shall resign and recontest my position. This will permit Caucus members at earliest moment and with least embarrassment to endorse their leader or substitute another. Letter will follow.

Few actions by Whitlam have been more widely condemned; to this day the most frequent description is 'petulant'. In fact, few of his actions have been more deliberate. Most of his friends dismiss it as an aberration. In fact, it was wholly consistent with his career and purpose in the Labor Party. It is still regarded as a needless interruption in Labor's steady recovery. In fact, it was an essential part of the recovery. It has always been put down as Whitlam's greatest failure and worst mistake before 1972. In fact, it succeeded in all its essential aims.

The 1967 Senate result had confirmed Whitlam in his determination to reform Victoria. Once again, Victoria achieved the worst result in the nation being 5 per cent below the national average and 8 per cent below the New South Wales vote. In February and March 1968, he undertook a number of visits to Melbourne to consult with the controlling group and the dissident groups. His message was clear: broaden the base of representation, or face intervention.

The meeting of the Federal Executive which began in Sydney on 17 April was the first significant test of the reforms worked out at the Federal Conference, with the four Parliamentary office bearers now there as full Executive members by right. The compromise inclusion of the two Senate leaders had been deliberately designed to keep the existing factional balance intact. However, Whitlam went to Sydney with high hopes. Since 1963, Federal Executive meetings had become a nightmare. The weakness of the Parliamentary Party confirmed the Executive as the most powerful unit within the Party structure. Each meeting became an exercise in asserting the power of the dominant faction and deepened the antagonism between the factions. Bloc voting by States predetermined almost every issue. Whitlam believed that the presence of the Parliamentary leadership would change the orientation of the Federal Executive. He was to be promptly disillusioned.

At 10.30 a.m. on 17 April, an incident began that became known as the Harradine Affair. In one form or another it was to bedevil the Labor Party for the next eight years. One of the two delegates from Tasmania, Brian Harradine, had been a member of the DLP in South Australia. Clyde Cameron believed that Harradine had gone to Tasmania to build a new base for the National Civic Council and he organized a campaign to prevent Harradine taking his seat at the Federal Executive. Harradine issued a circular stating: 'When I go to the meeting of the Federal Executive, the friends of the communists intend to try and silence me. I have been informed that they will try to exclude me from the Federal Executive meeting, so that there will be one vote less in support of Gough Whitlam.' The terms of prophecy guaranteed its fulfilment. Harradine was declared 'not a fit and proper person' to serve on the Executive unless he retracted and apologized. It was to be a long wait. Whitlam had no doubt that Harradine should have withdrawn the 'friends of communists' reference. But what appalled him most was the vindictiveness of the proceedings. The investigation into Harradine's credentials degenerated into an inquisition into his background and beliefs. Murphy took up the role as counsel for the prosecution. The interrogation continued for a day and a half, with increasing bitterness.

At this meeting, the new Secretary of the South Australian branch, Michael Young, first showed his calibre as a man of solid common sense with an ability to reduce highly emotional issues to a more reasonable perspective and to reduce temperatures. He drafted a skilful resolution that Harradine's credentials be accepted but that he be 'censured for his action in collecting and distributing a document, some sections of which have been interpreted as an attack upon members and branches of the ALP and that the issue and distribution of this document be referred back to the Tasmanian branch for their further consideration.' This called for some courage on Young's part because the drive for Harradine's destruction came from his friend and mentor, Clyde Cameron. Young's motion was defeated on a tied vote, 8–8. It was the intransigence against Young's compromise, which would have ended the 'Harradine Affair' once and for all, that finally determined Whitlam to resign. Young's emergence on the national stage was to prove the most hopeful and constructive event to come out of all this turmoil. In the atmosphere of recrimination, sectarian bitterness and vindictiveness in Sydney in April 1968, it passed unnoticed.

Having disposed of Harradine, the Executive then turned to Whitlam. A Victorian complaint against Whitlam's continuing criticisms against the performance and composition of the VCE was upheld. So was Calwell's complaint against Whitlam. It became clear that the new Executive was as bitter and damaging as ever. Whitlam was absolutely certain that the Harradine issue was simply a front for the real issue, the VCE.

By Thursday night, Whitlam had virtually decided on resignation. He told Barnard what was in his mind. Barnard questioned its wisdom but assured Whitlam of his support. During the lunch-time adjournment on Friday, Whitlam, Barnard and Wyndham met in Whitlam's Sydney office on the

8th floor of the Commonwealth Bank in Martin Place. Barnard agreed that Whitlam's course was correct and suggested that he too should resign and recontest. Whitlam rejected the offer. He was reasonably confident of his own re-election, even if there were a contest. He was not so certain of Barnard's chances. Wyndham, who arrived when the telegram was already being despatched, agreed with the course decided upon.

During the weekend, Whitlam drafted one of the most important letters of his career. A revised version, written in close collaboration with Wyndham, was sent out on 23 April. An elaborate exercise, involving the use of Commonwealth cars in every capital, was devised to ensure that members received the letter before they read it in the press. It began:

> As I undertook in my telegram of last Friday afternoon, I now set out the reasons which obliged me to take the unprecedented course of throwing open the leadership of our Party although my leadership has never been challenged inside Caucus. The proceedings of Federal Executive on Wednesday morning, Thursday morning and afternoon and Friday morning had produced a thoroughly damaging situation for the Party as a whole. A direct and deliberate affront had been offered to the Tasmanian Conference, the Tasmanian Executive and the sole remaining Labor Government. The proceedings created a well and deservedly publicized impression of intransigence, factionalism and bitterness.
>
> The Federal Executive still has no Rules for laying or hearing charges. It can at any time invoke the emergency powers adopted thirteen years ago. The Parliamentary Party bears the direct brunt of any loss of public confidence or esteem. Its leader was in the minority at every stage of the issue and in all attempts to find a reasonable solution. Nearly half the majority against me was provided on each occasion by four of our colleagues, including two of my fellow office bearers. All these factors created a situation which I could not ignore if I am to retain any public credit or credibility as Leader of our Party. Nor could the Caucus ignore them if it is to retain public credit or credibility as a future government of this nation.
>
> I do not contest the propriety of any member of the Caucus voting against me on the Federal Conference or Executive. On the

contrary, as you all know, I have resisted the whole concept of bloc voting at all levels of the Party. I do believe, however, that all members of Caucus have a special responsibility to promote the interests of the Caucus. In particular, they should promote our chief function, which is the formation of a Labor Government at the earliest possible moment . . .

The avowed intention of the Federal Executive majority was to exclude one Tasmanian representative. The predictable result was to exclude both Tasmanian representatives. Equally predictably, the Tasmanian Branch has been thrown into needless turmoil. The issue is not Mr Harradine personally, it is whether any delegate from any State is to be tried by the Federal Executive without notice and without charges and whether Caucus delegates are to back procedures which do not comply with the spirit of our Civil Liberties platform.

It is well known to you that those who excluded Mr Harradine did not aim simply to exclude Tasmania or even Mr Harradine. They aimed to preserve, in terms of voting strength on the present Executive, the position of the Victorian Central Executive. Because, as is equally well known to you, I have tried to secure reform of the present VCE, it was aimed at me. Therefore I am obliged to test whether my colleagues repudiate my efforts to secure reforms in Victoria . . .

I wish to make it clear beyond all doubt that, if you re-elect me, you will approve my efforts to secure a Special Conference to deal with the foregoing matters. I could accept the position on no lower terms. To do otherwise would make a mockery of all my efforts to fulfil the trust you reposed in me in February 1967 to do all I could within the framework of the Platform and Rules of the Party to lead us into government . . .

Beyond his letter, Whitlam did very little on his own behalf. Apart from Barnard, his most active supporter in Caucus was Reginald Francis Xavier Connor (Cunningham, N.S.W.). It was an unlikely combination. Connor, a massive shambling hulk of a man, already in his sixties, had spent years of frustration and resentment on the back-benches of the New South Wales Legislative Assembly. Totally out of sympathy with the right-wing controllers of the New South Wales organization and Parliamentary Party, he was scarcely more compatible with the New South Wales 'left', organized around the Miscellaneous

Workers' Union. In 1963, when Calwell heard that the New South Wales Party had smoothed Connor's transition to Canberra he complained to the New South Wales Labor Party Secretary William Colbourne, 'You want to get rid of the thorn in your side by putting it in mine.' He seemed a man rooted in the bitterness of the thirties and the Great Depression. Few guessed that he had a grand vision for Australia's future as the powerhouse of the Southern Hemisphere.

Gruffly monosyllabic with those of whom he was suspicious (a substantial company), he was capable of genuine and original eloquence in defence of his vision. He had nothing in common with Whitlam in terms of temperament, style, attitudes or background, but in 1967 he saw Whitlam as the coming man, the one through whom his vision might be translated into reality. He managed to secure a place on the Executive (Shadow Cabinet) at the Caucus elections. Whitlam had established a system of specific allocation of portfolios. Written off by most as a spent force in 1967, he was gratified to be appointed by Whitlam as Labor's spokesman on fuel and energy matters, a field big enough for his vast designs. This was the basis of his alliance with Whitlam in 1968.

On Anzac Eve, James Ford Cairns called a televised press conference to announce his challenge to the leadership. Its impact was somewhat diminished by the fact that because he was suffering from laryngitis his wife had to announce his challenge. In his own letter to Caucus, he put the issues in highly personal terms. This document encapsules almost all the criticisms of Whitlam's style of leadership ever made from 1968 down to this day:

> A contest for leadership is not my choice. It is the result alone of Mr Whitlam's decisions. To permit Mr Whitlam to contest the leadership unopposed could leave the incorrect impression that the Parliamentary Party unanimously supports his stand.
>
> His resignation and conduct have endangered our party because they brought completely into the public arena matters which should have been settled elsewhere within the Party's constitutional procedures.
>
> They have raised the question just how far Mr Whitlam can go

in defying majority decisions of the party authorities of which he is a member or with which he is associated.

They raise the question: Whose party is this — ours or his . . .

I am opposed to any attempt by any man to centralize power or to dominate his party colleagues. I cannot accede to the leader's demand that parliamentarians must agree with him in the various party conferences and executive. Disguise it as he may, this is the demand he makes.

This is intellectual arrogance and dangerous folly. Just this is what is involved in the present crisis. Unless harmony is established, it must be obvious that the party will be in a continuing state of crisis which will place in danger the seats of many of our present members whose majorities will not stand this state of affairs.

This means that we will be pushed further into the political wilderness . . .

The assumed appeal of Mr Whitlam to the middle-class electorate can never offset the devastation done by conflicts within the party which seem inevitable from his continued refusal to accept majority decisions when they go against him. I do not think Mr Whitlam has proved himself a stable leader and an unstable leader means an unstable party. Such a leader should be the last man to be given greater powers. It would be disastrous if Mr Whitlam could claim a mandate to continue what was this week called 'his war against the party'.

Cairns' question was: 'Whose party is it — ours or his?' It was half right, for Whitlam had indeed raised the question of whose party it was. Equally, Cairns had missed the point; a more accurate question would have been 'Theirs or yours?' — the party of those who controlled the Federal Executive by their grip on key positions in the State machines or of the Parliamentary Party and the wider membership. It was absurd to suggest that Whitlam sought dictatorial power by placing in jeopardy the post he already possessed. After all Cairns, not Whitlam, was the challenger. If he had defeated Whitlam, who would have benefited? Certainly not the members of the Parliamentary Party to whom he addressed his appeal.

Jim Cairns distrusted power and wanted power. He challenged authority and associated himself with the authoritarian 'left' of the Labor Party. He delivered speeches of burning passion in a cold monotone. In striving for a world view of human conduct, in seeking a framework of universality, he was equalled by only one other Australian political philosopher — his arch-enemy, B. A. Santamaria. Like Santamaria, he was essentially a product of political and intellectual Melbourne of the late thirties and forties. Like Whitlam, his political activity began only after the war.

Connor's counterpart as manager on behalf of Cairns was Clyde Robert Cameron. Cameron, member for Hindmarsh (S.A.) since 1949, bore the scars and bitterness of the Split and the years in the wilderness less comfortably than most of his contemporaries. The iron had entered his soul. First-rate talents were wasted in the years of frustration and failure. In 1968, he was altogether unimpressed by Whitlam's promise and performance. He did his work on Cairns' behalf well, but not quite well enough.

Taking the chair for the balloting on 30 April, Barnard ruled that the only matter before the meeting was the leadership, not the other matters canvassed in Whitlam's letter. This ruling was to lead to doubt whether Whitlam had any mandate from the Parliamentary Party to continue his campaign for the reform of the Victorian branch. Given a mandate he was; but was it a mandate merely to retain the leadership or a mandate for the style, methods and objectives of his leadership? Whitlam's margin over Cairns was narrower than expected — 38 to 32. Whitlam interpreted the mandate in the broadest sense; and although the special federal conference which he sought was never called, he continued his campaign for reform.

*

Was it worth all the turmoil? Part of the answer lies in asking the alternative. If Whitlam had done nothing and, as he himself put it, 'Was I just to cop it?', what would have happened? The Federal Executive would have been confirmed in its conduct. The Victorian Central Executive would have known that it could continue unchallenged in its conduct. The small dissident group in Victoria would have been further discouraged and probably would have disintegrated. The reforms of the Adelaide Conference would have been nullified. The hope of further reforms in the party's organization would have disappeared for all time. The 'faceless men' charge would have been revived. The Parliamentary Party would have been further demoralized.

Short of the extreme step of resigning, there was one other possibility open to Whitlam. He could simply have sought a vote of confidence from Caucus. This is the tactic of leaders on the defensive. It is always a sign of weakness, for it risks nothing and settles nothing. A leader under attack can usually secure a vote of confidence; it is a cheap victory which buys nothing, except a breathing-space. In April 1968, Whitlam was not in such a situation and he was not such a leader.

The big achievement of the resignation was that it put the Federal Executive on notice. The fact is that before April 1968, every meeting of the Federal Executive was a source of danger, damage and embarrassment to the Australian Labor Party. After the April 1968 trauma, the whole tone and nature of meetings changed. Thereafter, the Federal Executive began to fulfil its proper purposes of administering the general affairs of the Party to the public advantage of the Party. To assert that there was no cause and effect is to accept a higher degree of coincidence than normally occurs in politics. If it was a catharsis for Whitlam, it was a catalyst for the Federal Executive.

Above all, Whitlam had shown the Labor Party how not to split. Evatt split the Party by causing a confrontation from weakness, not risking a thing; Whitlam kept it together by confrontation from strength and risking all.

9
'I did it my way'

'You have just seen the beginning of the end of the Gorton Government.' The speaker was Clyde Cameron, in the Parliamentary office of the Opposition Leader on 20 March 1969 at 12.03 a.m. Two years less nine days would pass before his prophecy was fulfilled. With the insight vouchsafed only to the shrewdest of practitioners, Cameron had divined the fatal flaw which would bring down Gorton: he had offended the conservative Establishment; sooner or later they would destroy him. The incident of 19–20 March was trifling; its huge consequences sprang from the depths of Gorton's style and character. Given his nature and the nature of the Liberal Party, there was a Greek inevitability about his fall.

After a year as Prime Minister, Gorton was still enormously popular. But under this surface, he had undermined the basis of the Liberal hegemony he had inherited from Menzies and Holt. He did not understand its essential elements. He did not see the importance of the American alliance and clear unswerving support for American policies in Asia; he did not acknowledge the crucial role of the DLP; he did not understand how easily business could be unsettled, by even the smallest and most distant threat to its perceived interests; he did not realize how much conservative government depended upon a loyal, stable and secure bureaucracy; he did not know the power of Liberal State Premiers in the Liberal Party structure, particularly in its home base, Victoria; and he did not grasp the absolute necessity of maintaining a public facade of private respectability. By the end of his first year, Gorton had made the inner circle of the Liberal Party uneasy about all these matters. If he knew, he did not care.

In the heady emotion of the first days, he believed he had forged a lasting bond with the Australian people. He was convinced he could by-pass the Establishment and get straight to the heart of the people; he would by-pass established procedures and get straight to the heart of the peoples' affairs. As for his private life, 'John Grey Gorton will behave as John Grey Gorton bloody well wants,' he told the DLP Leader, Senator Gair, who replied with equal bluntness, 'Bloody John Grey Gorton will bloody remember that he is Prime Minister of Australia.'

He might still be Prime Minister of Australia if he had held an election in 1968. His failure to do so was a political blunder of the highest order. His popularity was high. The Labor Party was still recovering from the shock of Whitlam's resignation and its aftermath. He was entitled to seek a fresh mandate. 'The press was squared and the middle class quite prepared.' His refusal lay in his nature: everyone was expecting it and he delighted in the unexpected, and the DLP had put his back up.

Even in the great Liberal sweep of 1966, a quarter of their seats had been decided by DLP preferences. Since Gorton's statements on Vietnam during the Tet offensive, the DLP and its philosopher, B. A. Santamaria, had become increasingly uneasy. Gorton began developing a new defence doctrine which, despite his denials, was dubbed 'Fortress Australia' — a reversal of 'forward defence', described by Holt as 'fighting in other peoples' backyards'. To question this doctrine was to strike at the heart of the DLP's obsession — the ever-present threat from the North. Vietnam itself was the perfect expression of that doctrine. Visiting Singapore and Malaysia, Gorton had seemed to qualify Australia's support for new defence arrangements involving Britain and New Zealand. Visiting Djakarta, he had responded to a question about SEATO with a counter-question: 'Who is this General Seato?' He even appeared casual about the Ark of the Covenant, ANZUS itself. In Washington, he was asked whether ANZUS applied to Australian forces in Malaysia:

> I don't know I can give you any definite answer to that. ANZUS is a treaty. I think it applies in certain defined areas. But I would want to check this with the External Affairs people before I was sure that was correct. But by and large, I think it has been, what

shall I say—I cannot think of the exact words—a matter never spelled out whether it applied in the Malaysian and Singapore area or not ... you are asking really the sort of questions which one can pursue to the point where it is the whole sort of subject of discussions. And I do not think I am free to do that.

In all these matters, Gorton added to his difficulties by his clouded, convoluted use of English. Often, the confusion of language reflected simply confusion of thought. Yet there was also a degree of honesty in it; he could not be precise on matters which he had not yet reached a precise view. For the DLP, that was exactly the point—there were matters which brooked no doubt, no hesitation, no self-questioning. This was a steep decline from Menzies' grand simplifications and Holt's eager simplicities.

The DLP opposed an early election. They were short of funds and worried that Gorton would get a commanding personal mandate before he had demonstrated his soundness on basic dogma. They hinted that DLP preferences could no longer be taken for granted. When Gorton, in Whitlam's words, 'so archly and coyly' announced in October that there would be no election in 1968, he was doubly perverse. He was pleased to prove everybody wrong and shrugged off the disappointment of his followers. He told the press, 'If there were anything that would have led me to have an election, it was the statements made recently by the DLP. But I realized that was an emotional reaction, not a rational one. I believe the DLP would have been completely ineffective. I would have loved to see what happened. My sorrow is that I missed that fight.' He had missed more than a fight; he had missed the chance of a great victory and an impregnable prime ministership. He had further alienated the DLP, which had underwritten the Liberal Government since 1955. He had given his supporters grave concern about his judgement. And he had boosted Whitlam's confidence that he could get the measure of this erratic, enigmatic man.

Gorton's failure to seek his own mandate was all the more remarkable in the light of his determination to set his personal stamp on the whole process of government. He failed to see that before he could safely tamper with the Menzies inheritance he needed his own political capital. Only a successful election could provide that. The Liberals would forgive much if he could prove himself a vote-winner. Instead, Gorton appeared ready to drive the Liberal Party in new, uncharted directions before he had established his credentials. The Liberals would never pose Cairns' vulgar question: 'Whose party, his or ours?'; but in 1969, they might ask themselves, 'Whose party—Gorton's or Menzies'?'

Gorton deluded himself in thinking that popular support for particular actions guaranteed the political acceptability of those actions. An incident in September 1968 epitomized his style and its dangers. He learnt that a British corporation was buying up shares in one of Australia's largest insurance companies. In the absence of Treasurer McMahon overseas, and against the advice of the Treasury, and without consulting any of his Cabinet colleagues, he announced, in the middle of a tour of Western Australia, that any take-over bid would be blocked. Such intervention was only in his power because the company involved, M.L.C. Limited, happened to be registered in the Australian Capital Territory. Unquestionably, Gorton's action was popular. He had already identified increasing public concern at the growing international ownership of Australian companies and resources. In London, he had spoken at the Dorchester Hotel about the end of the 'tickle my tummy' approach to foreign investment. But behind the general applause, there were doubts in the places that mattered to a Liberal leader—doubts in the Cabinet, which had not been consulted, doubts among businessmen, who wondered where intervention of this kind might end, doubts in the public service, who had seen the Treasury by-passed and overborne. The MLC incident was small in itself, but it was a portent. Uneasiness stirred among powerful interests. If Gorton could intervene so suddenly and independently on this matter, where might it end? For such groups, it was no justification that his action was popular; on the contrary, its popularity made it all the more dangerous. There could be no greater threat to their entrenched interests than that a Prime Minister should use popular support to avoid 'proper channels'. Gorton did nothing to reassure them when he said in an interview in December 1968: 'A Prime Minister can make a decision and say that's it. Then if the rest of your cabinet doesn't like it, they can say, "Well, we won't go along with it." So that is just too bad for the Prime Minister. So he can get out, I suppose.' In saying that, he was speaking a kind of blunt truth, but in Australian politics, that is risky. For a Liberal leader to define the Prime Ministership in such extreme terms, with its implication of permanent confrontation, was to run a double risk. For a Prime Minister still without a public mandate to place his office on a 'take it or leave it' basis was running a triple risk. And in the MLC case, Liberals, even as they applauded Gorton's show of decisiveness, recalled how differently Menzies had handled a similar matter in 1961 — discreetly, quietly, smoothly, successfully; a few telephone calls, no confrontation, no fuss.

Part of Gorton's trouble as a Liberal Prime Minister was that he became interested in a liberal 'philosophy'. He offered in 1968 to hold a 'Liberal philosophy conference'. Menzies men were horrified to see Menzies' successor dickering about the philosophy. Wasn't Menzies' philosophy good enough? Wasn't a philosophy which had won eight elections good enough? What was the man about? Even when Gorton was trying his best, he was doing himself most harm. He was making everyone uncomfortable, and his Party saw not the slightest need for it. The worst thing about it was that people might begin asking questions. Questions were things that Labor people asked; they should not be asked about the Coalition Government, the Party of Menzies or the Party of McEwen.

Gorton was the first Prime Minister of the television age. Menzies was essentially a pre-television politician. A natural actor, his stage was Parliament and the platform. He was, in fact, a superb television performer, but disliked the medium and suffered bad nerves before his infrequent appearances. Australian television was nearly ten years old when Menzies retired, but as a political medium it was still in its infancy. The war in Vietnam, Johnson's visit, Holt's disappearance and the contest for the succession helped give television a new political importance. Television also influenced newspaper style. As the public relied increasingly on television for basic news, newspapers encouraged their political correspondents to supplement straight reporting with comment and background. A new generation of correspondents brought a new style and independence to the Canberra Press Gallery. The older journalists had depended heavily on the Labor Party to provide their most sensational stories. Among these, Ian Fitchett of the *Age* and then the *Sydney Morning Herald* was unequalled for his ability to get genuine 'scoops' from the Government rather than the Labor Party. With Menzies gone, the Government parties became almost equally newsworthy. By 1968, the Gallery doyens whose experience went back to Curtin and Chifley — Ian Fitchett, Alan Reid of the Sydney *Daily Telegraph*, Harold Cox of the Melbourne *Herald*, Jack Allsop and Neil O'Reilly of the Sydney *Sun*, Kevin Power of the *Daily Mirror* and Wallace Brown of the *Courier-Mail* — were being challenged by a new breed — Laurie Oakes of the Melbourne *Sun*, Allan Barnes of the *Age*, Maximilian Walsh of the *Financial Review*, Eric Walsh of the Sydney *Daily Mirror*, Alan Ramsey of the *Australian*, Jonathan Gaul of the *Canberra Times* and Peter Samuel of the *Bulletin*. The Canberra Press Gallery had long had a herd instinct, with its known herd-leaders. But the old herd-leaders, like Reid, were being challenged by new ones like the two Walshes, Oakes and Barnes. Mavericks also appeared, like Mungo MacCallum, rebellious scion of one of Australia's oldest families, the Wentworths. Subscription newsletters were being published, notably Don Whitington's 'Inside Canberra' and Maxwell Newton's 'Incentive', providing for their subscribers, mainly businessmen, the delights of 'inside knowledge'. For the first time, the public service itself became fair game for newspaper scrutiny. For the first time, the existence and role of ministerial staff, who had hitherto enjoyed absolute anonymity, was acknowledged. This flowering coincided with the arrival of Gorton. The combined impact of two such eminently newsworthy, controversial and contrasting leaders as Gorton and Whitlam added a new dimension to political reporting.

Significantly, Australia's most popular newspaper, the Melbourne *Sun*, with a saturation circulation in Victoria of more than 600,000, began to treat politics as seriously as Australian Rules football. Reading the *Sun* became as obligatory for politicians as the *Age* or the *Sydney Morning Herald*, Australia's traditional papers of record. The manner in which Gorton became Prime Minister gave the press a new sense of involvement and self-importance in the political processes. Thereafter, political journalists saw themselves, accurately enough, as participants as well as reporters. And in each event leading to Gorton's decline and fall, journalists were to play a direct part. Their role was more than professional; quite often it was altogether personal. The role of the press was always important and in Gorton's case it was to prove devastating. He had used the press and television in his drive for the leadership; in the end, he was to be brought down by the same instrument.

In the Australian Parliament, the debate on the motion 'That the House do now adjourn' is a convenient form for members to raise matters of special interest to themselves or their electorate which would not easily be raised in general debate. Most back-benchers use the adjournment debate to put something in the permanent record of Hansard to send to their electors and their local newspapers as proof of their hard work. Others use it as a platform for their more extravagent or eccentric flights of fancy. Coming at the end of the day's sitting, it occurs near the deadline of most daily newspapers. A speech on the adjournment is seldom reported. Under the Standing Orders of Parliament, the debate cannot be broadcast by the Australian Broadcasting Commission, and although it can be the most amusing and fascinating part of the proceedings, it is the least public and the least publicized. The adjournment debate is very much Parliament in its shirt sleeves.

When, at 10.45 p.m. on 19 March, Bert James (Hunter, N.S.W.) raised a certain matter on the adjournment, he was acting very much in the accepted knock about tradition of adjournment debates. James, a former policeman from New-castle holding his father's old seat, was 18 stone, gentle as a cooing dove yet rough as they come in politics, as easily moved to tears or rage, and held a deep contempt for the little hypocrisies of respectability. From his police experience he knew something about the links between crime, politics and business and enjoyed occasional muck-raking. Australia's professional muck-raker for two decades had been a journalist, Frank Browne, the proprietor of a newsletter, 'Things I Hear', a production in a fairly long Australian tradition of vituperative journalism. In 1955, Browne had been the victim of a shameful abuse of power by Parliament when the House, acting as Prosecutor, Judge and Jury in its own case, had sent him to prison for libelling one of its members. It was predictable that the whispers about Gorton would first surface in Browne's newsletter. It was also predictable that James would raise them in Parliament. James' speech touched briefly and vaguely on allegations by Browne that the Central Intelligence Agency had paid the London 'Private Eye' magazine to suppress stories about Gorton's behaviour. James' speech was mischievous and was meant to be.

The matters raised by James were not new. In late October 1968, President Johnson had informed Gorton that highly secret negotiations were under way with Hanoi and that he hoped to be able to announce cessation of the bombing before the Presidential election polling day. Returning to Parliament House after a theatre party, Gorton held an impromptu press conference with two or three journalists still in the building after midnight and revealed Johnson's secret. Johnson made known his understandable rage. On 1 November, just too late, as it turned out, to prevent Richard Nixon narrowly defeating Hubert Humphrey for the Presidency, Johnson announced the bombing halt. That night, Gorton was guest of honour at the Parliamentary Press Gallery annual dinner. The American Ambassador, William Crook, asked Gorton to call in at the Embassy after the dinner to discuss Johnson's announcement. By this gesture, Crook intended to smooth over any ruffled feelings lingering from Gorton's October indiscretion. Despite repeated messages inquiring when the Prime Minister might be expected at the Embassy, Gorton dallied on at the dinner until about 1.30 a.m. He finally arrived in the company of a young woman journalist and stayed until about 3.00 a.m. Versions of this incident circulated around the Canberra gossip circuit for several months. The matter became public property only because of the publicity given to James' speech.

The Minister for the Navy, Jim Killen, rose to the bait and the attack. With great eloquence and greater folly, he reminded Calwell of his denunciation of Browne and all his works in 1955. He challenged Calwell, who was in the House, and Whitlam, who was not, to repudiate James and Browne. Then, even more forcefully, T. E. F. Hughes, Q.C., repeated the demand that the Opposition repudiate James: 'Let somebody on the other side rise to disown him. If nobody does so, we all will know that the Labor Party in its present state of desperation is muck-raking in the bottom of the political bucket . . .' The veteran Labor member of the House, Fred Daly (Grayndler), enjoying himself immensely, cheerfully accepted the challenge. But when he sought leave to speak, the Government Leader in the House, Dudley Erwin, said 'No'. Calwell moved that the House should hear Daly. The House divided, and fifteen Government supporters voted with the Opposition on the motion that Daly be heard. The House divided 42-all. (The Government's normal majority was 36.) A footling piece of Parliamentary mischief had been escalated into a first-class Parliamentary row by the folly of Gorton's friends, Killen, Hughes and Erwin. McMahon and Fraser then intervened to ensure that James and Browne received the widest possible publicity. Barnard, in charge of House business on behalf of the Opposition, came into the Chamber, accepted the foolish challenge of Killen and Hughes, and moved that the whole matter be referred to the Privileges Committee of the House of Representatives. This moved the whole matter onto a new plane. Barnard's motion was ruled out of order for technical reasons. Calwell, drawing upon half a lifetime of parliamentary experience, moved the same motion formally. Faced with this formal motion, purporting to be about the highest of all parliamentary questions — the question of its own privilege — the Speaker, Sir William Aston, adjourned the House. On the Government side, Fraser cursed Erwin for his folly in preventing Daly from speaking which was the Government's last hope of finishing the affair. 'Leave Dudley alone, you long streak,' shouted McMahon, scurrying for the doors, leaving his colleagues aghast and the Press Gallery agog. It was then that Clyde Cameron delivered his remarkable prophecy.

Debate resumed as soon as the House met in the morning. Gorton, without any flourishes, defended himself and tried good-naturedly to defuse the debate. McEwen expostulated and protested, moving into the attack with a characteristic self-observing indignation which signalled to all who knew him well that he was on the defensive, embarrassed and uncomfortable. McEwen was the sort of man who rehearsed his outbursts of anger while shaving. Whitlam, who was embarrassed and uncomfortable in being involved in such an absurd matter as the original allegations, stated the political, distinct from the personal issue.

After McEwen gagged the debate, Gorton summoned a special Party meeting for lunch-time. He peremptorily announced: 'I am going to leave this room in thirty seconds and I want you to show by your words and your action that you approve of my conduct.' St John said, 'Sir, I don't agree.' Gorton snapped, 'I don't care whether you agree or not. If you disagree you can remain silent.' He then strode from the room to a standing ovation.

On the adjournment debate that night, St John crossed the Rubicon and as he wrote in his book, *A Time to Speak*: 'There was really no turning back now.' In his speech he canvassed the whole matter of Gorton's conduct on 1–2 November 1968, using information he had collected from a number of journalists. Gorton dismissed the whole thing as 'an interesting exercise in how something perfectly reasonable and proper can be twisted, turned and slimed over'. In any case, he said, he had left the American Embassy at 12.30 a.m., not 5.00 a.m. as St John had claimed.

At a press conference next day, St John kept up his campaign, putting Gorton's conduct in the context of the Australian-American alliance. 'That day', he said, 'was the first day on which there had been any public falling out between the United States and the South Vietnamese Governments as to the conduct of the war in Vietnam. It was a day when the Prime Minister and the Ambassador had been in contact with one another and there had been a little coolness developing between them on the flow of information. This was the day or, rather, the night, when our Prime Minister chose to linger on, as I would think, quite unnecessarily, into the small hours at a social occasion, stays until 3.00 a.m., not talking about these important developments, not reading the cables on this day, so we are told, but chatting with a nineteen-year-old journalist ... This was the Embassy of the United States, our most powerful ally, on this crucial day. I cannot think that this is private life.'

The end for St John came quickly. He scored one last devastating triumph over Gorton, who was forced to concede that the record showed that he had misled the Parliament about the times of his arrival and departure from the Embassy. It was not to be thought that the Liberals could excuse one who had so humiliated their leader. At the next Party meeting he was asked to withdraw from the Parliamentary Liberal Party. The Minister for Defence, Alan Fairhall, summed it up for his colleagues when he said that St John was 'too independent and too pure of heart to remain in the Liberal Party'.

In this tragi-comedy of errors, Killen stands out as the chief well-intentioned culprit. He paved Gorton's path to destruction with the best intentions. By defending his leader and his friend from a little bit of gossip, he ensured that the gossip would receive the widest possible publicity.

In 1969, Denis James Killen was doing what he had always done best — enjoying life and politics enormously. He was the last of the Regency men and would have been at home in the House of Commons of Burke and Fox, and above all, with Richard Brinsley Sheridan, seeing in the dawn at Brooks's Club. But he was profoundly Australian, a give-it-a-go jackaroo turned lawyer who had 'swum bare-arsed in the Condamine'. His addiction to the florid cadences of Burke led him into pomposities and absurdities. His sense of the ridiculous, not least about himself, saved him from becoming a bore. With a gift for friendship, he enjoyed a few good hates: at that time, chiefly St John, and later McMahon and Fraser. Alone of all the Liberals, he had a genuine affection for Whitlam, which was genuinely returned. Some of the best unpublished jokes in Australian politics are contained in the series of notes and telegrams they have exchanged over the past fifteen years. He was a true romantic in Australian politics, and on this score much has been forgiven him by his political opponents. Gorton and Killen could not help but be mates and Killen and St John could not help but despise each other. This meshing and clashing of personalities, with politicians like Gorton, Killen and St John, produced the events of 19-20 March.

If Killen, Hughes and Fraser had let the matter drop, there would have been no follow-up. Not one newspaper in Australia would have touched James' speech. By forcing a debate and a division, Killen and Fraser forced the press to report it. A speech that would have been buried for all time in the dead, unread pages of Hansard became a significant event. A newsletter with derisory circulation and authority became nation-wide public property. The embarrassment of the Liberals was enshrined in the division lists in Hansard. Further, their tactics forced St John into open rebellion. All this was done to Gorton by his loyal friends.

Killen's motives were friendship and loyalty. The motives of Calwell, whose swift parliamentary move divided the House and the Liberals, were more complex but equally clear. He had no wish to harm Gorton and less to help Whitlam who must inevitably benefit from any damage to Gorton. As an old parliamentary war-horse, he scented a good brawl, but he was aiming at the author of the newsletter, Frank Browne. He saw the chance of gloating over an old enemy; a repeat performance of the 1955 outrage. Browne, not Gorton, was his target. The motives of Fraser and McMahon are not so clear. Between them, they ensured that the debate would continue and would do the utmost damage to Gorton. Exactly two years later, they destroyed Gorton. It is, however, probably unreasonable to suppose that in that night McMahon and Fraser shared Clyde Cameron's clairvoyance.

*

The immediate beneficiary was Whitlam. For most of 1968, he had found it difficult to understand Gorton's character and to come to grips with him. Gorton now became vulnerable. Yet the opinion polls still showed Gorton's and the Government's popularity as high as ever. In the first half of 1968, the Labor

Party's support stayed at 40 per cent, the level of the 1966 elections. Speaking in Melbourne on 17 May, St John said, 'I confidently predict this will be a record year for Independents and the DLP.' Dr Alan Hughes, lecturer in political science at Melbourne University, wrote in the June issue of 'The Australian Quarterly': 'On present indications, Labor will suffer another electoral catastrophe in November.' The June Gallup Poll gave the Government 50.5 per cent, Labor 40 per cent, the DLP 7 per cent and the rest, 2.5 per cent. In May, the sole remaining Labor State Government fell in Tasmania. This had never been achieved, even in the Menzies years. The conservative sweep was complete. Yet, a by-election in Curtin (W.A.), caused by the appointment of Hasluck as Governor-General, recorded a 10 per cent swing against the Government and a by-election in Gwydir (N.S.W.) showed an 8 per cent swing against the Government. A by-election in the Labor-held seat of Bendigo provided the first campaign contest between Gorton and Whitlam. While Gorton did not make Holt's Capricornia error, he campaigned with breezy self-confidence. Bendigo was the See of Bishop Stewart, perhaps the most bitter critic of Labor in the hierarchy. The Labor candidate, David Kennedy, a young school-teacher, had none of the local credentials of the retiring member, Noel Beaton. Gorton failed to erode Labor's narrow majority. Fairly enough, Gorton dismissed these results as irrelevant to a national election. Yet it was not quite what his supporters had hoped for. They were uneasy at the contrast in the opinion polls and actual results. In the Labor Party, these results were enough to keep up hope.

10

Phoenix

For Labor and for Whitlam, it all began to come together at the 1969 Federal Conference in Melbourne at the end of July. From this Conference, Whitlam emerged the indisputedly dominant figure in the Labor Party and, in a way he had not before achieved, its true and accepted leader. The State aid issue was settled, the Parliamentary leadership established its authority over the Party organization, Chamberlain's grip was broken forever, the new Federal Secretary, Michael Young, emerged a bright new force in the Party's affairs and the old web of State alignments and factional intransigence was broken.

It was the first Conference since the Labor Split when the 'winner takes all' principle did not apply. It was also the first Conference in living memory where genuine debate influenced votes. If the parliamentarians tended to have most success, it was partly because they were more experienced debaters. In the week the Conference was held, the astronaut Armstrong took his 'giant step for mankind' on the moon. President Nixon said it was 'the greatest week in human history since the birth of Christ'. If Labor's concerns were more mundane, it was still a good week.

Mick Young's rise to the Federal secretaryship came about through a series of chances and misfortune. The resignation crisis of 1968 marked the high point of the collaboration between Whitlam and Wyndham. Thereafter, the relationship deteriorated. The main cause was Whitlam's unswerving dedication to the principle of equal representation. Proposals for a redistribution of electorates in 1968 failed to uphold this principle, although they were not manifestly unfair to the Labor Party. Whitlam wished the Party to oppose the redistribution in

Parliament and challenge it in the High Court. Wyndham believed that the proposals were the best Labor could reasonably expect and would, in fact, produce a better result for Labor than the existing electoral boundaries. In the event, the 1969 result proved the correctness of his practical judgement. He failed, however, to gauge the depth of Whitlam's commitment and the breach caused a marked cooling between the two. Wyndham, never outgoing, became taciturn and increasingly frustrated as the opinion polls showed Labor apparently headed for another disaster. Unhappy in Canberra, he successfully sought appointment to the secretaryship of the New South Wales branch. But on the eve of his transfer, antagonisms he had aroused among members of the Federal Executive, mainly because of his association with Whitlam but partly because of a certain prickliness of character, brought him down. The Queensland Secretary, Tom Burns, alleged certain discrepancies in the Federal Party accounts. In fact, such a charge was made possible only by the very meticulousness of Wyndham's accounting. But for Wyndham it was the last straw. Exhausted and embittered, he put himself out of reach of friends and enemies by disappearing into Western New South Wales. *In absentia*, he was censured by the Federal Executive and dismissed by the New South Wales Party. His political career was finished.

When the Federal Executive came to choose Wyndham's successor as Federal Secretary, the two candidates, Young and Chamberlain, each received seven votes. Chamberlain broke the deadlock by withdrawing in Young's favour. Throughout 1969, the Harradine matter had been threatening to disrupt Federal Conference. That it did not was mainly due to Young's finesse in defusing a potentially disastrous problem. He generated an atmosphere of determination that the Conference would be a success. He worked skilfully to isolate the Harradine issue as an irrelevancy and effectively isolated Harradine himself.

Whitlam's critics have always used Wyndham's fall and Harradine's isolation as evidence of weakness and even disloyalty. It is true that he chose to avoid a confrontation. However, the conditions for confrontation did not exist in this issue, nor was there cause or justification for it. He tried to protect

Wyndham, but Wyndham left himself defenceless by his disappearance. It should also be emphasized that Whitlam owed Harradine nothing. Whitlam's resignation in 1968 was not undertaken on Harradine's behalf but on behalf of the Parliamentary Party. By 1969, the balance of obstinacy and intransigence had passed from the Federal Executive to Harradine himself. For Whitlam to have plunged the Party into another crisis on the eve of this crucial Conference, itself on the eve of a critical election, would have been the ultimate betrayal.

It would also have been treachery against a great opportunity in Australia's history. The moment was at hand when seven generations of sectarian bitterness over State aid might be brought to an end. As the debate drew to its close, it became more intense than it had ever been. Throughout 1969, groups representing the opposing views organized a series of rival meetings throughout Australia, particularly in Melbourne and Sydney. These meetings produced the first opportunity for confrontation between Whitlam and Malcolm Fraser, to whom Gorton had given his old portfolio of Education and Science as a mark of favour and gratitude for his crucial support in the struggle for the Liberal leadership. In these debates, both Whitlam and Fraser were under some inhibition. Whitlam's key proposal — the establishment of an Australian Schools Commission — still lacked the imprimatur of Conference. Fraser's proposal to give non-government schools a direct grant for each student lacked final Cabinet approval. Fraser argued against the Schools Commission concept on two grounds: a Schools Commission would mean a vast bureaucracy prying into the needs of 10,000 Australian schools and it would mean the centralization of the education system controlled by Canberra. The first argument was pitched to the Catholics and the second to the State Education Departments. Fraser's per capita grants proposal was at that time the official policy of the Catholic Schools. At the meetings organized by the Catholic Schools, Fraser usually had the better reception. Whitlam used his opportunities to explain the proposal, directing his arguments not only towards Catholic parents and State teachers, but towards the Party itself. For instance, speaking on 6 July to the Catholic Schools Committee in Goulburn, which had dramatized the plight of Catholic Schools in 1969 by threatening a 'strike', he said:

> Both Mr Fraser's arguments are false. The Schools Commission would exist to recommend how the Commonwealth can and should best help schools, as the Universities Commission recommends how the Commonwealth can and should best help universities. The Schools Commission would receive claims from the State Education Departments, the Catholic schools and the various existing science blocks and libraries committees. It would make recommendations on these claims in the light of the findings of its own expert staff. This would be a continuing process. The Commission itself would be fully representative of both government and non-government schools. For the first time, teachers would be represented at the decision-making level. A Schools Commission could no more dominate or displace the State Education Departments than it could dominate or displace the Catholic system. It could and would have no powers of compulsion over government or non-government schools. It could and would have great means of assistance for both systems. It would not engage or pay teachers. It would give teachers a voice in determining the quality of education in Australia.
>
> Mr Fraser's opposition to a Schools Commission highlights the difference between the Liberal and Labor approach to education. Mr Fraser is content to continue a system of spasmodic slabs of Commonwealth aid, granted in an electoral and political context. Labor stands for a competent, continuing, comprehensive Commonwealth commitment for all schools. A Schools Commission provides the only possible framework in which the Commonwealth could make such a commitment in a non-political context.

The debates gave Whitlam his first direct insights into Fraser's style and attitudes. They were also the beginning of their mutual distaste. In all these debates, Whitlam tried to emphasize that they should be the last of their kind because the State aid debate side-tracked the real debate on the needs and quality of Australian education. Replying to an accusation in answer to an editorial in the *Age* in June, he wrote: 'It is divisive and destructive of our educational opportunities as individuals and as a nation if the State and Catholic systems appear to be in conflict. The pupils of both systems will be the sufferers in the event of such a conflict. It enriches and consolidates the educational opportunities of us all if the two systems understand each other and work together.'

Whitlam had no doubts about the stakes involved in the success or failure of the 1969 Conference. Important issues had to be settled and for the first time, television stations had arranged direct telecasts of proceedings.

He knew his opening speech on 20 July would be as important as any he had ever made, with the tone even more important than the substance. His aim was to establish a bond with the Conference and the Party. In a way he had not done before, he tried to reach out personally to the principal delegates, complimenting lavishly and chiding gently, more avuncular than fraternal.

'I know you all here so well, probably the only person in the room who on average knows each of you so well. I suppose nobody in Australia knows your capacities as well as I do, or, in some cases, your capers . . . I would like to testify that no man could have had a more loyal, effective and co-operative adjutant than I have in Lance Barnard. He is a man of some courage. He was wounded at Alamein and he has flown in the F-111 . . . Can anyone doubt Jim Cairns' superior economic qualifications as Trade Minister over McEwen . . . I will pitch Frank Crean as Treasurer against McMahon, not only in economic knowledge but in his compassion and wisdom . . . Senator Murphy has established for our party a proper regard and reputation for civil liberties and for human rights . . . I have done my very best to see that Clyde Cameron has an opportunity to write a great chapter in the history of this country. I don't want to see him end up just as a footnote in a book by Alan Reid.'

It was all very disarming and it worked well in creating an atmosphere of goodwill and relaxing tensions. There was, of course, steel beneath the velvet. Anticipating criticisms that he had gone beyond authorized party policy on State aid, he said:

> My job is to put the whole of the Party's policy . . . Am I to remain silent on topical issues . . . on education. There have been this year a great number of gatherings, the largest for a quarter of a century on this subject and so I put what the Party has put. I have repeated

Labor's promises. I have put them in the context of what we have
urged in the Federal Parliament, on 14 occasions in the House of
Representatives since 1957, when the comprehensive, continuous
commitment was made by the Commonwealth following the
Murray Committee's report for universities. Our general attitude
is that in Australia, we must do as much for schools—all schools—
as we already do for universities. Yet there are suggestions that I
am not putting the Party's policy . . . it would be an intolerable
situation if the Party spokesman had to say on every issue, 'I can't
say what our policy is until Federal Conference has met'.

Despite the spirit of compromise Whitlam and Young had
established, and despite his brave assertion that State aid had
been implicit in Labor's policy since 1957, the debate was as bitter
as ever. Cameron fought to the last in a speech which left even
Chamberlain a little shaken. 'You could hear the faggots
crackle,' he said. Cameron's effort to re-inject a little good old-
fashioned sectarian bitterness into the issue misfired. For
Conference to have endorsed his stand by rejecting Whitlam
would have been an affront to Australian Catholics which
Conference could never have supported. The very strength of his
speech weakened his case beyond saving. The resolution adopted
by Federal Conference included:

Australian Schools Commission
(a) The Commonwealth to establish an Australian Schools
Commission to examine and determine the needs of students in
government and non-government primary, secondary and technical schools and recommend grants which the Commonwealth
should make to the States to assist in meeting the requirements of
all school-age children on the basis of needs and priorities.

In making recommendations for such grants to the States, the
Commission shall have regard to (i) the primary obligation of
governments to provide and maintain government school systems
of the highest standard open to all children; (ii) the numbers of
students enrolled in the various schools; (iii) the need to bring all
schools up to acceptable standards; (iv) the need to ensure
optimum use of resources in the establishment, maintenance and
extension of schools.

The most troublesome part of the debate was not about the

Schools Commission, but about a Whitlam proposal for an emergency grant to all schools, the distribution of which was biased in favour of needy Catholic parish schools, particularly those in the newer and suburban areas and poorer country centres. Conference reasserted the primacy of the secular principle by insisting that any such grant be distributed on the basis of the numbers of pupils at all schools, not on a fifty-fifty basis between government and non-government schools as Whitlam had proposed. Whitlam quickly accepted the new arrangement. When the dust had settled, two things emerged. All the significant Australian political parties were now committed to a system of Commonwealth aid for all schools, and the Labor Party, if elected, could establish a comprehensive Commonwealth commitment to all forms of education. Whitlam, who had been within one vote of expulsion from the Labor Party on this issue in 1966, was entitled to say at a press conference after the debate, 'The Conference has facilitated and in fact promoted everything I have been saying on behalf of the Party in the field of education.'

For the rest, it was largely Whitlam's conference. The general policies he had been expounding on health, social welfare and urban development were incorporated in the platform. Without serious argument from Whitlam, Chamberlain successfully insisted on a specific timetable for withdrawal from Vietnam — six months after the advent of a Labor Government.

The week and its outcome justified Whitlam's explanation of the workings of a Federal Labor Conference to a slightly puzzled audience of Methodists in Wesley Church, Melbourne, on the eve of the Conference. Quoting John Wesley about the difficulties of keeping an organization together he said: 'In 1749, Wesley provided quite an apt description of the 1969 Federal Conference of the Australian Labor Party: "It frequently happened that one affirmed what another denied and this could not be cleared without seeing them together. Little misunderstandings of various kinds frequently arose among brothers and neighbours, effectually to remove which it was needful to see them all face to face. Advice or reproof was given as need required — quarrels made up, misunderstandings removed. It can scarce be conceived what advantages have been reaped from this little prudential regulation."'

The 'little prudential regulation' of 1969 left most of the 'brothers and neighbours' of the Labor Party reasonably well content. But there was no apparent reason to believe that the electoral position of the Labor Party had been significantly strengthened. All the things which might raise Labor's hopes or Liberal unease were vague, shadowy, unquantifiable. The substantial matters on which election results turn all seemed overwhelmingly in favour of the Government. There seemed no significant issue on which the Government could not outbid or at least match the Opposition. If the Federal Conference had patched over or side-stepped most of Labor's quarrels for the time being, the Government still seemed strong and united around Gorton. If Labor had at last settled on a State aid policy, then the Government's actual record and new proposals were still more acceptable to official Catholic spokesmen. If the United States was getting out of Vietnam, so was Australia; and if the Australians were now having doubts about the whole thing, so was their Government and their Prime Minister. If Gorton was a bit unpredictable compared with Menzies, well then, Whitlam was equally unpredictable. Such were the common arguments in the Spring of 1969.

The Government had every reason to believe it could coast to victory on its economic record. In 1969, the basic economic aims of conservative government were being met as perfectly as the system would allow. Inflation was about two per cent, unemployment was less than one per cent, and the annual growth rate about six per cent. As the election approached, a stock exchange boom, led by a frantic demand for Poseidon nickel shares, was under way. It was to end in disaster, but in September and October 1969, the burgeoning boom added to the general impression of confidence, buoyancy and prosperity. Thousands of middle-class Australians were experiencing for the first time the frisson of a flutter on the stock exchange. This year was to be the last of the long years of relatively untroubled economic growth characteristic of the 1960s and so basic to the Liberal hegemony of the decade. The Poseidon 'bust' six months later was to signal the end of the 'economic miracle'. But the troubles ahead caused little concern in 1969. When McMahon spoke in his Budget of an 'economy finely balanced', it was more a boast than a warning. The dictum attributed to Chifley that the most tender nerve in the elector's body is the hip-pocket nerve gave authority to the belief that the Gorton Government would be quite untroubled at the polls.

With a $1 increase in pensions, a slight easing of the means test, free health insurance for families with incomes of less than $39 a week, and most important electorally, it implemented Fraser's proposal for per capita grants to independent schools, allowing $50 per secondary school pupil and $35 per primary school pupil.

Yet there were storm warnings. St John, now out of the Liberal Party, published his book, *A Time to Speak*, which powerfully brought together all his misgivings, moral and political, about Gorton. Through newspaper serialization and television interviews, his charges reached a large audience. Gorton shrugged it all off, secure in his conviction that the Australian electors rather liked a bit of the larrikin in their Prime Minister. He took scarcely more seriously the rage of the DLP about a reference to the Soviet naval presence in the Indian Ocean by the Minister for External Affairs, Gordon Freeth. Freeth's appointment to replace Hasluck, had been regarded by the DLP as a downgrading of foreign affairs. For a speech on international affairs on 14 August, he innocently accepted, with Gorton's knowledge and approval, a departmental draft which stated: 'We need not panic every time a Russian ship appears in the Indian Ocean.' Equally disturbing to the DLP was the decision of the Minister for Defence, Alan Fairhall, to quit politics. The links with the certainties and stability of the Menzies era were snapping one by one.

Whitlam entered the campaign with a confidence unwarranted by the opinion polls. His policy speech was televised live from Sydney Town Hall and established a style for all future campaigns, Liberal as well as Labor. The full document brought together all the work of three years: 'The campaign of the Labor Party will have one dominant theme — the theme of opportunities, the taking of opportunities, the making of opportunities. We wish to renovate, rejuvenate, reinvigorate and liberate.' He took the opportunity of a national audience to put his personal philosophy:

> We of the Labor Party have an enduring commitment to a view about society. It is this: in modern countries, opportunities for all citizens — the opportunity for a complete education, opportunity for dignity in retirement, opportunity for proper medical treatment, opportunity to share in the nation's wealth and resources, opportunity for decent housing, the opportunity for civilized conditions in our cities and towns, opportunity to preserve and promote the natural beauty of the land — can be provided only if governments, the community itself acting through its elected representatives, will provide them. And increasingly in Australia the national government must initiate those opportunities.

This is the essence of Whitlamism.

Gorton was convinced that his popular Budget was sufficient to carry the election. He might have done better to have stuck to his earlier intention to 'run on the record'. Impressed by Whitlam's performance and anxious about reports from the electorates, Gorton's advisers and supporters urged on him the need for fresh policies. His response was characteristically half-hearted and half-baked. As a reply to Whitlam's general proposals for tax reform, Gorton undertook to forego $200 million in increased taxation over the next four years, about 10 per cent of the increase expected on the existing rates. In response to Labor's health scheme, he came up with a proposal that no operation should cost the patient more than $5. This proposal was so hurriedly devised in Gorton's own office that it was too late to be included in the printed policy speech document. The television version of this took the form of a closed studio presentation, with Gorton extolling the Government's virtues to his Cabinet colleagues and their wives. McEwen, stiff and sour, registered approval only when Gorton went down the line for high protection. McMahon was apparently asleep. It looked rather like an old-fashioned family will reading, and not a very happy family.

Writing in the Melbourne *Sun*, Laurie Oakes described

Whitlam's campaign as the 'politics of laughter'. There were two reasons why Whitlam deliberately set out to use the language of laughter. In general, the issues did not easily lend themselves to the rhetoric of high passion. It was not a year for anger, real or simulated. Laughter was the best bridge between Whitlam and his audiences. These were new audiences for Labor meetings, younger than usual, apparently more thoughtful, much quieter than in 1966, and giving some sign of coming to be convinced rather than stirred. Whitlam's issues of education, health, cities, housing and environment were very much issues of interest to the younger urban middle class. Their counterparts of fifteen years before had largely crossed to the Liberal Party of Menzies. Whitlam was the first Labor leader to make a deliberate effort to identify with them, and the mixture of humour and pedagogy rather than the old-style passion and demagoguery seemed the best, and for Whitlam the most natural, way of reaching out to them. The campaign of 1969 restored the public meeting as a relevant part of the Australian political process.

Whitlam also deliberately used laughter to attack Gorton on his strongest ground—his possession of all the authority that derives from being Prime Minister. As Gorton, then McMahon and Whitlam himself would discover, ridicule is the most brutal weapon in the political armoury. A Prime Minister who can be made to look ridiculous is finished. In 1969, the weapon against Gorton was his own words. For a generation brought up on the grand simplifications of Menzies, Gorton was a shock. Whitlam simply exposed Gorton's confusion. It was most effective when used to show that Gorton's policies were as confused as his language. On his proposal for health insurance rebates Gorton explained, 'On the other hand, the AMA agrees with us, or, I believe, will agree with us, that it is its policy and it will be its policy to inform patients who ask what the common fee is and what their own fee is so that a patient will know whether he is going to be operated on, if that is what it is, on the basis of the common fee or not.' In quoting this around Australia, it scarcely needed Whitlam's punchline, 'So now you know', for his audience to get the point. St John's high morality and Whitlam's low comedy worked in the same direction to raise deep doubts about Gorton's fitness to be Prime Minister. He increased the damage by his own response. He became morose and showed his bitterness. He continued drinking heavily and showed the signs. He described some journalists as 'slimy white things that crawl out of sewers'. He complained that Whitlam was able to attract better staff. 'I have to write all my own speeches.' Whitlam made the obvious retort, 'That is the one thing he has said in this campaign I can fully believe.'

On the polling night, 25 October, there were a few minutes when the more enthusiastic and less experienced supporters gathered at Whitlam's home in Cabramatta believed that a narrow Labor victory was possible. Whitlam knew it could not be. Nevertheless, the results were remarkable. The national swing to Labor was 7.1 per cent, the largest swing against a government since 1931. The swing occurred in all States, ranging from 11.8 per cent in South Australia to 2.2 per cent in Tasmania. Labor won seats in all States: 6 in New South Wales, 3 in Victoria (Benson did not stand for the seat he held as an Independent after 1966), 1 in Queensland, 5 in South Australia (including a new seat created by the 1968 redistribution), 3 in Western Australia and 1 in Tasmania. The coalition majority shrank from 39 to 7.

Election night 1969 was full of drama worthy of a play and accordingly has been made into one. *Don's Party*, by Australia's tallest and best playwright, David Williamson, hinges on the mounting optimism among a group of Labor supporters and their disintegration as the late count saves the Gorton Government. If Whitlam and his closest associates shared their disappointment, there was no despair. The political significance of the play is that Williamson perceives that in 1969 a significant number of Australians experienced a new interest and involvement in the fortunes of the Labor Party in a way which had not been felt since 1954 and which had never been felt by the post-Split generation at all. This was the real achievement of the 1969 campaign. How that flicker of interest was nurtured into a steady flame is the story of the years 1969 to 1972.

*

The 1969 election was the most significant Australian election since 1949. As a genuine political watershed it is more important than 1972, and may yet prove more important than 1975. Since 1949, Labor had done better in terms of votes in 1954 and better in terms of seats in 1961. However, the 1954 result represented the pro-Labor coalition of interests and ideals put together by Curtin just before and during the war, and consolidated by Chifley in the post-war reconstruction period. The 1954 election was essentially the last in the series beginning with the 1940 election. These two elections bracket the period of Labor unity under Curtin and Chifley. That is the only period in Australia's history when it was not altogether absurd to speak of a natural Labor majority. One of the reasons for the Split was that the men involved really believed that the natural majority existed and that it was a thing with its own existence, to be delivered to and controlled by whoever could control the official machinery of the Labor Party. The 1961 election result was due wholly to economic circumstances. In that sense, it was a copy-book election, even though its real result was to entrench Menzies. The 1969 election was the first of the present series.

There have been five 'series' of Australian elections since Federation. The elections between 1901 and 1914 gave Australia a definite two-party system, with Labor the firm challenger and established alternative. The series between 1917 and 1937 confirmed the basically conservative nature of Australian nationalism. The series between 1940 and 1954 established the power of modern laborism as kept together by Curtin and Chifley. The series between 1954 and 1966 established the power of modern liberalism, as expressed by Menzies. Within these great cycles, events like the landslide to Labor in 1929 and the near defeat of Menzies in 1961, can be seen as accidents which later confirmed the norm. The 1969 election began a new series, of which the elections of 1972, 1974 and 1975 are part of an ongoing pattern, in which the new and essential element is a greater volatility in the electorate than Australia has previously experienced.

The odd thing about 1969 was that Whitlam was actually helped by the buoyant economy. This is what marks 1969 as the point of departure from all before. It was precisely because the younger middle-class electors felt content and confident that Whitlam's appeal attracted them. Whitlam raised expectations, and those electors could see no reason why their expectations could not be met. Gorton could have done one of two things: take the classic conservative line of dampening expectations, emphasizing the coming difficulties of 'a finely balanced economy' or out-crusade Whitlam on confidence, hope and future expectations. He did the opposite: he conceded, in a muddled way, conservative failures while preaching doom in an equally muddled and unconvincing way.

The 1969 elections were the first since 1946 in which the coalition was unsuccessful in calling the issues. It was the first since 1949 in which communism was not effective as an issue. It was the first since 1954 in which international and defence issues did not dominate to Labor's overwhelming disadvantage. In the last week of the campaign, Gorton tried to get the campaign back to familiar grounds but by then it was too late. In any case, the attempt lacked credibility coming from a Liberal Prime Minister who had brought into question forward defence, placed limits on Australia's commitment to the United States in Vietnam and allowed his foreign minister to scoff at the Russian threat. He was no more successful than McMahon, who spoke about the Liberal tradition of fighting in the cause of freedom 'in Vietnam, in Korea, in the Boer war'.

The central question is why a government was brought so close to defeat in a time of such economic buoyancy? Gorton had broken away from the old certainties. He saw that the Menzies simplifications could not work much longer but he had not thought through the implications. He saw clearly enough that the Nixon doctrine, the end of Vietnam and the British withdrawal east of Suez meant a change in Australia's defence posture. His reward was to be described by the DLP as 'frivolous'. He believed Australia needed new arrangements between the Commonwealth and the States. In practice, he was no more centralist than Menzies. But the Liberal Premiers of Victoria, New South Wales and South Australia refused to campaign on behalf of the Gorton Government. He made business uneasy by coyly refusing to deny rumours that McMahon would remain Treasurer after the election. And, as he struck off the props to his inherited power, the confidence of the business community, the support of the State machines and the DLP, he undermined his own source of power — the personal authority derived from the Prime Ministership.

At the declaration of the poll in his own seat of Higgins, Gorton undertook to 'identify and rectify' the causes of the Liberal disaster. He beat off a challenge by McMahon who remained Deputy Leader and ceased to be Treasurer. As part of his campaign, McMahon had announced that he was a Federalist, which Whitlam said reminded him of Molière's Monsieur Jourdain who was delighted to discover he had been talking prose all his life. But McMahon was really offering a return to Liberal orthodoxies and aligning himself with the State Premiers against Gorton. The 1969 result destroyed Gorton's original and almost only claim to the leadership — that he was a winner, that he was the man to beat Whitlam. From 25 October 1969, it became almost certain that Labor would win the next election; and almost as likely that Gorton would not lead the Liberal Party at that election.

11

Reconstruction

Vietnam had not played a prominent part in the official campaigns of either side in 1969. Yet it remained an issue of critical importance. It had played its part in undermining Gorton by weakening public confidence in the infallibility of Liberal foreign policy and in restoring Labor's self-confidence as events in Vietnam vindicated its early opposition to involvement. Events in Washington made its opposition credible and even respectable. Electorally, Vietnam had provided the solvent to detach a significant number of the younger middle class from the Liberal Party. For the first time since 1946, Labor increased its support in the crucial 21–35 age group. A significant, if imponderable, factor was the impact of university students on their middle-class, Liberal-voting parents. Their opposition to Vietnam may not have translated directly into votes for Labor, but it at least had the effect of conditioning their seniors and moderating their hostility to Labor.

Yet the Vietnam issue had not exhausted its capacity to cause conflict within the Labor Party. As Australia's involvement drew to a close, Whitlam was more concerned than ever that the Labor Party should not be seen as a single-issue party at the very time that the issue was losing its force as incomparably the greatest issue of its time. He took the opportunity of an exchange of messages with Chamberlain to set out a careful justification of his handling of the Vietnam issue since 1965.

The Western Australian Party had sent a message to Whitlam asking him to put himself at the head of a national movement to end the war in Vietnam. Whitlam took great care with his reply and wrote on 18 December:

Mr Chamberlain,

I thank you for your telegram and letter conveying the State Executive's resolution. I completely endorse your desire for a constant mobilization of Australian public opinion to help bring an end to the war in Vietnam and I am also emphatically of your opinion that the Labor Party is the appropriate organization to develop and lead public opinion in this regard.

The great question for our Party as a whole and for each of us as members is how public opinion can most effectively and rapidly be translated into national action. In the United States, the burden of opposition to the war has been borne by individuals acting outside the established party organizations. This has been so even in the case of members of Congress. The majority of Democrats and the substantial minority of Republicans who oppose the war have had to oppose Administrations formed by their own parties. The honour due to these Americans, from the Senate Majority Leader to the millions who have marched, should never be forgotten in any assessment which outsiders make of American attitudes to the war.

In Australia individual members of our Party have not had to make this agonizing choice between party and principle. Our problem has been of a different, if no less difficult, kind: namely, how to give practical effect to our principled opposition to the war and Australian participation in it. Inasmuch as all Australians must accept blame for what we have done to Vietnam, the guilt of the Australian Labor Party lies not in our policies but in our failure to form a government which would have applied those policies.

No fundamental change can ever be made in Australian foreign policy except by a change of government. Had there been a change of government in October, the 8th Battalion would not have gone to Vietnam last month and no Australian troops would have been in Vietnam within six months. If, however, the present Parliament runs its full course it is almost certain that there will be no Australian troops in Vietnam by the time of the next House of Representatives elections. This will have occurred not because of any real change in the policy or attitudes of the Gorton Government but because withdrawal will have been forced on it by the inevitable and accelerating withdrawal by the Nixon Administration.

As a Party and as individuals we must now set out to achieve national redemption for the policies hitherto pursued in our name as a nation. This applies not only to what we have allowed to happen to Vietnam but what we have allowed to happen to the United States. In times to come that great people will feel no gratitude towards governments and nations which have willed their prolonged embroilment in this physically, politically, socially and morally debilitating and divisive war. There will, however, be gratitude for those friends who have recognized America's agony and have tried to use such influence as they have to help her end it. It is in this light that I view the further mobilization of Australian public opinion as the beginning of an act of national redemption. Otherwise there will be an era of national and international recrimination.

Your Executive refers generously to my own role. Since I became Leader I have chiefly tried to bear in mind that the wider the support we can command for our Party and policies as a whole, the more effective will be our advocacy of any particular aspect of those policies. We maximize our opposition to the Vietnam war by maximizing support for our Party. I do not believe that I should single out any one of half a dozen major issues for my exclusive attention. Support for us on one issue is strengthened by support for us on other issues.

In the Senate elections in 1967 I concentrated on the need to end the Vietnam bombing and to negotiate with the N.L.F. and in the House of Representatives elections in 1969 I presented specific proposals to end the Australian commitment by June 1970. Between those elections I have used every appropriate forum and principally the national Parliament, the greatest forum available to our Party, to explain and expound those proposals and the Party's policies which lie behind them.

As Leader, I have not thought it proper or prudent to sign statements or to appear with persons expressing a less complete view than our Caucus or Conference or presenting a different emphasis. The traps into which one can fall were highlighted last Monday. [A reference to a recent rally in Melbourne which had called for a military victory by North Vietnam.] The President-elect of the ACTU [Robert Hawke] briefly attended and the Victorian President of the ALP [George Crawford] chaired a meeting which carried a Vietnam resolution which is not and never could be ACTU or ALP policy. Members of the Party should not give the false and damaging impression that under a

Labor Government foreign policy would be determined at mass meetings or by public petitions. For this reason I concentrate my own actions in party and parliamentary channels. I will readily and fully participate in any further activity your Executive may have in contemplation under party auspices.

Our principal electoral task in the coming year will be to organize for a Senate victory. There can be no doubt that a massive vote for Labor in those elections will be the most effective contribution we can make towards ending not only the Gorton Government but the disastrous war which it endorsed and is most unwilling to end.

That letter marked the end of the Labor Party's disputes on Vietnam. From the beginning they had been only about the method and timing of withdrawal. The leadership of the anti-war movement beyond the official Party remained with Cairns. In May 1970, after Nixon unleashed the South Vietnamese army and destroyed Cambodia in the name of saving American lives, the Moratorium movement reached its peak as Cairns led 70,000 marchers down Bourke Street, Melbourne. Profoundly moving in itself, with a deep emotional significance for Cairns, a vindication of his past and a pointer to the future development of his thinking, this greatest success of the Moratorium movement came too late to influence events in Vietnam or Australia.

Whitlam's December letter marks the change in his relations with Chamberlain. For nearly a decade the relations between these two men dominated political Labor. They divided the Labor Party because they both held strong views on great issues, such as the role of the Parliamentary Party, State aid and Vietnam. It is not often that a personal dispute can be maintained over so long a time and kept at such a high level.

The key to Chamberlain was his suspicion of parliamentarians. The essence of Whitlam's challenge to Chamberlain was his insistence of the idea of the Labor Party as a Parliamentary Party seeking its aims through parliamentary means. Their conflict was inevitable. The issue was whether the Parliamentary Party would be altogether dominated by the machine, according to a very narrow interpretation of the rules. Chamberlain was, in fact, doomed to failure because his interpretation of the rules was altogether too narrow. Had he made the issue the politicians against the rank and file, he might have won. But he made it the Federal Executive against all the rest of the Party. In the end it came down to the issue of Chamberlain against all the Party. There was only one more battle between them—the reconstruction of the Victorian branch. But after 1969, the personal bitterness between them disappeared.

The last conflict was perhaps the most important of all, for its outcome decided the result of the 1972 election. It was the final struggle over the control of the Victorian branch of the Labor Party. The outcome is known as 'reconstruction'. Labor won in 1972 because it won seats in Melbourne. It won seats in Melbourne in 1972 because of the reconstruction of the Victorian branch in 1970. In the final battle, Whitlam and Chamberlain were the opposing generals, as they had been in the battles of 1966, 1967, 1968 and 1969. But in 1970, Whitlam, for the first time in his contests within the Party, found allies in Cameron and Young, who were as persistent as he, and who were better tacticians than Chamberlain.

In his letter to Chamberlain, Whitlam referred to the activities of the Chairman of the Victorian State Executive, George Crawford. This was a little-noticed signal for renewed warfare between Whitlam and the Victorian branch. This time it was to be carried through to a finish. Unlike the previous confrontations, Whitlam this time worked from a well-prepared position of great strength and had this time powerful allies on the 'left'. In this final and successful assault, the challenge came on the same issue as in 1966—State aid. It also rested on the same principle—the primacy of the Parliamentary Party.

Nothing proves the political folly of the controllers of the Victorian branch better than the manner in which they provoked this final, fateful confrontation. They chose the wrong time, the wrong issue, the wrong tactics and the wrong enemies. The Victorian contribution to the 1969 result should have won the VCE a reprieve, if not immunity, from attack until 1972. Three seats had come to Labor, although only one of these, Maribyrnong, was a clear gain, Batman being recovered from Benson, and Cairns' new seat of Lalor having become a clear Labor seat in the 1968 redistribution. A swing in the State of 5.6 per cent, though below the national average of 7.1 per cent, was respectable enough. During the campaign, Hartley had effectively squashed an attack by Calwell on Whitlam's handling of Vietnam. A State by-election in outer Melbourne in December was won by Labor, continuing the trend of October. All the basic weaknesses of the Victorian branch remained; but Whitlam saw no prospects for reform. At the beginning of 1970, the conditions for Federal intervention in the affairs of the Victorian branch did not exist.

The Victorian Premier, Sir Henry Bolte, called an election for May. The deterioration in his relations with Canberra, the public disunity of the Federal Liberal Party, a stock exchange bust after the Poseidon boom, seemed to give Victorian Labor its best chance since the 1955 Split. The Leader of the State Opposition, Clyde Holding, encouraged by the October national result and his personal success in the Dandenong by-election, looked forward to a vigorous and united campaign. These hopes were torpedoed by the Executive. The Executive drew up its own election manifesto, a kind of rival policy speech, which included an assertion that a State Labor government would 'phase out' State aid to non-government schools. To emphasize its humiliation of the Parliamentary Leader, the Executive, through Hartley, arrogantly offered Holding the option of including the 'phasing out' proposal in his own speech. Whitlam warned Hartley that he would repudiate the proposal, a clear breach of the policy so painfully established by the 1969 Federal Conference.

Whitlam, Barnard and Dunstan went to Melbourne for a show of leadership unity around Holding at his opening meeting. Dunstan was in the middle of his own State campaign in South Australia, where a split in the State Liberal Government over electoral reform had given him an unexpected chance to return to power after his defeat in 1968. Central to his election programme was an education proposal along the lines of the 1969 Federal policy. Dining together in Dunstan's hotel room, the three leaders learnt from a reporter that the Executive manifesto with its version of Labor's education policy had already been released to the press. Less than an hour before they were to be at St Kilda Town Hall, they discussed a boycott of the meeting. Of the three, Dunstan was the most outraged. 'If this is the policy, then in Adelaide I am a fool or liar.' Accepting Barnard's advice to attend the meeting, Dunstan made it clear that he was acting only out of consideration for Holding. 'But after the elections...' Dunstan left the threat unsaid. At St Kilda Town Hall, Crawford shepherded his group of reluctant leaders into the hall and said cheerfully, 'Clyde's doing all right.' 'He would have been if you bastards had left him alone,' said Whitlam. Speaking briefly in support of Holding, Whitlam confined himself to a general restatement of education policy. Four nights later, at Bendigo, he denounced the Executive manifesto as 'spurious'. The VCE then withdrew any further invitations to Whitlam to take part in the campaign. On 30 May, the Bolte Government increased its majority. Dunstan became the only Labor Premier in Australia.

The questions raised by the conduct of the leaders of the Victorian Executive were profoundly important. They asserted the right of a State branch to set aside national policy; either Victoria was in breach of Federal policy or each State could interpret Federal policy according to its particular whim. They asserted in a way never so brutally attempted, the power of the paid officials and the machine to override the Parliamentary Party. They stripped the Parliamentary Leader of his traditional responsibility for the election programme. Beyond the principles involved, their conduct and their timing brought into play personal and political factors full of danger to them. They had given their declared enemy, Whitlam, spectacular evidence of all his allegations against them at a time when his prestige was at a peak and in circumstances which made plain to everybody that there was nothing cranky, obsessive or idiosyncratic about his long and often lonely crusade. Even more dangerous was Dunstan's new enmity.

Without knowing it, the VCE had changed fundamentally the balance of numbers against them in the Federal Executive. Their immunity against Whitlam's attacks lay in the support they had hitherto received from South Australia. Once South Australia moved in favour of intervention, they were doomed. That happened when they made Dunstan their active enemy.

Dunstan's anger was shared by Clyde Cameron and Mick Young who, in 1970, was acting both as South Australian Secretary and National Secretary. Young had joined Whitlam in warning Hartley that the 'phasing out' proposal breached the policy established at the 1969 Conference. Cameron, the most bitter speaker against that policy, knew what had been decided and that his defeat on that issue was final. One of the toughest and most tenacious fighters in the Party, he was determined that the battle he had lost on State aid would not be reopened. No one in the Labor Party knew better how to sustain a fight or a grudge, or when a fight had been irretrievably lost. He had taken to heart Whitlam's call to destiny — to be a great Minister for Labour or a footnote in history. He had been surprised and impressed by Whitlam's success in the 1969 elections. After the elections, Whitlam appointed Cameron as the parliamentary spokesman on industrial relations, the 'shadow Minister' for Labour. This was what he had always wanted, and the prospects for 1972 gave substance to that 'shadow ministry'. He also valued the appointment as a slap in the eye for his arch-enemy, the Secretary of the Australian Workers' Union, Tom Dougherty. The great days of the AWU, when it had dictated to Labor Governments in Queensland and New South Wales, were past; its relative decline reflected the declining importance of rural industries in the Australian economy. It remained Australia's largest and wealthiest union, an independent empire within the Labor movement. Tom Dougherty was its king. For more than a decade, Cameron had waged unrelenting and resourceful war through ballots and in the courts against what he regarded as Dougherty's dictatorship and what he believed to be the lack of political and industrial principle of the AWU. Offering him the appointment, Whitlam insisted, 'You must promise me that you will not use this just as part of your vendetta.' The bargain, thus struck, was crucial in 1970 and as it turned out, crucial to Labor's victory in 1972. Whitlam was to describe Cameron as 'my Carnot — the architect of victory'.

Since the 1955 Split, the left-wing controllers of Victoria had been protected against Federal intervention by the right-wing controllers of New South Wales. It was a mutual non-aggression pact, or as Whitlam put it, a 'knock-for-knock' agreement. The pact ended in 1970 with new leadership in New South Wales under John Ducker, an astute Yorkshireman blending real toughness with a degree of sentimentality. Cameron had long advocated intervention in New South Wales, not least because he regarded it as being dominated and therefore corrupted by Dougherty's Australian Workers' Union. Ducker's succession to the New South Wales presidency, replacing Charles Oliver, New South Wales Secretary of the AWU, opened the way for a new deal. Anticipating a change of mood and the shifting power balance on the Federal Executive, Ducker made it known that the New South Wales branch would not implacably resist a Federal investigation into its affairs. The 'knock-for-knock' agreement was broken.

A working alliance between Cameron and Ducker was forged at a fateful meeting of the Federal Executive at Broken Hill in August. It was at once apparent that there had been a decisive shift in the balance of power. Victorian complaints against Whitlam's conduct during the May State campaign were promptly dismissed. An appeal by the Leader of the Labor Party in Victoria's Upper House, the Legislative Council, John Galbally, against his suspension for supporting Whitlam on State aid was upheld. A motion by Chamberlain endorsing the VCE's 'phasing out' policy on State aid was defeated eleven to six. Finally, Cameron produced the South Australian plan, worked out by Dunstan, Young and Cameron himself, for a Federal investigation into the affairs of both the New South Wales and Victorian branches. Ducker announced New South Wales acceptance — in Cameron's words, submitting to rape and enjoying it. The Victorian delegates resisted bitterly. In doing so they made the further mistake of venting all their rage on Cameron personally, insinuating treachery and a sell-out. William Brown warned that the Party would split in Victoria and made it a threat not a prediction. It is a measure of the frustration and fury of the old 'left' forces supporting Victoria, their shock at the sudden change in their fortunes, that Chamberlain accused

Cameron of 'using numbers blatantly'. Cameron took it as the highest compliment from one of the great masters of the numbers game, now suddenly weakened but still formidable. At that stage Cameron made one of the most adroit manoeuvres in the Byzantine annals of Federal Executive politics. He retreated. He withdrew his motion in return for an undertaking that the Federal officers were welcome to visit Victoria at any time. What Cameron had done, and what probably only Cameron knew at the time, was to give the Victorians more rope with which to hang themselves and more time for rank-and-file pressure in Victoria to rise against the VCE.

This tactic may not have succeeded had the VCE controllers not behaved with predictable folly. They did not know when to quit while still in front. Just before leaving his motel at Broken Hill, Cameron overheard Brown telephoning Melbourne and boasting how they had called the bluff. In Melbourne, George Crawford went on television to say that the proceedings of the Federal Executive 'were reminiscent of the Mad Hatter's Tea Party'. And Hartley asserted that the phasing-out of State aid still stood as far as Victoria was concerned. By their intransigence and indiscretion they made a show-down inevitable. They had also converted Cameron into an implacable enemy.

It was Cameron, the ex-shearer, rather than Whitlam, Q.C., who perceived the absolute necessity of a water-tight legal case for Federal intervention in Victoria. And it was Cameron who compiled the case. Whitlam cheerfully admitted that, in this matter, he was acting as Cameron's junior. In the 1955 Split, the legal case had been botched, enabling the breakaway group to lay legal claim to the Party's assets and property in Victoria. This oversight intensified and deepened the Split. Cameron remembered all this; in addition, his long legal battles against Dougherty had taught him the importance of proper legal preparation. To assist him in compiling a voluminous case, in preparing the ground step by step and in framing a series of water-tight resolutions to underpin intervention, to pre-empt challenge in the courts and, if any challenge occurred, to ensure the case would stand up in court, he called in three Queen's Counsel — Richard McGarvie, Xavier Connor and John Sweeney.

When, after a preliminary meeting in Sydney, the Federal Executive assembled in plenary session at the Travelodge, St Kilda, the basic outcome was a foregone conclusion—there would be Federal intervention and there would be reconstruction of Victoria. But the form of reconstruction, its extent and depth, was by no means predetermined. That this nine-day meeting, the longest in the history of the Labor Party, was no rubber stamp for what Whitlam, Cameron and Young had already decided is shown by the defeat, amendment or withdrawal of several of the motions. But the marathon meeting went on to its inevitable conclusion. The grip of the old controllers of the VCE was broken at last. A set of complicated procedures for re-establishing rank-and-file representation, union representation and for the election of new officers was settled. Whitlam's long fight was over, triumphantly over.

In the reconstruction crisis, the new President of the ACTU, Robert Hawke, had a particularly difficult and unrewarding role to play. To the extent that the old controllers of the VCE claimed to represent and defend the trade union wing, he had to defend them. With his political base in Victoria, he could not afford a break-up of the Party, or a split in the industrial wing. In the position in which he was placed, it was impossible for him to please everyone and almost impossible to please anybody. He defended Victoria against intervention and led the case for the defence against Cameron's prosecution. But when the final decision had been made, it was Hawke himself who presided over the reconstruction in the months ahead and whose role was critical to the success of reconstruction.

Robert James Lee Hawke may prove to be the first completely modern Australian politician. He was the only Australian to have left Oxford more convincedly Australian than before he went there. Oxford had much the effect on Hawke as Cambridge had on Lee Kuan Yew a decade before; both learnt that there was nobody better than them there, but that their destiny lay absolutely at home. Hawke, in his generation, was the most significant of those who learned in England to patronize the English, as the English had patronized the Australians for six generations.

In 1963, Whitlam had opened Hawke's unsuccessful campaign for the Federal seat of Corio. Whitlam urged him to run again in the 1967 by-election. Hawke, then highly successful as the industrial advocate for the ACTU, had staked a clear claim to succeed the wily, but ageing, ACTU President, Albert Monk. He told Whitlam that he wanted to see if he succeeded Monk; if he did, he wanted six years in the job; if he did not, he would enter parliamentary politics immediately. As he then put it to Whitlam: 'If I become President of the ACTU, I shall always be involved in decisions. If I get into Parliament, I'll be involved in decisions only when we get in.'

In a sense, the reconstruction of the Victorian branch of the Labor Party was the last chapter in the long, agonizing story of the Split. But in no sense did it represent a readjustment in favour of the old 'right'. Its principle purpose was to break the control of a narrow group and to open the Party to greater rank-and-file representation. It did not represent, and was not intended to represent, a massive move to the 'right' in Victoria. The Victorian Party as a whole remained oriented to the 'left'. Reconstruction left the old sources of power virtually untouched but, by creating new sources of influence, tilted the balance of power to the 'centre'. Critics of reconstruction, in and outside the Labor Party, sneered that it was only a 'cosmetic job', a mere public relations exercise. This is true to the extent that reconstruction was designed to improve the Party's standing with the public by giving it a more representative, more democratic appearance. But those who asserted it was all a fraud were really wanting expulsions, sackings and political bloodshed. The managers of the reconstruction—from Whitlam and Cameron to Young—were not out for blood. The most remarkable achievement of reconstruction was precisely its lack of recrimination and enduring bitterness. For the smooth and orderly process of reconstruction, Cameron must take chief credit. For the lack of vindictiveness in the aftermath, the credit belongs to Whitlam and Young, for whatever their failings, they are not vindictive men. As Whitlam emerged from the Conference on the last triumphant day, the VCE organizer, Robert Hogg, hissed at him, 'We'll get you, Gough.' Seven years later, Hogg, who became Secretary of the Victorian branch, was one of Whitlam's firmest champions in the leadership battle of 1977.

As a test of public response to reconstruction, the 1970 Senate election was not encouraging. It was a poor campaign, attended by bad luck, poor organization and unenthusiastic audiences. At the author's urging, Whitlam opened his campaign in Melbourne instead of Sydney, to symbolize a fresh start in Victoria. Any boost this might have given fizzled out in the half-empty Dallas Brooks Hall. Senator Murphy refused to attend because he was not given equal billing with the Leader in the House of Representatives. An attempt to place a bomb in Whitlam's R.A.A.F. aircraft at Brisbane turned out to be a break-in by R.A.A.F. guards in search of beer. A plan to take the campaign to rural districts ended limply in Adelaide because of fog over Hamilton in the Western district of Victoria. A powerful plane in search of a destination — it seemed to sum up the campaign.

The press judged the Prime Minister to have run a reasonably good campaign, at least compared to 1969. The results were discouraging for the Labor Party but disastrous for Gorton. The Government and the Opposition each lost two seats and the DLP won a fifth seat, reaching the high-water mark of its representation. In apparent rebuke to the major parties, two independent senators were elected. But, for Gorton, the critical figure was the coalition's aggregate vote — 37 per cent across the nation. It was the lowest vote for any major grouping in any national election since 1943. Gorton had achieved the impossible. All analyses of Australian voting habits, in a two-party system, rest on the theory of an irreducible minimum of 40 per cent for each of the major groups. This hard-core support is the base from which the parties fight over the remaining 20 per cent. In falling below the magic forty, Gorton delivered a shocking blow to Liberal morale. When those numbers went up, Gorton's number was up. Should any incident provide opportunity for a challenge in the Party room, he was bound to be defeated. All that was required was for someone to manufacture an incident and provide the opportunity.

12

The Destruction of Gorton

The identification and rectification of problems Gorton had promised after the 1969 election took a curious form. One of the charges against him was that he had scant regard for Parliament and its forms. To underline the truth of the charge, he called Parliament together for a farcical one-day sitting on 29 November, humiliated Hasluck as Governor-General by giving him a speech of 100 words to deliver, and fuelled Whitlam's contempt by gagging debate and closing down Parliament after five chaotic hours. It was an ominous start. Whitlam was not alone in concluding that Gorton was incorrigible and that, despite his electoral battering, this was still the John Grey Gorton who had shrugged off Liberal concern at his methods with the defiant claim: 'You ain't seen nothing yet.'

Whitlam's enmity he could afford. The enemies he was making in his own ranks he could not afford. He removed McMahon from the Treasury. After that, Sir Frank Packer, proprietor of the Sydney *Daily Telegraph*, campaigned to get rid of Gorton with the same determination with which hitherto he had hounded the Labor Party. Gorton then broke up the partnership which had worked to get him elected leader. Dudley Erwin lost his Ministry. Explaining his sacking in a press statement, Erwin said: 'The political manoeuvre used to get me out of office wiggles, it's shapely and its name is Ainsley Gotto.' The school-boy peevishness of Erwin's outburst only emphasized the pitch to which Menzies' estate had come.

Central to all Gorton's difficulties was his attempt to redefine relations between the Federal Government and the States. It was on this issue that he made the most formidable enemies. It is frequently claimed though scarcely true that State Governments are closer to the people and command their first loyalties. It is true, however, that the loyalties of the party machines are State-oriented. The strength of the parties lies in their State machines and the best opportunities for influence, manipulation and, occasionally, corruption are in the State capitals. In all parties, the hearts of most Party officials are in their State organizations, the source of their influence and income. In seeking to express the national sentiment, Gorton underestimated the State loyalties of the Liberal Establishment.

In practice, Gorton was no more centralist than Menzies. He was less skilful in disguising it. Like Menzies, he thought that the pretensions of the States to sovereignty was humbug; unlike Menzies, he said so. And thus he made enemies of the two most powerful Liberals, Sir Henry Bolte, Premier of Victoria, and Robin Askin, Premier of New South Wales. Their power to harm him rested on their control of their State parties and the fact they could heavily influence the metropolitan newspapers in Melbourne and Sydney. The State-orientation of the metropolitan press is as significant as the State-orientation of the parties. Those who want to exercise political influence will seek its nearest source. For metropolitan press proprietors and Party officials, that is usually State Parliaments. The best that Gorton could bring to bear against entrenched State interests was a vague national sentiment; and as he was so spectacularly unsuccessful in national elections, he had no authority with which to embody or express this sentiment. In any case, he had no concrete policy to apply his own nationalism on the national sentiment—with one exception.

The power of State hostility to harm a Prime Minister was demonstrated in a matter in which Gorton had logically, historically and, as it was to be proved, legally an incontrovertible case—the matter of off-shore legislation. On his side he had overwhelming public opinion, the Labor Party, the tide of history and, in this case, the press. Yet it was an issue which nearly brought him down.

In May 1970, the former Minister for National Development, David Fairbairn (Farrer, N.S.W.), who had refused to serve under Gorton in 1969, provoked a first-class parliamentary row. He claimed that records of meetings held between Federal and State Ministers in 1969 showed that the Federal Government had promised further discussions before bringing in legislation to assert the Commonwealth's authority over the continental offshore. Gorton maintained there had been no prior commitment for further negotiations. On 15 May, the Opposition moved a no-confidence motion, charging Gorton with dishonouring an agreement with the States. This was defeated only by some smart footwork by McEwen. He framed a face-saving motion which asserted that there had been no breach of faith, but conceded justification for Fairbairn's complaint. The motion said that the proposed off-shore legislation covered different matters from those under discussion with the States and continued: 'The House is of the opinion that it is this fact which has led to the member for Farrer feeling justified in believing that an undertaking that there would be further consultations, which he gave the States, had been dishonoured.' It was all highly specious, but enough to remove the threat of the Government's defeat on the floor of the House. But the affair worsened Gorton's relations with the State Premiers and opened a further rift with his colleagues — notably, fatefully, Malcolm Fraser.

The difficulties Gorton encountered in his relations with the State Premiers, with the mining corporations, with sections of manufacturing industry and with the Sydney press proprietors, foreshadowed many of those Whitlam would face. Without pushing parallels too far, the manner of his fall bore some noteworthy similarities to Whitlam's: the thumbs-down signal was given in both cases to the same agent, Malcolm Fraser, by the Melbourne Establishment and the Sydney press; like Whitlam, Gorton was brought down at the very time when he was asserting his leadership more effectively than he had done for some time past and when he appeared to be recovering public approval; and the fatal move against him was made by a man whom he had every reason to trust.

Just before his fall, Gorton, ever resilient, seemed to be making a recovery from the Senate election débâcle. In January 1971, he made his only really successful appearance abroad as Prime

Minister at the Commonwealth Heads of Government meeting in Singapore. In February, he confronted the Australian Medical Association, with considerable public support, to withstand its demands for a 15 per cent rise in medical fees. In March, he was deposed.

The long-expected retirement of Sir John McEwen in February transformed the power balance within the coalition and indeed within the Liberal Party itself. Gorton had alienated McEwen by resisting McEwen's last-ditch effort to castrate the Tariff Board. Although McEwen had formally withdrawn his veto on McMahon as Liberal leader after the 1969 elections, his mere presence, with all the bitter memories it embodied, remained a formidable barrier against McMahon's leadership hopes. Once McEwen's presence was removed, McMahon automatically became a valid alternative and a potential challenger for the leadership. It was because McMahon was now available that Malcolm Fraser made his devastating move. From the days of their alliance, which helped deliver the Liberal leadership to Gorton in 1968, their relations deteriorated throughout 1970. Gorton believed that Fraser had not supported him loyally in the Party room over the off-shore legislation. Fraser believed Gorton had not supported him loyally in his moves to reorganize the defence forces. But Gorton was not aware of the deep animosity against him and the deep distaste for his style of government which was mounting up in the heart of his old friend, ally and confidant. Only a capacity for a very special kind of hatred can explain Fraser's actions of 7–10 March 1971. Only a capacity for a very idiosyncratic code of honour can excuse them.

In February, the ABC correspondent in Saigon filed a story stating that the Australian Army in Vietnam was ending its civil action programme. Fraser pointed out that this was contrary to Government policy. He then gave a briefing to Peter Samuel, Canberra correspondent for the *Bulletin*, whose report, based on this briefing, told of serious tension and lack of mutual trust between the Minister for Defence and the armed forces. Similar stories appeared in the *Daily Telegraph* and the *Canberra Times*. Gorton's response to these reports was to summon the Chief of the

General Staff, Lieutenant-General Sir Thomas Daly, to assure him that the Army had the Prime Minister's complete loyalty and trust. A version of this meeting was written by Alan Ramsey, the political correspondent of the *Australian*. Before filing his story, Ramsey showed it to Gorton. Gorton made no comment on it, even on Ramsey's suggestion that in the interview between Gorton and Daly, the General had accused Fraser of disloyalty to the Army. Fraser was later to claim in Parliament that Gorton could have stopped the Ramsey report 'with a single sentence... but that word was not given'. This is nonsense. Ramsey, like any good journalist, might have modified, altered or added to his story in the light of any comments Gorton may have made, but there was no way he would have scrapped an essentially true story. The Ramsey story appeared on 4 March. Fraser made no attempt to discuss his grievance with Gorton. On Monday 8 March, he sent Gorton his resignation and, by the simple expedient of taking his telephone off its hook, made sure that Gorton could not contact him. When Parliament assembled on the afternoon of Tuesday 9 March, Fraser was given leave to make a statement of explanation. He charged Gorton with disloyalty for failing to repudiate the Ramsey article.

Fraser gave his version of the incidents leading up to his resignation, and said:

> On Wednesday, 3rd March, journalist Alan Ramsey saw the Prime Minister. He told him he had a story that General Daly had accused me of disloyalty to the Army and to [the Minister for the Army] Mr Peacock. In his own words, the Prime Minister did not comment on this report, on this essential part of it. He refused to comment. The Prime Minister claims he does not comment on reports of private conversations even though he knew it would lead to a report damaging both to its target — myself — and to its alleged spokesman, General Daly. In plain words, the Prime Minister would prefer to allow a false and damaging report to be published about a senior Minister. Later, both he and General Daly denied the report. Which principle is more important — silence about a conversation, or loyalty to a senior colleague? One sentence would have killed the report. The Prime Minister, by his inaction, made sure it would cover the front page. As I have

> indicated in my letter of resignation, I found that disloyalty intolerable and not to be endured.
>
> It should not be thought that this act alone has brought me to this point. Since his election to office, the Prime Minister has seriously damaged the Liberal Party and cast aside the stability and sense of direction of earlier times. He has a dangerous reluctance to consult Cabinet, and an obstinate determination to get his own way. He ridicules the advice of a great Public Service unless it supports his view.

Fraser claimed that Gorton had acted high-handedly, and 'possibly illegally', in securing a Cabinet order to call out troops in Rabaul during a crisis on the Gazelle Peninsula in July 1970. Fraser had stored this grievance — his only other charge against Gorton — for nine months. But as he was to say in another context in 1975: 'The timing must be impeccable.' He concluded:

> The Prime Minister fought to prevent Cabinet discussion. If such a discussion had not been held despite my insistence, I would have resigned then. I have now done so as a result of what I have regarded as the Prime Minister's disloyalty to a senior Minister. The Prime Minister, because of his unreasoned drive to get his own way, his obstinacy, impetuous and emotional reactions, has imposed strains upon the Liberal Party, the Government and the Public Service. I do not believe he is fit to hold the great office of Prime Minister, and I cannot serve in his Government.

Gorton was clearly stunned by the venom of Fraser's statement. His rambling, stumbling reply was enlivened by an interjection from Alan Ramsey in the Press Gallery.

Hansard records:

> The matter on which I would not comment at all was Mr Ramsey's suggestions as to what General Daly might or might not have said. I believe it wrong to do this, to make comments or affirmations or denials in cases like this. I think this is so whether the person who is a third party is a General, an Admiral, a politician, a civil servant or a businessman. I therefore replied to that question: 'Had General Daly said what it was claimed he did say?' by saying that I thought it wrong to discuss or comment with Mr Ramsey on what a third party had said and Mr Ramsey replied: 'Fair enough'.

A voice: You liar.
Mr Speaker: Order!
Mr Calwell: Why don't you deal with the animal?

A prompt apology by Ramsey headed off by a move by Whitlam to have him called before the Bar of the House to explain just what he meant. The real drama then moved to the Liberal Party meeting room.

In the subsequent Party room battle of tactics that night and next morning, 10 March, Gorton was once more, this time fatally, harmed by the good intentions of his friends. Instead of forcing Fraser or one of his supporters to take the initiative, they made the mistake of seeking a vote of confidence. The vote was tied, 33 to 33. Gorton, as chairman, declared the motion lost. It is claimed that Gorton nobly used his casting vote against himself. In fact, he had no choice. At least four Liberals — Fairbairn, Jess, Howson and Bate — had said that they would cross the floor to vote for the Opposition's inevitable no-confidence motion if Gorton remained Prime Minister. This would have brought down the Government.

The drama and surprises of the day were not quite exhausted. McMahon easily defeated Snedden for the leadership and the Prime Ministership. Gorton nominated against Fraser and Fairbairn for the deputy leadership — and was elected. When the news was announced to the waiting journalists, King's Hall echoed with disbelieving laughter. To add the spice of mischief to the day, McMahon's first action as the new Prime Minister was to ask his deputy to choose his portfolio; Gorton chose Defence. The McMahon Prime Ministership was thus deeply embarrassed and flawed from its outset. It never really recovered from its first day attempt to mask its profound and traumatic disunity and recrimination by such a patent and phoney display of unity and reconciliation. When the inevitable came and McMahon disposed of Gorton, the effort cost him his last hopes of presenting a front as a secure and confident Prime Minister in control of a united party.

It needs no conspiracy theory to explain why the Liberal Party rid itself of a leader who had proved himself an electoral disaster. The 1969 and 1970 election results contain in themselves sufficient reason for his fall. That is not to say there was no conspiracy in the actual means used to bring about an inevitable end.

Whitlam believed that Fraser, McMahon and Sir Frank Packer wanted to bring down Gorton and used the trumped-up story about civil aid to do it. There was, in fact, nothing intrinsic to this issue to require either the resignation of a Minister, much less the destruction of a Prime Minister. If a Minister were to resign every time a Prime Minister refused to repudiate damaging press articles, no government would survive a month. The matter could never have become a momentous crisis unless Fraser was determined to make it so. Whitlam believed that the curious assertion, 'one sentence would have killed that story' came from Alan Reid. Certainly it was not a true statement about the situation. Ramsey would not have killed the story had Gorton denied it; he might have modified it, but his paper would not have killed it. On Tuesday 9 March, Reid did something he had not done since Whitlam became leader; he sought an interview with Whitlam to give him some helpful advice. He pointed out that if the Opposition moved a motion against Gorton, the government would perforce close ranks; Gorton would be saved; better leave things to see what would develop. Reid's advice was gratuitous. The Opposition was quite capable of determining its own tactics which, in any case, were irrelevant to the central drama and its dénouement in the Liberal Party room. Whatever his motives, Reid certainly was not known for his desire to assist the Labor Party.

What was it about Gorton's actions which made enemies? The Gorton paradox is that taken one by one, most of his decisions were popular; yet when taken together, they added up to deep unpopularity and distrust for his Government. The decisions on Vietnam, the five-power agreement, support for Israel, his defence of Australia's immigration policy, his attitude to foreign investment, the Esso-BHP oil price arrangements, the off-shore legislation, protection of the Great Barrier Reef, his defiance of the A.M.A., his anti-inflationary measures, each received the apparent support of his Party, most of the press and whenever separately surveyed in opinion polls, of the public. His posture of a staunch nationalist, reflected in his attitude to Commonwealth-State relations and foreign investment, was in tune with prevailing sentiment; he was entitled to believe that he was reflecting an attitude he shared with most Australians. Yet each little plus seemed only to add up to one big minus.

Some, like Fraser, purported to explain the Gorton paradox — and justify their own conduct — by putting it down to the Gorton style — the alleged 'shooting from the hip' style of decision-making, the breach of established forms. Yet when it came down to specifics, how thin was their case. Fraser's own philippic of 9 March ended with its final denunciation: 'This man is unfit to be Prime Minister'.

It is not particularly helpful to say that Australia was not ready for Gorton, though that is the explanation for his failure favoured by his friends. It is truer to say that he was not ready for the Australia he was called upon to lead, in that he had not formed his ideas clearly enough to make the most of the popular support potentially available to him.

It can be seen with hindsight that there were bad omens for the future Whitlam Government in Gorton's fall. Beyond Gorton's manifest shortcomings, his Prime Ministership posed daunting questions: would the power grouping in Australia tolerate any departure from their perceived norm; would State governments tolerate any alteration in existing arrangements; would businessmen tolerate anything which might in the slightest degree make them uneasy; would the electorate tolerate anything which, in the personal style of the Prime Minister, differed from Menzies, even when such a departure accorded with its wishes expressed in the opinion polls?

To the limited extent that Gorton tried to bring about change, he did a great disservice to the cause of change and reform — not because he tried, but because he failed. His failure and destruction further strengthened entrenched interests, who were delighted to discover how easy it was to get rid of anybody, even a Prime Minister, who disturbed them.

Outside Gorton's immediate following, the people most worried by his sacking were Labor members holding marginal seats. William Morrison, with a majority of 69 in 1969, was barometric in his moods; on a good day, he felt his majority had gone up to 72 for 1972; he could measure Labor's lows and highs to the last voter. Since the Senate election, the common conviction within the Labor Party was thus expressed, 'Gorton is our best asset'. Some were aghast at seeing Labor's 'best asset' lost. Even Fred Daly, neither new nor marginal, was impressed by McMahon's first press conference as Prime Minister. It was a competent performance, during which only a slight tremor of the upper lip and later, to an observant journalist, a wet patch on the table where his hands had rested, betrayed his nervousness. Watching it on television with Whitlam in his office, Daly said, 'Don't underestimate him. We all know he's hopeless, but the people will think they've got a real change without having to go through an election . . .' Whitlam sought to calm these anxieties by putting them in this perspective: 'Now listen. If there had been an open contest after Holt went, who would we have wanted? Not Gorton, certainly not Hasluck. We would have wanted them to choose Bill McMahon . . .'

13
Whirlwinds of Change

For nearly a century, China and New Guinea were linked, dimly but deeply, in the Australian consciousness: China the threat, New Guinea the bastion. Actually, the threat from the North materialized in the shape of Imperial Japan; but this merely proved the reality of the 'yellow peril'. When, after the Chinese revolution, the 'yellow peril' and the 'red peril' coalesced, the threat became total. Race, religion, ideology and recent history all combined against any rational debate on China. By the 1960s, Vietnam had become the first line of defence against 'the downward thrust of China between the Indian and Pacific Oceans'. New Guinea remained the last and most important line. Under the apparent indifference of Australians to the political future of their colony of Papua-New Guinea lay a deep concern about Australia's security.

To place these matters on the agenda of current political controversy involved going against some of the deepest and often the darkest instincts of Australians. This is what Whitlam decided he had to do in 1970. In part, it was a deck-clearing operation. The 1969 election result convinced him of the certainty of a Labor victory in 1972. To keep questions like the independence of Papua-New Guinea and the recognition of China out of the public debate was the safe course, but this would have meant wasting another three years before Australians could be brought to face realities. To keep the issues on ice meant that Australia's postures would have become frozen even further. Whitlam's conduct on New Guinea and China in 1970 and 1971 bear the classic elements of his style and purpose. In both cases, the starting point was a position long held, in advance of the established Party position. In both cases, he enunciated general principles in terms of an accumulation of specific details. In both cases, he could focus his passion on a private perception of evil — in the case of China, his contempt for Dulles; in the case of Papua-New Guinea, his inherited contempt for colonialism and racism. In both cases, he took his own course against more cautious counsel. In both cases, the reaction of his opponents was fierce and vindictive. And common to both ventures was a high element of chance and luck. Without great good luck, Whitlam would never have been so quickly and dramatically vindicated. But, like all risk-takers in politics, war and love, he had good luck because he worked hard for it.

Whitlam broke through the wall of silence on New Guinea affairs by which Australian politicians had conspired to retard any movement towards independence. Hasluck has testified that of all members of Parliament, Calwell was the most 'helpful and encouraging' in securing bipartisan support for his policy of enlightened paternalism. Hasluck's policy was enlightened in that during the building years of his long ministry from 1950 to 1963, he skilfully thwarted the political dominance of the economically dominant class, the Australian managers, planters and officials. By keeping all groups in political tutelage to Canberra, he prevented the expatriate Australians ruling from Port Moresby. He is entitled to his claim, which he makes in his book *A Time For Building*, that he prevented the emergence of Papua-New Guinea's Ian Smith. The price paid, however, was the retarding of political development at all levels, indigenous as much as expatriate. Calwell's co-operation was based on the concept of the islands of Papua-New Guinea as Australia's defence bulwark, the maintenance of which required Australian control. This was the rationale of the 'consensus'.

The political consensus on the future of Papua-New Guinea was really a conspiracy of silence to retard independence. After Hasluck's long paternalist rule, the Territory became a Country Party fief under the Ministry of C. E. Barnes, a Queensland grazier and racehorse owner. The Country Party's interest in New Guinea was entirely commercial, mostly in a negative sense. For example, the Country Party was anxious to ensure that New

Guinea did not develop its own sugar industry. In 1960, Menzies had startled some of his supporters by hinting at independence 'sooner than later'. Returning home from a Commonwealth Prime Ministers' Conference, the first after Harold Macmillan's *Winds of Change* speech on colonial Africa in 1958, he said at a press conference on 6 June:

> Whereas at one time many of us might have thought that it was better to go slowly in granting independence so that all the conditions existed for a wise exercise of self government, I think the prevailing school of thought today is that if in doubt you should go sooner, not later. I belong to that school of thought myself now, though I didn't once.

Hasluck put both Menzies and Macmillan in their place on 22 September 1960:

> The lessons of Africa today are not that we have to gallop madly along a path of political change but that we have to choose wisely and carefully a path of change which really satisfies the needs of and provides opportunities for, the people and the country with whom we are concerned.

In October 1961, in a speech to the Economic Society of Australia and New Zealand, Hasluck said that the necessary conditions for independence, industry and financial self-sufficiency 'would appear unlikely under twenty years' and indeed 'anything up to fifty years'. 'It is possible', he said, 'to peddle political dreams and fantasies, but the economic reality is that Papua and New Guinea is a dependent Territory and will continue to be dependent . . . for very many years to come.' In his turn, Barnes envisaged independence in half to three-quarters of a century. It was Barnes' deliberate policy to retard independence by appealing to the conservatism of the tribal leaders, particularly in the Highlands, against the stirrings of the emerging younger educated groups in the coastal towns, particularly Port Moresby and Rabaul. At the Australian Institute of Political Science Summer School in Canberra on 28 January 1968, he said:

> I urge that Australians who think about Papua-New Guinea

should admit the possibility that Papuans and New Guineans in general are conscious of the shortness of time of their development and are not anxious to push into self-government or independence . . . I myself see little need for or advantage to be gained in forcing the pace of constitutional change.

It would be thoroughly misleading to deny that there had been no constitutional progress under Hasluck and Barnes. A partly-elected Legislative Council had been opened in 1960. By 1969, a fully-elected House of Assembly, with limited powers, provided a training ground for local politicians. But despite a fair record, especially in education, health and aviation, Australian policy and Australian concern was still overwhelmingly governed by the ideas set out by Menzies' first Minister for External Affairs, Percy Spender, who said in the House of Representatives on 1 June 1950: 'The purpose will be to ensure that they are administered and developed in a way best calculated to protect the welfare of the native inhabitants and, at the same time, to serve Australia's defence interests.' The political consensus which Calwell helped create by his approving silence was deeply rooted in the Australian belief that New Guinea was its last defence bastion. The Australians were interested in staying in New Guinea for the same reasons that they were persuaded to go into Vietnam.

When Whitlam deliberately chose in January 1970 to break the consensus, he was striking at one of their most deeply-cherished beliefs, linking them with two World Wars, and some of their deepest fears and proudest memories. In the whole of his career, there is nothing that teaches more about Whitlam than his New Guinea initiative. For no better reason than that he believed it was the right thing to do, Whitlam set out to change the history of two nations in January 1970.

Like most things in politics, it was not a coldly planned venture. Whitlam did not anticipate the anger, fear and hatred his visit would create. Flying to Port Moresby, he outlined to four accompanying journalists his general ideas, including the setting of target dates—self-government immediately upon the election of a Labor government and independence by 1976. He was only repeating, or in fact modifying, a view he had expressed at

Goroka in 1960 that independence should come by 1970. It was a measure of Australia's lack of interest in New Guinea that this statement by a mere Deputy Leader of the Opposition passed unnoticed. It was Whitlam's position as Leader of the Opposition after a nearly victorious election campaign which gave new force to everything he said and did in New Guinea in January 1970. Whitlam had intended his briefing to be off the record, but the journalists, fairly enough, immediately reported the substance of his remarks. The premature disclosure sent a shock wave through the white community of the Territory, particularly the three groups with the deepest interest in indefinite continuance of Australian rule: the middle-range Administrative officers, the plantation owners and the private business interests from the octopus Burns, Philp and Co. Ltd trading company to service station operators. The expatriate economic monopoly was probably more complete in Papua-New Guinea than in any former British colony.

Whitlam's timetable for independence was not the only cause for the expatriate hostility. Just as menacing was his refusal to take them at their own value estimate of their worth as pioneers and builders in a harsh, hot land. Whitlam saw them as a privileged, philistine élite, blind to the hatred they were breeding around them. Above all, Whitlam believed that colonial rule corrupted both the rulers and the ruled. But Whitlam had not gone to New Guinea to talk to the expatriates. The importance of his visit was his contact with the emerging leaders. Whitlam was the first significant Australian politician to take them seriously as leaders who would shortly assume political responsibility for a self-governing nation. He was the first to assert that they were ready for leadership, not at some distant date, but now, in 1970.

The Administration and the Australian-owned newspaper ensured maximum publicity for any of the tribal leaders most hostile to Whitlam, particularly in the Highlands. The most vocal political leaders, later to form a loose conservative alliance called the United Party, expressed absolute hostility to any timetable for independence. The infant Pangu Party alone expressed support from the beginning. This proved decisive. In 1970, Pangu seemed weak and unrepresentative, drawing its support mainly from the coastal towns. But in less than two years, it had become the strongest party; in less than five years, it formed the first government of an independent Papua-New Guinea. The Pangu leaders who risked supporting Whitlam were Michael Somare, then member of the House of Assembly for his home district of Wewak; Albert Maori Kiki, then leader of the rudimentary trade union movement; and John Guise, Speaker of the House of Assembly. Somare became the first Prime Minister of Papua-New Guinea, Maori Kiki her first Foreign Minister and Guise her first Governor-General.

The great significance of Whitlam's 1970 visit was its impact on the development of political parties. It gave a boost in self-confidence to the Pangu leaders. Their own political programme gained a relevance it could never have had as long as the Australian political consensus made it unrealistic for any Papuan to talk about independence. By placing independence squarely on the political agenda in Australia and in Papua-New Guinea, Whitlam placed Pangu on the wave of the future. Just as significant was the impetus the visit gave to the political organization of the independents and conservatives opposed to Pangu. As Don Woolford points out in his book *Papua-New Guinea —Initiation and Independence*:

> Few members then (before Whitlam's visit) seemed to think the forming of a party was particularly urgent. But a month later, with the arrival of Gough Whitlam for his first visit, the situation changed... About eighteen months earlier, the independents had reacted to the threat of Pangu; now they reacted to the threat of Whitlam.

From moves which began as soon as Whitlam departed emerged the conservative coalition called the United Party which, until the House of Assembly elections in 1972, was regarded by most observers (and by Pangu itself) as the majority party and the group most likely to form a government in a self-governing Papua-New Guinea. Whitlam was the catalyst by which, within a year, Papua-New Guinea emerged with a relatively coherent party system.

Three years later, when Whitlam was Prime Minister and

Somare was Chief Minister of a country only seven months away from full independence, Whitlam recalled with relish:

> We have both come a long way from the lounge of the Sepik Hotel at Wewak in January 1970, when we were both virtually under siege, and under surveillance from the Territory's security police, and when we were both deemed in certain quarters to be committing political suicide by daring to talk about early independence for Papua-New Guinea.

The passions Whitlam unleashed reached a peak when he visited Rabaul. The Gazelle Peninsula, home of the proud Tolai people, who were the most politically aware of all the New Guineans, was already in a state of ferment over an Australian decision to establish a multi-racial local government council for the area. The Tolai suspected that this was a device to preserve white supremacy. They had boycotted the new council and formed a co-operative society, the Mataungan Association, to collect their own taxes and run their own affairs. Three of their leaders were in jail. The Mataungans knew little of Whitlam, but they knew that his was an important voice saying new things. His reception at Rabaul airport by more than 1,000 Tolai and then at a rally of more than 10,000 at the nearby Matupit oval was tumultuous. It was a deeply emotional moment, and it needed considerable self-control for Whitlam, Kim Beazley and Bill Hayden, who also addressed the meeting, not to be swept away by the almost religious fervour of the meeting. While they both tried to give encouragement and sympathy to the Mataungans the key to their message was a plea for discipline and an appeal against violence. The Gazelle Planters' Association sent a telegram to Gorton charging Whitlam with stirring up violence. Gorton responded immediately in an attempt to fix upon Whitlam responsibility for any violence or bloodshed that might occur in the area. For months after, the fear did indeed exist; it was a fear which haunted Whitlam. He reacted vigorously to Gorton's charge. In a statement from Port Moresby on 11 January, he said:

> Mr Gorton's first statement ever on New Guinea affairs strengthens elements whose chief fear is any discussion or

investigation which would expose to the people of Australia the injustices and indignities being perpetrated in their name.

The difference between Mr Gorton's action and mine is this: he has supported and encouraged those who, some through incompetence and some through malevolence, have created a perilous situation in the Gazelle. I have spoken directly to the people to deplore violence, to encourage peaceful and orderly processes and to reassure them that the people of Australia will listen and will care. There is no surer path to violence than by allowing these people to believe that no-one will listen. It is not by speaking the truth, but by suppressing the truth that the real damage is being done. There is a deep-seated fear among these people. Hate, the companion of fear, cannot be far behind. Emotions like this will not be suppressed by mechanical appeals to law and order. Unless people have faith in the law, or believe that unjust laws can be changed by orderly processes, there can be neither law nor order.

The Tolai fear of the multi-racial Council is not in fact a radical emphasis on race but a conservative fear of losing land. This fear was aggravated by Government action in attempting to force the people of Bougainville to sell their land. Tear gas and batons were used in that attempt. Mr Gorton and his colleagues, a section of expatriate Australians and a section of the Administration are knowingly or unknowingly combining to create in New Guinea the classic pattern of disruptive colonialism which ultimately destroys both the ruled and the rulers. I will not be silenced or intimidated on these matters. I am not prepared to allow Australia's name to be besmirched, dishonoured and disgraced, even under the cloak of technical legality.

When moving the Papua-New Guinea Independence Bill on 20 August 1975, Whitlam recalled:

It is hard, in retrospect, to understand the obloquy, the hatred, the contempt, and even worse, the ridicule, which we brought upon ourselves in those far-off days — a whole five years ago — when we first stated that the independence of Papua-New Guinea was imminent and inevitable. It was not easy in those days to live with the accusation — as in particular, the Minister for Education, the Treasurer and I had to live with it in the early months of 1970 — that we had stirred up and should be blamed for possible bloodshed. It was said of me — in the newspapers and in this Parliament — that by my efforts to point out the inescapable fact that Papua-New Guinea was already providing more leaders than Australia could ever impose or be willing to provide – I would have 'blood on my hands' – the words of the then Prime Minister, the Right Honourable Member for Higgins.

The Mataungan meeting was to have a curious aftermath. By a strange twist it was to play a part in Gorton's fall. His ill-advised and opportunistic reaction against Whitlam led to a series of misjudgements culminating in his authorization for the use of the Army to control disturbances on the Gazelle. This was one of the grievances which Fraser harboured until he was ready to strike in March 1971.

On the eve of his departure, Whitlam issued a statement to bring together his views:

> New Guinea is already rich in leadership. The time when that leadership will assume its full and proper responsibilities cannot and must not be long delayed.
>
> We quickly found that our most urgent and difficult task was to reassure the people and their leaders, and even expatriates, about Australia's relationship with New Guinea after self-government and after independence. We have been appalled to discover how widespread and deeply rooted is the impression that independence or even self-government means the end of Australian concern and Australian help for this country. I make no comment about the origins of the falsehood. Any Australian who propagates it does a grave disservice to the people of Australia and the people of New Guinea.
>
> Our second task has been to clear away fundamental misunderstanding about the reality of Australia's present relationship to New Guinea. Australia would not have been permitted to remain in New Guinea as trustee had she not promised to prepare New Guinea for independence. The Governments of 112 nations have just called on Australia to transfer full executive and legislative powers to elected New Guineans. The fact of independence is just not negotiable.

Whitlam set out to counter the prevailing orthodoxy that the pace and timing of independence was exclusively a decision for the Papuans and New Guineans:

> The Australian Parliament has responsibilities beyond New

Guinea. Its primary responsibility is to the people of Australia. It has the responsibility of protecting the reputation and the relations of the nation with all countries. These are the responsibilities of the elected persons of the Australian Parliament and the elected government, and of no others, elected or non-elected. The Australian Parliament cannot escape or share this responsibility. Therefore it is either misleading or meaningless to assert that the decision for independence is one for the people of New Guinea alone. The form of independence is certainly for them to decide for themselves. The fact of independence has already been decided.

We have tried to impart a feeling not only of urgency but of self-confidence in the ability of New Guineans to make their own decisions. There are New Guineans who are well-equipped to fill the highest political and administrative positions in their country. Indeed, no Australian could claim the contrary without reflecting on Australia's record of administration at its crucial point.

The House of Assembly and its members should now be taken seriously as the representatives of their people and treated by all Australians with proper respect and seriousness.

The only thing in which New Guinea is really unique among the countries of the world is that alone among significant populations its people make no final decisions on any matter affecting their welfare.

It is not unique in its economy, in the difference of economic standards between sections of the country, its educational or social standards, its need for economic aid from abroad, its need for advisers, the diversity of its local customs, or even the multiplicity of its languages. All these matters present complex and difficult problems for any future government of New Guinea. None of these problems require colonial rule for their solution or easing. In fact, many of them will worsen if foreign techniques, methods, laws and customs continue to exclude local custom, knowledge and experience. An outside administration cannot teach or impose unity. It can by errors unite a people against it. This is the very situation which Australians at home will not permit, and Australians in New Guinea must most avoid.

Whitlam's 1970 visit was a catalyst for political change within Papua-New Guinea. It transformed the political debate within Australia. For the first time in Australia's history as a colonial power, the Australian Parliament began seriously to debate its role. Gorton toured Papua-New Guinea in June. He had planned his visit as a mission of reassurance that Whitlam need not be taken seriously. He found Whitlam's proposals had already taken root. By the end of his visit, he had committed the Australian Government to a programme not discernibly different from Whitlam's, except for continuing to be vague on specific target dates for self-government and independence.

Exactly a year after his 1970 visit, Whitlam made a return visit and was able to say:

> In the past year the political climate of Papua-New Guinea has been transformed. A year ago proposals for early self-government were met with official hostility and public dismay. Some elementary truths about the early and inevitable end of colonialism in Papua-New Guinea held the terror of the novel and the unknown. Now the most significant leaders of Papua-New Guinea and significant sections of the population accept that they must shortly come to terms with their own future as a self-governing nation.

It was no more than the truth. What had seemed an audacious proposal in 1970 was already becoming commonplace. A new Administrator, Leslie Johnson, cheerfully predicted he would be the last Australian Administrator.

New Guinea had a deep hold on Whitlam's emotions and instincts, perhaps deeper than any other issue in which he was involved. Gorton's cry that blood would be on his hands was more hurtful than Gorton could have credited, for it was a prophecy that could easily have become self-fulfilling. It was while Whitlam was under this cloud that he heard of the death of the son of the former Labor Leader in the Senate, Nicholas McKenna; unaccountably, except to those who suddenly realized the pressure he was under, he broke down. Years later, a few minutes before the midnight hour at which Papua-New Guinea would become independent, Michael Somare, at a private dinner for a few colleagues who had been through the brief struggle, singled out Whitlam as 'his mate Gough'. When Whitlam rose to reply it was many moments before he could command his voice. But then, it was a time — 16 September 1975 — when few were disposed to profess mateship for the embattled Labor Prime Minister of Australia.

Whitlam always saw the issue of Australia's role in Papua-New Guinea in terms of Australia's international reputation, particularly as a racist nation. In his farewell declaration at the end of his 1971 visit, he said:

> All Australians must now realize how damaging and dangerous a reputation Australia's present policies produce. We are a European nation on the fringe of the most populous and deprived coloured nations in the world. What the world sees about Australia is that we have an aboriginal population with the highest infant mortality rate on earth, that we have eagerly supported the most unpopular war in modern times on the ground that Asia should be a battleground of our freedom, that we fail to oppose the sale of arms to South Africa, that the whole world believes that our immigration policy is based on colour and that we run one of the world's last colonies. We may rightly profess our good intentions and feel that we are merely the victims of special circumstances, but the combination of such policies leans heavily indeed on the world's goodwill and on Australia's credibility. The true patriot therefore will not seek to justify and prolong these policies but will seek to change them.

*

Since 1949, the 'true patriot' had been judged by his soundness on the 'China question'. After Mao Tse-Tung's final victory in October 1949, the British Labour Government suggested to its fraternal governments in Australia and New Zealand that the three governments should recognize the People's Republic simultaneously. Australia and New Zealand both faced general elections on 10 December 1949 and therefore, decided to delay recognition until after the elections, in which communism — domestic and international — loomed as a major issue. Britain went ahead alone. In the first months of his new government, Menzies was by no means implacably opposed to eventual recognition, but the outbreak of war in Korea in June 1950 brought tentative moves to a halt. The French defeat in Indo-China and the enshrining of the Dulles doctrine in the South East

Asia Treaty Organization in 1954 settled Australia's attitude for the next decade. Menzies, however, always resisted pressure from the United States and some of his colleagues to recognize formally Chiang Kai Shek's claims to be the Government of China by appointing an Ambassador to Taiwan. In 1966, Holt, for no better reason than a personal friendship with the Taiwanese Ambassador in Canberra and against the advice of the Department of External Affairs, reversed this policy and established an Embassy in Taipei.

The Labor Party did not include a plank to recognize China until after the Split in 1955. Whitlam had taken an independent view as early as 12 August 1954, when he told Parliament:

> We have to face the fact that the countries of South-East Asia, and the Colombo Plan countries in particular, do not regard the Communist Government in China as being hostile to them. In these circumstances, they do not wish to align themselves with either of the two power blocs as they regard them. A still more serious phase of our policy is that we say not only that the Communist Government of China is not, and should not be, the Government of China. We must recognize the fact that the Government installed on Formosa has no chance of ever again becoming the Government of China unless it is enabled to do so as a result of a third world war. When we say that that government should be the government of China, we not only take an unrealistic view, but a menacing one.

Australia's refusal to recognize the People's Republic and the official line that China was the real enemy in Vietnam did not prevent substantial trade between the two countries, particularly in wheat. Early in 1971, however, the Chinese Government failed to renew its contract with the Australian Wheat Board and, through the British Embassy in Peking, made it known to the Australian Government that its decision was influenced by political considerations. The new Leader of the Country Party and Deputy Prime Minister, Douglas Anthony (succeeding McEwen, who retired early in 1971), had announced that he 'would not sell his soul for a mess of potage', although it had not been previously thought that the Country Party was the keeper of the Australian conscience, particularly in trade matters. About the same time, hostile references towards the People's Republic by the Australian Ambassador in Tokyo, Gordon Freeth, had got back to Peking and had caused great offence there.

Dr Rex Patterson, in his role as Labor's spokesman on primary industry, skilfully exploited the Government's embarrassment. Australian interest in the China matter was further stimulated by the cordial welcome given in Peking to an American table tennis team and an apparently off-hand remark by President Nixon that he hoped sometime to visit Peking. In the climate of early 1971, that the heir of Dulles should even joke about such matters was an omen of considerable portent. In fact, Nixon was not joking at all. This was the beginning of 'ping-pong diplomacy'.

In April, the Federal Executive met in Adelaide. The Federal Secretary, Mick Young, who had been deeply stirred by his experiences during a three-month visit to China before the upheaval of the Cultural Revolution, suggested an initiative designed to advance Labor's policy and take advantage of the Government's embarrassment. Accordingly, Whitlam sent a cable to Premier Chou En-lai suggesting a visit by an Australian Labor Party delegation for discussion on 'matters of mutual concern'. Whitlam's Private Secretary, Richard Hall, suggested that the cable refer to a 'delegation' rather than any particular individual. After the despatch of the cable to Peking, Hall contacted an Australian academic, Dr Ross Terrill, then lecturing at Harvard University. Terrill was a recognized 'friend of China', personally known to Premier Chou En-lai. Terrill pressed Labor's case, and vouched for Labor's *bona fides* with the Chinese representatives in Canada. Terrill's intervention was crucial in overcoming the opposition to the proposal from Peking's contact in Australia, the Melbourne lawyer E. F. Hill, leader of the pro-Chinese Australian communist faction. It can now be seen in retrospect that a visit from such a delegation from such a country — a non-recognizing country, trading with China but allied with the United States — meshed perfectly with the policy of seeking an opening towards the West which Chou En-lai was then developing. Yet, as the weeks passed, and Peking maintained its silence, hopes faded. The cable seemed to have sunk without a trace in the vastness of China. The initiative was set fair to become an embarrassment. McMahon sought a tortured parallel with a letter Dr Evatt had sent to the Soviet Foreign Minister, Mr Molotov, in 1954 during the Petrov Royal Commission. 'What a strange thing it is', he told Parliament, 'that in one case the letter should have been written to a country that was a communist country and in the second place a similar kind of application was made too. What they are attempting to do, when the political interests of both those countries were involved, was that they sought the support of those countries on the one hand in order to interfere with legal processes in Australia that were taking place in a Royal Commission; and on the other to try and interfere in the operations of the Australian Wheat Board which is appointed by the Government and which genuinely represents the wheat interests in this country.'

On 11 May, Whitlam received from the Chinese People's Institute of Foreign Affairs (a body used by Peking for contacts with non-recognizing nations) a telegram: 'We have learned about your cable to Premier Chou En-lai dated 14 April. Our Institute will welcome an Australian Labor Party delegation to China in mid-June or the latter part of June for discussions on questions concerning the relations between the two countries.' In *Whitlam P.M.*, Laurie Oakes states accurately:

> Whitlam phoned Freudenberg at home and asked, 'Do you think I should go?' Freudenberg, thinking of the political risks, answered 'No' — but on the way to the office he changed his mind. He reasoned that the potential advantages of Whitlam being seen taking a major diplomatic initiative in playing the international statesman justified the domestic political dangers which would be involved. When he got to the office, he found that Whitlam had decided to make the trip anyway, and was already discussing arrangements.

The delegation Whitlam and Young devised was: Whitlam, Tom Burns (by then Federal President of the Party), Patterson, Young, Dr Stephen Fitzgerald and the author. Fitzgerald had left the Foreign Affairs Department in protest at the Government's Vietnam policy, thus depriving it of one of its few Chinese speakers and almost the last of its officers with any claim to knowledge of Chinese affairs; he was therefore a nuisance and the Department was glad to see him revert to a more suitable career at the Australian National University. Patterson, who felt that he had started the whole thing, was miffed that the delegation had been expanded both in personnel and purpose. Until he actually arrived at Sydney airport to fly to Hong Kong with his colleagues, it was not altogether certain that he would consent to go at all. The lure of China overcame all. On the day that Whitlam announced the delegation, McMahon also made an announcement: 'The Government has decided that we will now explore the possibilities of establishing a dialogue with the Government of the People's Republic of China.'

The Labor delegation (augmented by Australian journalists who had been given last-minute approval by the Chinese) entered China from Hong Kong on Friday 2 July. After a kind of acclimatization period in Canton, a dishevelled, exhausted group reached Peking near midnight on Saturday 3 July after a turbulent five-hour flight. At the Peking Hotel, the spokesman for the People's Institute suggested a rest day for Sunday. Whitlam, already slightly uneasy at the Chinese refusal to indicate any clear itinerary said firmly, 'No. We have come here to work.' It was the right comment at the right time. 'Then there will be a meeting with a responsible person at 9 a.m.,' said the spokesman. It turned out to be the Acting Foreign Minister, Chi Peng Fei. Next morning there was another interview, this time with the Minister for Trade, Pai Hsiang Kuo. On the day of the Australian delegation's arrival in China, a Canadian trade mission had received an assurance that recognizing countries, like Canada, would receive trade preference from China. The formula used by the Trade Minister to Whitlam and Patterson was: 'Political relations cannot but affect trade relations.' Patterson was invited for separate talks with Chinese trade officials. But the silence on further arrangements and appointments remained.

Just before lunch on 5 July, the Australians were told by the spokesman for the People's Institute: 'Please remain in the hotel.' In the afternoon, the official, scarcely suppressing his own excitement, announced that 'some time in the night, you will be taken to see the Premier'. Asked what form the interview with

Chou En-lai would take, he said it would be exactly as the other interviews: that is, in private. The summons came at 9 p.m. As soon as the delegation arrived at the Great Hall of the People, it was plain that the proceedings would be anything but private. Forty officials were already seated in a great horseshoe arrangement of chairs. Whitlam had to adjust immediately to the fact that his encounter with one of the world's most formidable personages would take place before an audience of nearly sixty, including the Australian journalists, and that every word would be sent back to Australia. Chou had the further advantage of speaking in Chinese but understanding English perfectly.

After an exchange of courtesies, Chou plunged into Vietnam.
Chou: I've heard that your party advocates the withdrawal of all United States troops from Vietnam and withdrawal of all forces sent to Vietnam by the Australian Government.
Whitlam: We have consistently advocated the withdrawal of the Australian troops in South Vietnam. We have said that everything should be done to end that war and Australian participation in it.
Chou: That is quite right. Does your Party have this general proposition that foreign troops should be withdrawn from other countries.
Whitlam: Yes, we have. And at our Federal Conference last month we once again stated that foreign troops should not be stationed in Czechoslovakia, Korea, Vietnam.
Chou. It is a correct proposition . . .
Whitlam: We have found that the attitudes of the Australian Labor Party are well known and understood and none of your Ministers and none of your officials questioned our right to have different assessments from those of the Chinese Government. In our interchange of views, when there were differences, they were understood and respected.
Chou: It is obvious there are bound to be different views between us . . . you mentioned in your discussions that you looked upon the ANZUS Treaty as preventing restoration of Japanese militarism. That is a fresh approach to us. I would like to ask you to inform us as to what points of that Treaty are directed against militarism.

Whitlam: That Treaty was made in 1951 by the Truman Government.

Chou: In the Truman Government, Dulles was already responsible for some parts of the work.

Whitlam: He was an officer of the State Department. Australia has only been attacked by one country in her history — Japan . . . So Australians in 1951 had a fear of the Japanese. They had the same fear of the Japanese then as I believe your people have now . . . The Americans were much more anxious to sign a peace treaty with Japan than were Australia or New Zealand and, to reassure the Australians and New Zealanders, the Americans entered into the obligations of the ANZUS Treaty. All political parties in Australia support the conclusion of the Treaty. I told your Acting Minister for Foreign Affairs that we in Australia regard the ANZUS Treaty as entirely defensive. It has never been used as justification for operating in Vietnam . . .

Chou: You are also a member of SEATO. You cannot call SEATO a defensive treaty.

Whitlam: It is moribund. I do not think anybody places much reliance on a treaty where Britain has so clearly determined never to undertake any military operations again, where France never turns up at the meetings, where Pakistan is quite obviously an embarrassment. Of the other members, The Philippines and Thailand are trying desperately to insinuate themselves into your Government's good graces.

Chou then very skilfully invited Whitlam to denounce ANZUS.

Chou: But in linking up the ANZUS Treaty with SEATO we can learn this lesson: that is, both of them have the United States as a principal member. That was the policy of John Foster Dulles. You may say it was his soul . . . his policy was by a whole series of alliances to encircle China. Now he has a successor to the north (the Soviet). We too had a defensive treaty, concluded in 1950 between China and the Soviet Union. The treaty was called the Sino-Soviet Friendship Treaty of Alliance and Mutual Aid. Its first article was to prevent the resurgence of Japanese militarism . . . Now the United States–Japanese reactionaries together are reviving Japanese militarism . . . And what about our so-called ally. They are in very warm relations with the Japanese Government and also engage in very warm discussions with the Nixon Government on so-called nuclear disarmament, while we, their ally, are being threatened by them together. So we feel that our ally is not very reliable. Is your ally very reliable? You see they have succeeded in dragging you into the Vietnam battlefield. How is that defensive? That is aggression.

Whitlam: I must say with respect I see no parallel in the Sino-Soviet Pact and the ANZUS Treaty. There has been no similar deterioration between Australia and the United States as between China and the Soviet Union . . . Yours has been a bitter experience and I understand your feeling. May I put this qualifying argument on behalf of the United States. I still deplore the destructive style of John Foster Dulles but his soul does not keep marching on. The American people have broken President Lyndon Baines Johnson and if Richard Milhous Nixon does not continue to withdraw his forces from Vietnam they will destroy him similarly. The Australian people have had a bitter experience in going all the way with LBJ. They know the American people made him change his policy and they will never again allow an American President to send troops to another country in this way.

Chou: I have similar sentiments with you. Such a very good appraisal of the American people. I do believe the American people will rise up and restrict the policies of the American President and overthrow him.

Chou was well-briefed for the occasion. 'Probably because your excellencies are here, the Australian Prime Minister declared yesterday that the establishment of diplomatic relations with China is far off now. They do not want to establish relations. He seems to be quite confident. It is probably because your party is in China.'

Whitlam: This may be. I must say even to the credit of my opponents that they are catching up with the realities of life on China to a certain extent. They know Dulles' policies have failed dismally and if President Nixon says he wants to visit China, can Mr McMahon be far behind?

Chou's amusement at this sally can now be seen to have been

that of a man who knew something nobody else knew. On that day, Dr Henry Kissinger, Nixon's National Security adviser, was in Bangkok, en route to New Delhi, Islamabad and Peking. Sharing the universal ignorance, Whitlam was puzzled by the relative mildness of Chinese references to the United States, at least compared to his fierceness against Russia and Japan. On 9 July, the day Kissinger arrived for his secret rendezvous, he wrote for the *Australian*: 'I did not detect the depth of animosity towards the United States I would have expected. It seemed to me that Chou En-lai was more sceptical about ANZUS than hostile towards it.'

The accompanying pressmen, aware of the disadvantages under which Whitlam had spoken and witnessing the subtle nuances of Chou's probing, were highly impressed by his performance and emphasized Whitlam's steadfast refusal to be drawn into a denunciation of ANZUS. At first, the view of the men on the spot was reflected in the editorial comment in their newspapers. On 8 July, the *Sydney Morning Herald* said in an editorial:

> It is easy to point to what is tendentious, question-begging or fallacious in Premier Chou En-lai's conversation with Mr Whitlam. There were, for instance, the far-fetched and irrelevant parallels between the ANZUS Treaty and the Sino-Soviet Pact ... which Mr Whitlam very properly denied ... Its interest will not be limited to Australia. Indeed, others may well ask why Mr Chou elected to give so much time and confidence to a politician who is, by world standards, obscure. The timing suggests part of the answer ... What is new is the rational tone of argument ... What seems clear is that China has the best motives for seeking to rally world moral support as one kind of limited insurance against gathering hostility. These motives will not be ours but Australia will be foolish if it does not make every effort to establish the relations in which this informal dialogue can be continued formally.

Yet, a week later, this 'rational tone of argument' altered completely. On 14 July, the same editorial writer commented:

> Mr Whitlam has not hesitated to seek Chinese smiles of approval at the cost of Australian interests ... Examples of Mr Whitlam's

servility are rife . . . In his most offensive passage about the United States Government, he declared that President Nixon would be destroyed by the United States people if he did not continue to withdraw his forces from Vietnam . . . If Mr Whitlam thinks that this wholesale selling out of friends to gain a despot's smile is diplomacy, then Heaven protect this country if ever he directs its foreign policy.

In the intervening week, the Department of Foreign Affairs, on instruction from the Government, had mounted an intensive effort to 'inform' the newspaper. Yet the original assessment of the *Sydney Morning Herald* proved to be immensely superior to the 'informed' Foreign Affairs view. There is no better proof that Whitlam had done well and was perceived to have done well than in the fury of the attack orchestrated against him between 6 and 15 July. If on any matter related to foreign affairs there is a sharp change of editorial opinion within a week, the explanation should be sought not in what has happened in the world but who lunched with whom in Sydney, Melbourne or Canberra that week.

At a DLP conference on 11 July, Senator Gair said, 'I feel sickened to see an Australian political leader fawn, praise and flatter Chou as Whitlam did.' On television, Santamaria said, 'Australia gained a Chinese candidate, if not a "Manchurian candidate", for the Prime Ministership when the Chinese communist leader personally backed Mr Whitlam's campaign for the next election.' On 12 July, McMahon told 400 cheering Young Liberals in Melbourne, 'It is time to expose the shams and absurdities of his excursion into instant coffee diplomacy. We must not become pawns of the giant Communist power in our region. I find it incredible that at a time when Australian soldiers are still engaged in Vietnam, the Leader of the Labor Party is becoming a spokesman for those against whom we are fighting . . . By accepting Peking as the sole capital of China, he is abandoning Taiwan . . . In no time at all, Mr Chou had Mr Whitlam on a hook and he played him as a fisherman plays a trout.' The *Australian* reported that McMahon's performance was 'accompanied by flag waving, banners, whistles, streamers, trumpet blasts and a pop band.' Next day, at another Liberal gathering in Devonport, Tasmania, he assured any doubters: 'China has been a political asset to the Liberal Party in the past and is likely to remain one in the future.'

On 15 July, President Nixon announced that, as a result of talks in Peking during 9–11 July, he had accepted an invitation to visit China before May 1972. 'I have taken this action because of my profound conviction that all nations will gain from a reduction of tensions and a better relationship between the United States and the People's Republic of China. It is in this spirit that I will undertake what I deeply hope will become a journey for peace—peace not just for our generation but for future generations on this earth we share together.'

The *Sydney Morning Herald* headlined McMahon's response— 'P.M. says: it's also our policy.' 'It makes an awful farce of Whitlam's visit,' he said. 'Whitlam didn't even know that Kissinger was there. That's how much the Chinese trust him. It makes a mockery of the man.' The Deputy Prime Minister, J. D. Anthony, said he was 'very pleased'. Barnard, holding the fort for Whitlam in Australia, invoked McMahon's own piscatorial metaphors and said the Prime Minister was 'like a stunned mullet'. In China, the senior official who brought the news to Young and Fitzgerald, who had extended their stay in China beyond the departure of Whitlam, Patterson and Burns on 13 July, displayed a canny grasp of Australian politics: 'Who's being played like a trout now?' There was quite a celebration in Wuxi that night.

It was true, as Whitlam claimed, that his visit made Australia 'look less slow-footed, less imitative' in the light of Nixon's coup. Yet McMahon's claim that the Australian Government had taken its own initiatives towards China was true enough. Throughout 1971, Australian diplomats had had exploratory discussions with their Chinese counterparts in Colombo, Bucharest and Paris. The less excusable, therefore, is the overblown rhetoric in denunciation of Whitlam, much of it originating in the Foreign Affairs Department in the days before the Nixon announcement. This slow, cumbersome exercise in discreet diplomacy was overtaken by events which caught the United States State Department as much off-balance as McMahon. The

Secretary of State, William Rogers, was as much in the dark as anybody else. The general outline of McMahon's embarrassing speech of 12 July had been referred to the American Embassy in Canberra and received its endorsement. Years later, McMahon complained to this author, 'We thought we were being helpful.'

Whitlam heard the news in Tokyo. In the morning he had called on Takeo Miki, a former Foreign Minister, later Prime Minister, but then in eclipse. Miki led the faction of the ruling Liberal Democratic Party which advocated normalization of relations with China. Whitlam was one of the few foreigners who maintained contact with Miki throughout his years in the shadows; most wrote him off as a back number. For Miki, as much as Whitlam, Nixon's announcement was an opening to the future. The Japanese Government, and Prime Minister Eisaku Sato most of all, were in a state of shock; the announcement of 15 July was the most severe of the Nixon *shokku* suffered by the Japanese during his Presidency. When Whitlam met Sato, he found himself, as the last Westerner to speak with Chou En-lai before Kissinger, in instant demand with the status of an acknowledged expert. Sato spoke of his sense of humiliation: 'The journalists told me about it.' There had been a mix-up in Washington and the White House message to the Japanese Embassy had been delayed. 'I have done everything they (the American Administration) have asked. They let me down.' There was a dignity about this old man, as tears welled in his eyes, admitting the truth of this tremendous loss of face to a foreigner.

This visit to Japan, following the visit to China, was the genesis of the Treaty of Friendship between Australia and Japan signed by Prime Minister Fraser in 1976. In Peking, Whitlam had been surprised by the depth of Chinese fears about Japan. At this time, there was an official formula in vogue, frequently repeated by the Chinese to the Australian delegation: 'The four enemies of the people of China are (in this order) Soviet revisionism; American imperialism; Japanese revived militarism; and the Japan Communist Party.' The JCP had such a high place in current Chinese demonology apparently because it was taking a strong pro-Moscow line against Peking. The Chinese made sure that the

Australian delegation was made aware of their genuine fears about both Russia and Japan. At the Great Wall of China, they emphasized its total defence irrelevance against Soviet missiles and aircraft, whatever its ancient military significance might have been. In Shanghai, the delegation was taken to the former site of the old European racecourse, not to be reminded of the degeneracies of our Imperial past, but to see the planning for the world's largest underground air-raid shelter, so that Shanghai would never again be vulnerable to air attacks from Japan.

Equally, in Tokyo, in the heightened atmosphere following Nixon's astonishing announcement, Whitlam was made deeply aware of the Japanese sense of insecurity, their sense of loneliness in the wake of the Nixon *shokku*. Whitlam believed that it was possible for Australia, as Japan's largest supplier of raw materials, to play a role, limited but significant, in reassuring Japan. The treaty, which was agreed to in principle between Prime Minister Whitlam and Prime Minister Tanaka in Tokyo in October 1973, was originally designed to give formal expression to this reassurance.

At home, the attacks on Whitlam's visit continued, severely depleted in their credibility and capacity to do damage by Nixon's announcement. The most vitriolic attack of all came from Malcolm Fraser. Speaking on 17 August from his self-imposed exile on the back-bench, he spoke in terms reminiscent of his onslaught on Gorton five months before. Whitlam, he said, had 'clearly become the Chinese candidate for the next Australian elections'. He had forsaken his right ever to claim to be the Prime Minister of this country, because Australia just could not run the risk of having as its Prime Minister 'a person who has in all likelihood secret agreements of one kind or another or who has given promises which Chou En-lai expects him to put into effect'. Fraser asserted: 'The Leader of the Opposition has ignored China's support for revolutionary warfare; he has ignored China's blatant support and pursuit of nuclear policies; he has ignored the fact that China is building strategic military roads in Asia that could be used to invade Thailand through Laos or through Nepal and which could be used for invasions in other areas; he has ignored the basic interests of Korea, Japan, Taiwan,

The Philippines, South-East Asian countries and Australia. What else has he promised in private if he ever becomes Prime Minister?' Just as in March he had declared Gorton 'unfit to be Prime Minister', he concluded: 'The man is a disgrace to Australia.'

In letters to newspapers, Fraser and W. C. Wentworth had been especially critical of Whitlam for his alleged 'desertion of a former friend'.

Throughout the China visit, Whitlam had tried to solve the mystery of the fate of the Australian journalist Francis James. Whitlam and James had been pupils together at Canberra Grammar School, although their relationship was, then or later, by no means as close as James tended to claim. Whitlam described James as 'the last of the Elizabethans' in tribute to his combination of literary skill, classical scholarship and love of swashbuckling adventure. He was also an Australian eccentric in the English tradition of Gordon, Lawrence and Biggles, part-hero, part-mystic, part-scholar, part-schoolboy. Upon the outbreak of World War II, he enlisted in the Royal Air Force, distinguished himself as a fighter pilot and suffered severe injuries to the face, eyes and body. As editor of the *Anglican* newspaper, he became the gad-fly of the Bishops over their silence on Vietnam (with some honourable exceptions, notably Bishop Moyes of Armidale). In 1969, he disappeared in circumstances of total mystery, having been last seen at Shumchun, China's point of entry from Hong Kong.

The whole purpose of Whitlam's inquiries in China was to signal his interest and concern to the highest quarters. He judged correctly that a formal and public inquiry would be the worst, possibly the most dangerous, course; to have courted an official denial of James' presence or existence in China could have been, literally, fatal. With the collaboration of the pressmen, Whitlam took the matter up, with calculated casualness, a dozen times with most of the officials accompanying him, confident that his concern would be relayed to the top. Their invariable reply was: 'We know nothing of this matter.' On the last day, on the train from Canton to Hong Kong, the farewelling official told him, 'You may give three messages to Mr James' family. He is in good health in China; he is under detention; he is charged with a breach of Chinese law.' On his return home, Whitlam for the first time raised the matter formally at the highest level. He wrote to Chou En-lai, setting out James' record 'in the war against Fascism', his early advocacy of normal relations with China and his opposition to the war in Vietnam. Whitlam told Chou En-lai that if James was in fact guilty of any breach of Chinese law, it could be explained by James' eccentricity, certainly not by any enmity towards China. He hoped the Chinese Government would free James 'as an act of clemency'.

Publicly, Whitlam kept steadfastly silent, refusing to be drawn by attacks by Fraser and Wentworth that he had 'abandoned a friend'. It would have been easy enough for Whitlam to end the attacks which, involving as they did charges of personal betrayal, were more hurtful and harmful than purely political attacks. He could easily have ended the matter; he might also have ended Francis James.

Early in 1972, the Australian Government learned from British sources that James was about to be released. In an attempt to win publicity and praise, the office of the Foreign Minister, Nigel Bowen, leaked the news and let it be known that elaborate arrangements were being made in Hong Kong for James' reception, hospitalization and interrogation. Affronted, the Chinese Government changed its mind. James was not released until after the election; the first fruits of Labor's normalization of diplomatic relations. Clumsiness or opportunism by Foreign Affairs cost James nine more months in prison.

14

'Tiberius with a telephone'

'He was determined, like other little Caesars, to destroy the member for Higgins (John Gorton). There he sat, on the Isle of Capri at Surfers Paradise, plotting his destruction — Tiberius with a telephone.' Thus Whitlam pilloried the manoeuvres leading to Gorton's dismissal as Defence Minister.

In August 1971, McMahon finally destroyed Gorton as a Liberal politician; in doing so, he also helped destroy himself. By sacking Gorton, McMahon effectively ended any chance he might have had of establishing himself as a credible or remotely popular leader. Whitlam's barb was no obscure classical reference. The Australian Broadcasting Commission had been telecasting a widely popular dramatized series on the life of the Caesars, with the Emperor Tiberius as the central character, and half of McMahon's colleagues had suffered from his addiction to the telephone. The point that Whitlam made, summed up in a single jibe, was that McMahon's action was the impossible combination — vindictive and ridiculous. The jibe was powerful and damaging precisely because McMahon's own followers felt it to be absolutely true. McMahon had inspired hate without inspiring fear.

Yet McMahon's sacking of Gorton was an act of frustration rather than vindictiveness. The Liberals had imposed on him an almost impossible task and then deprived him of the support needed to carry it out. His task was to restore unity, the key to the coalition's long success. The dissident Liberals of the Gorton faction scarcely bothered to conceal their disloyalty and their contempt. The election of Gorton as deputy leader, far from closing the ranks, merely exposed the crack in the Liberal Party from top to bottom. By July, McMahon desperately needed an excuse to get rid of Gorton. Gorton himself provided the excuse. It was not a very good one; a breach which would have been passed over as trivial if any vestige of goodwill had remained between the two men, but enough for McMahon's urgent purpose. In July, the journalist Alan Reid's book, *The Gorton Experiment*, appeared. Outraged by what he believed to be the book's gross unfairness, Gorton agreed to write a series of articles for the *Sunday Australian* to refute Reid's account of certain Cabinet decisions. Gorton wrote: 'One of the problems of Cabinet in recent years has been the difficulty of keeping anything under wraps. Even in Sir Robert Menzies' day, the leak and gossip continued to bedevil Cabinet, and the situation has not improved. From time to time Cabinet ministers have shown themselves so uncertain of their own opinions that they have chosen to canvass the value of impending legislation far beyond the Cabinet and the confines of Parliament altogether. Others are addicted to trying out ideas on their wives.' McMahon convinced himself—or accepted the urgings of Sir Frank Packer—that Gorton's articles constituted a breach of Cabinet responsibility. After days of indecision, using the telephone furiously from a holiday retreat at Surfers Paradise to sound out opinion, he screwed himself up to demand Gorton's resignation.

The disloyalty and disaffection within the Liberal Parliamentary Party remained as deep as ever; but now it spread to all levels of the Party. With his fall, Gorton reached a popularity he had never held as Prime Minister. McMahon lost the only claims to public goodwill he had held since becoming Prime Minister—sympathy for a man placed in a difficult situation, admiration for a 'little battler' who had won through after years of humiliation and unfairness.

The appearance of disarray was deepened by the dismissal from the Ministry of Leslie Bury, who had been Gorton's last Treasurer. In this move, McMahon did not enhance his credibility by denying it only days before he did it. McMahon said that the reason for the removal was Bury's failing health. Bury replied that he was in perfect health. Thus McMahon, who at the outset of his Prime Ministership had seen his task as that of a unifier, divided his Cabinet and his Party as deeply as Gorton had done.

In a rare message to the Party he had founded, Robert Menzies, in September, called upon the Liberals to 'recapture that first fine, careless rapture' of the late forties. Even supposing that to be an accurate description of the state of mind of the resurgent conservatives under Menzies, it was an impossible demand for the Liberals under McMahon. After the Gorton storm, the Liberals wanted a return to normalcy. They identified the Menzies era not with any 'careless rapture' but with orthodoxy and stability. But this yearning was not in tune with the dominant spirit of the time. Gorton and Vietnam had broken the mould. McMahon was not the man to put the pieces together.

McMahon's misfortune was that he could never make anything he did look Prime Ministerial. He appeared as a living example of a man promoted above the level of his competence. Paradoxically, his reputation as the most successful Treasurer since Sir Arthur Fadden diminished his prestige as a Prime Minister, for it seemed to underline the case 'a fine Treasurer, yes; Prime Minister, no'. With great unfairness, the contrast of his age and that of his wife was used by his critics to create an impression of a man barely below senility. The contrast between his height and that of Whitlam was used to portray him as hardly more than a dwarf. It was all very unfair, but by then he had no reserves of Party loyalty or of public goodwill, to counter the denigration to which he was daily subjected.

More than any Prime Minister, McMahon suffered from the Menzies syndrome, the deeply-imbedded folk-memory (by no means limited to Liberal Australians) that Menzies was the measure and the model of an Australian Prime Minister. With that august shadow looming behind him, McMahon just never looked like a Prime Minister, or the Australian cliché for one. In particular, he was an embarrassment abroad, at those times in which the Australian cultural cringe is especially potent. There is a tape extant which records McMahon's speech at the White House in October 1971. He had been given a speech by his Foreign Affairs adviser, Richard Woolcott, and Dr H. C.

Coombs, whom he had publicly designated as his 'guru'. As Richard Nixon made, or appeared to make, a neat off-the-cuff speech, McMahon decided to reply in kind and pocketed his prepared speech. On and on he warbled: 'I take as my text a few familiar words. There comes a time in the life of a man in the flood of time that taken at the flood leads on to fortune . . .' On the tape (not one of Nixon's notorious tapes), above McMahon's meanderings can be heard the groans of the Australian journalists listening to the speech piped through to the White House library and one agonized voice above the rest: 'O God, I wish I were an Italian.'

Such things, however, were only symptoms of the Liberal decline under McMahon. At the heart of the Liberal decline was the collapse of the two pillars of the Liberal hegemony — public acceptance of their superior management of the economy at home and foreign policy abroad. By 1971, these assumptions no longer held.

It was a measure of the turn-around in public attitudes that by 1971, Vietnam had become a disaster area for the Government in sheer political terms. Even the American alliance had lost its magic, as it became clear that the United States had suffered a massive defeat. The invasion of Cambodia in April 1970, the unleashing of the South Vietnamese Army upon its traditional Khmer enemy, and the destruction of Prince Norodom Sihanouk's little oasis of peace, however tenuous, however artificial, however ambiguous it may have been, marked a decisive change. For the first time, Australians in general began to acknowledge the immorality of the war in Indo-China. In coming to this conclusion after five years of approval, they were mightily assisted by the fact that the United States was clearly losing the war. After the Cambodian invasion and the killing of protesting students at Kent State University, Ohio, the Australian protest movement reached its peak strength when Jim Cairns led 70,000 Moratorium marchers down Bourke Street, Melbourne in May 1970. By 1971, McMahon's uncomfortable task was to preside over Australia's withdrawal in line with America's, and call it victory.

In his first Prime Ministerial statement of policy on Indo-

China on 30 March 1971, he said: 'The Australian Government will continue to assist the Republic of Vietnam, although the character of our assistance will progressively change.' The United States under Nixon had no intention of making it any easier for McMahon than Johnson had made it for Holt. On 30 September, the American Embassy in Canberra conveyed a request to the Defence Department for more assistance in the form of military advisers to train Cambodian troops. Incredibly, Gorton's successor as Defence Minister, David Fairbairn, neglected to inform McMahon or the Cabinet of this request. His failure, which can only be explained by the political embarrassment such requests now caused, is a measure of how completely the balance of political advantage of the Indo-Chinese question had shifted. A month later, McMahon was still in the dark. On the eve of his departure for Washington, he denied press reports that the request had been made and would be discussed with Nixon. With understandable asperity he told reporters: 'If such a request had been made, I, as Prime Minister, would certainly have been told of it. There has been no request.' In one sense, McMahon was lucky that he was leaving for the United States, so that his denial was given to the press and not to Parliament. He was lucky not to face charges of having misled Parliament. Fairbairn and Anthony, as Acting Prime Minister, had to face the Parliamentary attack and their humiliation and embarrassment overshadowed McMahon's Washington and London visit. The journalists accompanying McMahon felt quite cheated at finding themselves out of the real action; for the first time, a Prime Ministerial overseas visit became a sideshow to the main action in Parliament. This resentment may partly explain why they extracted the last drop of ridicule from McMahon's odyssey of 1971.

*

The other pillar of Liberal prestige was its claim to provide good economic management. In three disastrous days on the eve of Christmas 1971, the McMahon Government destroyed any remaining reputation it had for superior economic expertise and probably destroyed any remaining chance of winning the 1972 elections. Like most of the significant events in these years, the Cabinet convulsion of December 1971 was a by-product of Vietnam. The United States economy could no longer absorb the cost of the war. Early in December, Nixon devalued the American dollar by 8.5 per cent. On 20 December, Cabinet met to settle whether Australia should devalue with the United States dollar or revalue with sterling. The Country Party members insisted on devaluation; Treasury and Snedden were as vehement in favour of revaluation. By 4 a.m. on 21 December, Cabinet reached the decision to revalue by less than the full percentage required to keep the dollar in line with sterling. By the afternoon, Treasury returned, with new arguments, to its original line of full revaluation with sterling. Anthony thereupon threatened to take the Country Party out of the coalition. McMahon compromised. On 22 December, he announced: 'The Government has decided to establish the market rate for the Australian dollar on the United States dollar at an appreciation of 6.32 per cent on the previous parity relationship. This means an overall depreciation in relation to the parities of our trading partners as a whole of approximately 1.75 per cent . . .'

The economic consequences of this decision were largely ignored because of the absorbing spectacle of the Cabinet brawl, which had so nearly led to the destruction of the coalition. For the next year, the Australian dollar remained grossly undervalued and presented the Labor Government with a massive liquidity problem from the outset. But this was all in the future. What came across explosively in December 1971 was the image of a deeply-divided coalition, a brawling Cabinet and a weak, ineffectual Prime Minister. An opinion poll taken in January gave Labor a lead of 17 per cent over the Government parties.

McMahon tried hard to meet the two competing demands upon his Prime Ministership: to satisfy Liberal demands for a return to conservative orthodoxy after the Gorton turmoil, and yet make some response to the expectations for change which Gorton and Whitlam between them had raised in the electorate. He satisfied neither. He correctly identified the urgent need to restore peace between the Federal Government and the Liberal State governments; he abandoned the off-shore legislation and returned payroll tax to the States; but the deflationary strategy of the 1971 Budget left him with little scope for more than pious expression of a desire for closer co-operation with the States. He set up a national inquiry into poverty, back-tracking on his earlier refusal to do so, thus gaining credit neither for flexibility nor compassion. Recognizing Whitlam's success in exploiting urban issues, he established an Urban and Regional Development Authority. He tried to keep exploratory talks with China going; but as long as he insisted that the Embassy in Taiwan would stay, there was no chance that the Chinese would listen. Thus, he dissipated the strength of the old intransigence and won no praise for progressiveness. Perhaps his most adventurous and constructive action was to appoint Andrew Peacock as Minister for External Territories with a definite brief to advance self-government for Papua-New Guinea; but by 1972, the main political result of this initiative was to validate and endorse Whitlam's actions in 1970–71.

*

McMahon was correct in later complaining that the press were far harder on him than on Gorton. Until 1969 at least, the press campaign against Gorton was largely subterranean, unsupported by the proprietors. For all his life-long cultivation of proprietors, McMahon's press support was minimal and grudging. His last prop was removed in 1972 when Sir Frank Packer sold the *Daily Telegraph* to Rupert Murdoch. Murdoch had committed his papers to a change of Government. He telephoned McMahon in Sydney to tell him of his deal with Packer: 'I can promise, Prime Minister, that we will be as fair to you as you deserve.' In the background, Packer warned: 'If you do that, you'll murder him.' The significance of the defection of the *Daily Telegraph* was essentially negative; Labor was meant to make great gains in Melbourne, where Murdoch's influence was marginal. But the change of ownership removed a constant threat to Labor—the presentation of every story in the most unfavourable light possible. For example, just before the sale, the *Daily Telegraph* was gearing up for an 'exposé' of the New South Wales Labor Party for alleged irregularities in its pre-selection of candidates; after the sale, nothing more was heard.

The main weapon against McMahon was ridicule. He was an easy target. In fact, Whitlam never enjoyed scoffing at McMahon in the way he had with Gorton. McMahon was altogether too vulnerable. He suffered especially at the hands of the cartoonists and he was doubly unlucky that his Prime Ministership coincided with a great flowering of the art of the cartoon in Australia, notably Bruce Petty and Larry Pickering. In this context, McMahon was handicapped by his inadequacies as a television performer. The brilliantly and cruelly destructive cartoons were in no way countered by his television image. It reinforced the caricature.

McMahon's Prime Ministership is recalled both by supporters and opponents as an interim period of non-achievement. Yet, it had valuable elements as a period of transition. It is all too easy to see McMahon as a total failure and a rather comic failure at that. Had he been nothing more than that, the Labor victory of 1972 would have been vastly greater than it was, and the Liberal resurgence after 1972 would have taken much longer than it did. Indeed, by being the perfect scapegoat, McMahon saved the Liberal Party from going into deep, destructive shock when it lost power after twenty-three years. McMahon saved the Liberals from despair. It was precisely because they could convince themselves so easily that McMahon was not fit to be supreme ruler, that his defeat left undamaged their inner conviction that they are Australia's natural rulers, 'the men born to rule'. This was McMahon's legacy to the Liberal Party.

After the Gorton sacking, McMahon stood lower in approval rating as recorded in opinion polls than any Prime Minister has ever done since such polls were taken. Paradoxically, Australians generally seemed to enjoy one of their happiest periods during his Prime Ministership. He was seriously damaged by rising inflation and rising unemployment — the first manifestation of the intractability of the post-Vietnam world economy which was to be so ruinous for the Whitlam Government. But in 1972, there was no feeling at all that there was anything seriously wrong. There was no sense at all of any deep sickness, in either the Australian economy or world capitalism. There was no sense of irretrievable crisis. On the contrary, it was a time of high optimism. As Gibbon set the time of the greatest human happiness in the Age of the Antonines, because that was the period when more people felt happy than otherwise, we may well place 1972 as Australia's happiest year. There was in that year a brilliant balance between hope for better things and satisfaction with the present; between expectation and experience; between a desire for change and enjoyment of the present. It was a time of general good humour and general goodwill such as Australians have not shared since. On the Labor side, the high optimism, the certainty of a coming victory, the exhilaration of being on the side of the future after twenty-three years of defeat and humiliation; on the Liberal side, an acceptance of inevitability and a resignation to it. Labor men and women — and in 1972 there were more Australians willing to so identify themselves than at any time since 1949 — enjoyed that ineffable felicity, high expectations without responsibility. For Liberals, at the parliamentary, organization and electorate levels, the fact. McMahon commanded nothing more than nominal loyalty as Menzies' vestigial heir, eased their passage towards defeat. On the Liberals' part, it was never a question of genuine goodwill towards Labor. There was none. But there was, in 1972, a widespread absence of active ill-will, scarcely any of the hatred and fear which disfigured the fifties. For this condition, McMahon was largely, if indirectly, responsible; and though it was essentially a negative achievement, it is not the worst thing to be said of a Prime Minister that he presided over one of the happiest years of his country's history. There were times in 1972 when it seemed that there had indeed been a recapture of a 'fine, careless rapture', but it was in altogether the contrary sense that Menzies had desired and shared by the wrong sort of people altogether.

15
'It's Time'

After each of Labor's eight successive defeats since the 1949 election, Party leaders and officials had written solemn reports about the futility of trying to communicate to the public a three-year programme during a three-week campaign. The 1972 campaign was the first in which the obvious was acted on. It was the first campaign which involved a sustained co-operative effort, over more than a year, between the parliamentary leaders and their staff, the Federal Secretariat, the Federal Executive, and the Party's advertising agency. It was this long exercise in regular collaboration, rather than any exceptional brilliance in specific campaign techniques, which made the 1972 campaign the most successful the Labor Party had undertaken since 1943.

Nothing new in politics is without its dangers. Mick Young convinced Whitlam of the merits of having a rehearsal for the election campaign in October 1971. The idea was to introduce the public to a range of issues and policies and to introduce the parliamentary spokesmen on those issues, the 'shadow ministry', to the public. Whitlam chose the topics and the spokesmen. The contrivance was to launch a chosen topic in the appropriate capital — industrial relations in Sydney, the capital of the largest industrial state; urban affairs in Melbourne, the political target of Whitlam's metropolitan strategy; defence in Brisbane, generally regarded as the most defence-conscious State; immigration in Perth, where at that time an ugly race controversy was threatening because of a minor migration, mostly Eurasian, from India and Sri Lanka. Adelaide and Hobart were chosen for the topics of education and Federal financial relations, for no better reason than that they had to be discussed somewhere. The exercise was a mixed success. For all its essential artificiality, it did achieve press coverage for Labor's policies worth far more than the money paid for the operation. It gave a valuable impression of an energetic party actively raising problems for public debate. It gave publicity to prospective ministers and promoted the idea that behind Whitlam there was a team. It gave the younger people of Labor's long-standing advertising agency, Hansen Rubensohn-McCann Erickson Pty. Ltd, insights into the curious and subtle tricky nature of Australian politics.

For each plus, however, there was at least one minus. At the Sydney unveiling of Labor's industrial relations, the author foolishly allowed the press conference to drift into extended, uncontrolled chit-chat, during which Clyde Cameron casually dropped a suggestion — an old hobby-horse of his — that fines should be imposed on both employers and employees for breaches of negotiated agreements. He had revived a proposal which had been rejected by the 1971 Federal Conference in Launceston. The consequent squabble in Caucus nullified any publicity gains his exposition of Labor's general industrial relations policy might have achieved. Whitlam's denunciation of a White Australia caused a dispute with the spokesman on immigration, Fred Daly, which ended in Daly's removal from that area of responsibility in the 'shadow Cabinet'. Barnard's genuinely important announcement in Brisbane of a proposal to establish a Natural Disaster Agency, based on the Defence forces, was overshadowed by a casual remark by Whitlam. He told a television interviewer that he advised conscripts ordered to Vietnam to notify their commanding officer in writing that they refused duty on conscientious grounds. This provoked a press and parliamentary storm. Whitlam, it was claimed, was inciting mutiny. Nonetheless, the exercise provided valuable exposure and valuable experience for the real thing in 1972.

On 10 October, McMahon announced the election date, 2 December 1972. Whitlam joyfully responded: 'The Prime Minister has steadfastly adhered to the principle he announced for himself on this subject last March: "What I have never done is to fix a date until I have made up my mind what the date is likely to be." We have not only had speculation on the election date; we have even had just as much speculation about the day that the date would be announced. Whoever will be able to say that the right honourable gentleman cannot keep a secret? Who will ever say again that he cannot grasp the nettle? The date announced places the Deputy Prime Minister (Mr Anthony) in a quite extraordinary position. He appeared to intend to hold his own election on 25 November — the ultimate gesture of Country Party independence. Certainly one now has a magnificent example of the trust, the confidence, the comradeship between the two leaders of the coalition. As the Deputy Prime Minister said: "He (McMahon) told me he would not tell me the date so I did not ask him." So now we have the date, and I must say I think it is jolly decent of the Prime Minister to let us know officially. The second of December is a memorable day. It is the anniversary of Austerlitz. Far be it from me to wish, or appear to wish, to assume the mantle of Napoleon, but I cannot forget that the second of December was a date on which a crushing defeat was administered to a coalition — another ramshackle, reactionary coalition.'

In the weeks before the campaign, Whitlam would respond with some irritation to pleas from the agency or Party officials for advance information about the contents of the policy speech: 'It's written already. The policy is in the platform and the priorities are in Hansard.' The policy speech of 1972 was certainly the most thoroughly-prepared document of this peculiarly Australian genre ever produced in Australia, but the document itself was simply a summary of the work of the previous six years. To the extent that Whitlam had done or initiated so much of the work himself, and to the extent that the platform now embraced a great many of his ideas, formulae and priorities, the policy speech was a highly personal document. But no Labor policy speech has ever so completely embodied the collective work of so many people, through the three Federal Conferences from 1967, innumerable committees and the work of the principal members of the Parliamentary Party and the small Opposition staff, especially Race Mathews in Whitlam's office and Clem Lloyd in Barnard's office. Although an air of contrived secrecy traditionally surrounds the policy speeches of all parties before their delivery, this was almost wholly predictable. The physical task of its preparation, of which the author had the carriage, was simply one of organizing a mass of material worked over for years past into a coherent framework.

In September, Whitlam met the Federal Executive in Sydney to discuss the broad outlines of the speech. There was only one matter of serious argument — the question of taxation. Hawke and Cameron urged strongly that there should be no promise not to increase taxation. Hawke argued that during the campaign Labor would again be pinned to the question, 'Where's the money coming from?'; the only credible response would be, 'From increased taxation.' Whitlam pointed out that the Party was committed to tax reduction and Crean, as putative Treasurer, strongly supported Whitlam and pointed out that even at existing rates of taxation, revenue would increase automatically by at least $5,000 million. The formula ultimately inserted in the policy speech carefully reflected this debate. There was no hard and fast commitment against any tax increases, but there was a specific commitment to revise the tax schedules. The speech read: 'The huge and automatic increase in Commonwealth revenue ensures that the rates of taxation need not be increased at any level to implement a Labor Government's programme . . . The most pressing need in the tax field is to retard the trend by which inflation has forced lower and middle income earners into the high tax brackets.' The debate at the Federal Executive was to assume some significance when the Labor Government found its economic problems mounting in late 1973. Hawke was to claim that Whitlam's refusal to agree to a tax increase was the result of a commitment made without consultation with the Party. In fact, there had been considerable consultation and there was no doubt that the policy speech formula, though not absolutely precluding tax increases, faithfully expressed the clear opinion of the Executive meeting in 1972 at which the matter had been thoroughly canvassed.

In discussing the policy speech with the Executive, Whitlam had submitted to consultation in a way none of his recent predecessors had done. Scullin was the last Labor leader to submit his policy speech for the approval of Caucus and there had never before been any meaningful discussions between the leader and the Federal Executive. Evatt and Calwell in particular had jealously guarded their sole proprietorial rights over the policy speech. Evatt's unilateral inclusion in his 1954 policy speech of a promise to abolish the means test was made one of the complaints against his leadership by the dissident Victorians who later formed the DLP. The centre-piece of Calwell's 1961 campaign, a proposal to stimulate the economy with a £100 million deficit, was inspired by the management of the *Sydney Morning Herald*. By comparison, Whitlam's 1972 policy speech was a model of consultation and collaboration. The Australian policy speech is a uniquely Australian institution. It is not a manifesto nor a platform. The leader who delivers it assumes personal responsibility for it. In Australia, the policy speech has taken on a certain mystical quality, by which it becomes a compact of honour between the party leader and the people. It is this personal element which puts the Australian policy speech in a category all of its own. The manifestos of the British parties or the platforms of the American parties altogether lack the mandatory quality of the Australian policy speech: they are usually ignored during the campaign and mercifully forgotten after it. They are never regarded as the blueprint for government by the successful party. Campaign speeches and promises, other than the policy speech, do not have this quality of imposing inescapable obligations. In 1932, the candidate Roosevelt undertook in a speech in Chicago to balance his first Budget; President Roosevelt in 1933 asked his adviser, Judge Samuel Rosenman, how he was to explain that one away. Rosenman (who had written the speech) replied, 'Mr President, the only way to explain that speech is to deny you ever made it.' It would be impossible for an Australian Prime Minister to explain away a policy speech in that way. Ironically, it is nowadays the one speech which can never be denied. (Significantly, the 1975 election, when nothing was normal, saw the first real break from custom; it remains to be seen whether this is an aberration or a precedent.) The demands of television have reduced to total fiction the idea of a policy speech being a speech which is actually delivered. In the half-hour of free time conventionally allowed on national television, the speech is reduced to the barest outline of the proposed programme, with a few rhetorical flourishes; the effect — the entrance of the leader, the cheers, the contrived enthusiasm — is everything. What the newspaper reader reads next day as the policy speech bears only the sketchiest relationship to what he saw on television the previous night. This convention, the assumption of personal responsibility for carrying out the promises of the speech, makes a successful policy speech a document of exceptional importance and unique significance in Australian politics. The 1972 speech carried the practice to its highest pitch.

The Labor Party policy speech was delivered on 13 November 1972 at Blacktown Civic Centre, the heart of the new suburban electorate to which Whitlam had been speaking for two decades. It was not so much a public meeting as an act of communion and a celebration of hope and love. Touching the author lightly on the shoulder, a curious ritual that had developed between us before major speeches on which we had collaborated, an act not of superstition but of remembrance of things past, Whitlam said, 'It's been a long road, comrade, but we're there.'

He opened with the words used by John Curtin in 1943:

Men and Women of Australia!

The decision we will make for our country on 2 December is a choice between the past and the future, between the habits and fears of the past, and the demands and opportunities of the future. There are moments in history when the whole fate and future of nations can be decided by a single decision. For Australia, this is such a time. It's time for a new team, a new programme, a new drive for equality of opportunities; it's time to create new opportunities for Australians, time for a new vision of what we can achieve in this generation for our nation and the region in which we live.

It's time for a new government — a Labor Government.

My fellow citizens, I put these questions to you:

Do you believe that Australia can afford another three years like the last twenty months? Are you prepared to maintain at the head of your affairs a coalition which has lurched into crisis after crisis, embarrassment piled on embarrassment week after week? Will you accept another three years of waiting for next week's crisis, next week's blunder? Will you again entrust the nation's economy to the men who deliberately, but needlessly, created Australia's worst unemployment for ten years? Or to the same men who have presided over the worst inflation for twenty years?

Each rhetorical question met with a resounding 'No', so spontaneously that it seemed orchestrated. Snedden, as Leader of the Opposition, was to use the same set of questions with considerable effect eighteen months later. Whitlam continued:

> Can you trust the last-minute promises of men who stood against these very same proposals for twenty-three years? Would you trust your international affairs again to the men who gave you Vietnam? Will you trust your defences to the men who haven't even yet given you the F-111?
>
> We have a new chance for our nation. We can recreate this nation. We have a new chance for our region. We can help recreate this region.
>
> The war of intervention in Vietnam is ending. The great powers are rethinking and remoulding their relationships and their obligations. Australia cannot stand still at such a time. We cannot afford to limp along with men whose attitudes are rooted in the slogans of the 1950s — the slogans of fear and hate. If we made such a mistake, we would make Australia a backwater in our region and a back number in history. The Australian Labor Party — vindicated as we have been on all the great issues of the past — stands ready to take Australia forward to her rightful, proud, secure and independent place in the future of our region.
>
> Our programme has three great aims. They are:
> — to promote equality
> — to involve the people of Australia in the decision-making processes of our land
> — and to liberate the talents and uplift the horizons of the Australian people.
>
> We want to give a new life and a new meaning in this new nation to the touchstone of modern democracy — to liberty, equality, fraternity.

In its comment on the speech, the *Sydney Morning Herald* professed to find this revolutionary sentiment particularly objectionable: 'the slogan of the French revolutionaries who replaced the old order with mob rule and terror and paved the way for the dictatorships of Napoleon . . . a curious banner for an Australian political leader to campaign under in this day and age.'

The potted version of the speech for television set out the priorities of a Labor Government:

> We will make pre-school education available to every Australian child. We do this not just because we believe that all Australian children should have the opportunities now available only to children in Canberra, but because pre-school education is the most important single weapon in promoting equality and in overcoming social, economic and language inequalities.
>
> Under a Labor Government, Commonwealth spending on schools and teacher training will be the fastest expanding sector of Budget expenditure. This must be done, not just because the basic resource of this nation is the skills of its people, but because education is the key to equality of opportunity. Sure, we can have education on the cheap . . . but our children will be paying for it for the rest of their lives.
>
> We will abolish fees at universities and colleges of advanced education. We believe that a student's merit rather than a parent's wealth should decide who should benefit from the community's vast financial commitment to tertiary education. And more, it's time to strike a blow for the ideal that education should be free. Under the Liberals this basic principle has been massively eroded. We will re-assert that principle at the commanding heights of education, at the level of the university itself.
>
> We intend to raise the basic pension rate to 25 per cent of average weekly earnings. Australia did that in the late 40s. Does anyone say we cannot afford it now? The important thing is this: the present method of irregular, uneven and politically inspired pension increases has been a source of needless anxiety, insecurity and indignity to those who depend on pensions for their sole income.
>
> We will establish a universal health insurance system not just because the Liberal system is grossly inadequate and inefficient, but because we reject a system by which the more one earns the less one pays, a system by which a person on $20,000 a year pays only half as much as a person on $5,000 a year.
>
> We will exert our powers against prices. We will establish a Prices Justification Tribunal not only because inflation will be the major economic problem facing Australia over the next three

years but because industrial co-operation and goodwill is being undermined by the conviction among employees that the price of labour alone is subject to regulation and restraint.

Under Labor, the national government—itself the largest customer—will move directly and solidly into the field of consumer protection.

We will change the emphasis in immigration from government recruiting to family reunion and to retaining the migrants already here. The important thing is to stop the drift away from Australia. We believe that the Australian people rather than governments should have the real say in the composition of the population.

We will issue national development bonds through an expanded Australian Industry Development Corporation—not just because we are determined to reverse the trend towards foreign control of Australian resources, but because we want ordinary Australians to play their part in buying Australia back.

We will abolish conscription forthwith. It must be done not just because a volunteer army means a better army, but because we profoundly believe that it is intolerable that a free nation at peace and under no threat should cull by lottery the best of its youth to provide defence on the cheap.

We will legislate to give aborigines land rights—not just because their case is beyond argument, but because all of us as Australians are diminished while the aborigines are denied their rightful place in this nation.

We will co-operate whole-heartedly with the New Guinea House of Assembly in reaching successfully its timetable for self-government and independence—not just because it is Australia's obligation to the United Nations, but because we believe it wrong and unnatural that a nation like Australia should continue to run a colony.

All of us as Australians have to insist that we can do so much better as a nation. We ought to be angry, with a deep determined anger, that a country as rich and skilled as ours should be producing so much inequality, so much poverty, so much that is shoddy and sub-standard. We ought to be angry—with an unrelenting anger—that our aborigines have the world's highest infant mortality rate.

The printed version of the policy speech, a document of forty-two pages, gave the details of how these general promises would be implemented. In all, there were over 120 specific undertakings. They included: a Schools Commission and Pre-school Commission, a Hospitals Commission, a Fuel and Energy Commission and revival of the Interstate Commission; a Prices Justification Tribunal and a Parliamentary Standing Committee on Prices; strengthening of the restrictive trade practices legislation; a comprehensive health insurance system; a five-year school dental health programme; an Australian Assistance Plan; automatic quarterly increases in all pensions to raise them to 25 per cent of the average weekly earnings; abolition of the means test within three years; transferability of pensions to foreign countries; a Ministry of Urban Affairs; Land Commissions to buy, develop and sell residential land at cost; the development of Albury-Wodonga; tax deductibility of mortgage repayments; a national sewerage programme; local government representation on the Loan Council; Grants Commission direct grants to local government; transfer of State railways to the national government; direct grants to urban public transport authorities; a national disaster organization; acquisition of the Australian wool clip by the Wool Corporation; abolition of the wine excise; a Ministry of Northern Development; expansion of the activities of the Australian Industry Development Corporation; abolition of the industrial penal clauses; an expanded job retraining scheme; removal of the Post Office from Public Service Board control; equal pay for women; four weeks annual leave for public servants; a Freedom of Information Act; an Ombudsman; abolition of appeals to the Privy Council, a Commonwealth Superior Court; abolition of the death penalty; voting at 18; a Securities and Exchange Commission; a legal aid scheme; an Australian Council for the Arts, with a number of autonomous boards to control the allocation of public funds for the arts; a public lending right; a Department of the Media; increased Australian content on television; a Ministry for Tourism and Sport; a Ministry for Aboriginal Affairs; aboriginal land rights; an Aborigines Land Fund; anti-discrimination legislation. This was the legislative and structural framework of the Labor programme at home. Moving to foreign policy, Whitlam linked Australia's treatment of her aboriginals to her role abroad:

Let us never forget this: Australia's real test as far as the rest of the world, and particularly our region, is concerned is the role we create for our own aborigines. In this sense, and it is a very real sense, the aborigines are our true link with our region. More than any foreign aid programme more than any international obligation which we meet or forfeit, more than any part we may play in any treaty or agreement or alliance, Australia's treatment of her aboriginal people will be the thing upon which the rest of the world will judge Australia and Australians — not just now, but in the greater perspective of history. The world will little note, not long remember, Australia's part in the Vietnam intervention.

Even the people of the United States will not recall nor care how four successive Australian Prime Ministers from Menzies to McMahon sought to keep their forces bogged down on the mainland of Asia, no matter what the cost of American blood, treasure, no matter how it weakened America abroad, and even more at home. The aborigines are a responsibility we cannot escape, cannot share, cannot shuffle off; the world will not let us forget that.

We now enter a new and more hopeful era in our region. Let us not foul it up this time. Australia has been given a second chance. The settlement agreed upon by Washington and Hanoi is the settlement easily obtainable in 1954. The settlement now in reach — the settlement that 30,000 Australian troops were sent to prevent, the settlement which Mr McMahon described in November 1967 as treachery — was obtainable on a dozen occasions since 1954. Behind it all, behind those 18 years of bombing, butchering and global blundering, was the Dulles policy of containing China.

Until barely a year ago, to oppose this policy, even to question it, was being described by Mr McMahon — and even some other people — as treason. If President Nixon had not gone to China nine months after I did, Mr McMahon would still be denouncing me, just as he was on the very eve of President Nixon's announcement that he would go to Peking. This is the man, this is the party, which expects you to trust them with the conduct of your nation's international affairs for another three years. A Labor Government will transfer Australia's China Embassy from Taipei to Peking.

We will take the question of French nuclear tests to the International Court of Justice to get an injunction against further tests. We shall act in this matter on the same high legal advice which Mr McMahon has received — but failed to act upon. We will ratify the Treaty on the Non-Proliferation of Nuclear Weapons.

We will give no visas to or through Australia to racially selected sporting teams.

A nation's foreign policy depends on striking a wise, proper and prudent balance between commitment and power. Labor will have four commitments commensurate to our power and resources; First — to our own national security; Secondly — to a secure, united and friendly Papua-New Guinea; Thirdly — to achieve closer relations with our nearest and largest neighbour, Indonesia; Fourthly — to promote the peace and prosperity of our neighbourhood.

For his audience, the 4,000 in and around the hall, the millions on television, and for Whitlam himself, his peroration was a deeply emotional moment, for there were few who were not certain they were listening to Australia's next Prime Minister:

Will you believe with me that Australia can be changed, should be changed, must be changed, if we are to have for ourselves and our children a better Australia, with a better grip on the realities of living in the modern world, and in our region as it really is? And will you believe with me that a new government, a new programme, a new team, is desperately needed to provide that change? I believe it is, and I believe that most Australians in their heart know these things to be true. We just cannot keep going the way we have these past twenty months. We cannot afford the instability of a government which has had sixty ministerial changes in the six years since Sir Robert Menzies.

We are coming into government after twenty-three years of opposition. This programme is ambitious. I acknowledge that. It has to be so; it should be so, because the backlog is so great. And we cannot expect to clear away that backlog in three months or even three years.

Before this campaign is out, I shall have completed twenty years as a member of Parliament. The basic foundations of this speech lie in my very first speeches in the Parliament, because I have never wavered from my fundamental belief that until the national government became involved in great matters like schools and cities, this nation would never fulfil its real capabilities.

For thirteen years I have had the honour to fill the second highest and then the highest place my Party can bestow. Throughout that time I have striven to make the policies of the Australian Labor Party, its machinery, its membership, more and more representative of the whole Australian people and more and more responsive to the needs and hope of the whole Australian people. This at least I have tried to do, and will continue to do; and, supporting me, I have the best of colleagues and the best of friends.

But the best team, the best policies, the best advisers are not enough. I need your help. I need the help of the Australian people; and, given that, I do not for a moment believe that we should set limits on what we can achieve together, for our country, our people, our future.

The campaign was smoothly predictable. Whitlam rarely diverted from his main themes, the themes refined and rehearsed for so many years—national involvement and national responsibility for schools, cities, health and welfare. He announced that he would appoint Dr H. C. Coombs as a personal adviser and that Australia would choose a new national anthem. These were suggested to an enthusiastically receptive Whitlam by Rupert Murdoch. Whitlam's regard for Coombs as a public servant who had served five Prime Ministers since Chifley was enhanced by his refusal to accept a knighthood and his advocacy of the aboriginal cause. Coombs had drafted the arts platform presented in the 1972 policy speech. (It is not true, as some commentators asserted, that Murdoch provided the plan for the promise on tax deductibility for housing interest rates. This was done at Whitlam's request by an economist at the Australian National University. Murdoch did, however, guarantee publicity for such a proposal.)

At a national Press Club luncheon three days before polling day, he allowed himself to be drawn into an argument over the first-past-the-post voting system. Calwell had made the same mistake in 1963. The major concern of Labor Party officials was the possibility of rowdy demonstrations against McMahon of the kind which had helped Holt in 1966; but the passions of 1972 were very different from those of 1966. A criticism of Labor's education policy by the Catholic rural Bishops was more than countered by a speech by Archbishop James Carroll, who said that the policy of both parties should be acceptable to Catholics. Eric Walsh, Whitlam's public relations officer-designate, astutely ensured publicity for this statement. Fittingly, this was the same Bishop whose letter to Calwell in 1966 had played a part in Labor's crisis over State aid in February of that year. For McMahon, it was a deeply unhappy campaign. Badgered and pilloried by the press, poorly advised, ill-supported by his colleagues and the Liberal machine, he lashed around wildly in search of issues. He plagiarized the *Sydney Morning Herald*'s protest about the revolutionary implications of Whitlam's phrase, 'liberty, equality, fraternity'. He gave contradictory estimates of the cost of Labor's proposals and his own. In the last week, he raised the spectre of the permissiveness and pornography which would engulf Australia under a Labor Government. But nothing worked properly, not even the auto-cue device from which he read his speeches. If the campaign had been less one-sided, a turning point would have been a radio interview he gave in Brisbane on 25 November in which he appeared to criticize his Cabinet colleagues: 'I have learnt that I now must make more decisions than I had intention of making when I first became the Prime Minister. I wanted to be the head of a team. I wanted to delegate the authority to the relevant Minister . . . I couldn't get the work done quickly enough and I found frequently that the political approaches to it were not as good as I thought they should be.' But it was not a turning point in the campaign, just another step on the road to defeat.

If the leaders' campaigns were the decisive factor in elections, Labor's victory would have been more complete. It was indeed a substantial victory, but it was no landslide. McMahon was entitled to his consolation. In an aside after his prompt, brave and graceful concession of defeat he said: 'At least I didn't lose as many seats as Gorton.'

Labor won six seats in New South Wales, four in Victoria, one in Queensland and one in Tasmania. It lost one seat in Victoria (Bendigo, where Bishop Stewart had strongly attacked Labor's David Kennedy on abortion and education), one in South

Australia and two in Western Australia. The net gain was eight. The actual result partly disguised the very great gains made in Melbourne, where a few score more votes would have given three more seats. Reconstruction had borne its fruit. Aided by reconstruction, the coalition of urban home mortgage payers, the younger middle class and the disaffected Liberals which Whitlam had stitched together in 1969, held together in 1972, as it was still to do in 1974. Through this, and by maximizing Labor's traditional support for the first time since 1946, and by reversing the drift of Catholic support since 1954, Whitlam had become the first Labor leader to carry his Party into Government from Opposition since 1929.

16

The Duumvirate and After

In the forty-one years between 19 December 1931 and 2 December 1972, the Australians had only once changed their national government at an election—in 1949. Since Federation, there had been twenty Prime Ministers; only three—Cook, Scullin and Lyons—had not previously been Prime Minister when called upon to create a government after a successful election. In this crucial test of democracy, the orderly transfer of power, precedents and experience were limited and distant. In deciding how and when to take over, Whitlam had few settled rules to work by.

His first decision was almost instinctive. It was to get to Canberra as soon as possible. He delayed his departure from Sydney until 3 p.m. on 3 December only in deference to the anticipated hangovers of his staff. As it turned out, whether through tension, excitement or exhaustion, everyone stayed sober, however much they tried otherwise. For those who had been with Whitlam on the long campaign trail, it was all a curious anti-climax. Not until a crowd of about 500 swirled around Whitlam at the gate of the R.A.A.F. Base at Canberra did the shock of reality hit them; and then there were more tears than anything else. We were rather given to tears in the Whitlam Government.

At Whitlam's request, McMahon had arranged that key public servants involved in the transfer should attend at Parliament House. Their functions foreshadowed Whitlam's intentions: Sir John Bunting, Permanent Secretary of the Prime Minister's Department; Sir Keith Waller, Secretary of the Foreign Affairs Department; involved in subsequent meetings jointly and separately were Alan Cooley, Chairman of the Public Service Board; and Clarence Harders, Secretary of the Attorney-General's Department. The first meeting with Bunting and Waller lasted more than two hours. Staff were excluded. It was this meeting which determined Whitlam to have the new government installed as soon as forms would allow, and that the government would exercise immediately the full powers of government. Waller told him that some votes on race questions were imminent at the United Nations General Assembly; pending the swearing-in of a new government, the proper course for Australia would be to abstain. Whitlam said he would consider it; in fact, he was determined to seize the opportunity to signal to the world that a new government with new attitudes had taken charge.

Equally important was to signal to the Australian people that the change they had made was real. If Whitlam was to justify his own career and all the trials he had put the Labor Party through, he had to show that the electoral change was meaningful. The events of 5–18 December — the days of Australia's duumvirate — were implicit in Whitlam's whole conduct and character. The duumvirate was not an aberration; it was an inevitability.

Practical considerations also determined Whitlam's course. These involved Australia's electoral system, the rules of the Labor Party and the workings of the public service. The Chief Electoral Officer, Frank Ley, told Whitlam that counting of absentee votes and the distribution of preferences was unlikely to be completed until Friday 15 December and at least nine seats must be considered doubtful until then. Under Party rules, the Ministry would be elected by the whole Parliamentary Party. So Caucus could not be called together until 18 December. Thus, a fully effective Labor Ministry could not be sworn-in until a week before Christmas. On this timetable, there would be at the most three working days before the Christmas shut-down. Given Australian habits and customs, it would have been extremely difficult to begin serious, effective work before the New Year, a full month after the election. Any momentum generated by the election result might melt under the summer sun.

McMahon tendered his resignation to Sir Paul Hasluck at 11

a.m. on 5 December. At 12.30 p.m., Whitlam issued a statement drawn up in the Attorney-General's Department:

> I have today recommended to His Excellency the Governor-General that he swear my Deputy, Mr Lance Barnard, and me to all existing Ministerial offices. His Excellency has agreed that the swearing will take place this afternoon. This allocation of portfolios is, of course, temporary. It will operate only until such time as the full Ministry is elected and can be sworn. I have given His Excellency an assurance on this point.

At 3.30 p.m. on 5 December, Sir Paul Hasluck administered the oaths of office to Whitlam as Prime Minister with twelve other portfolios and Barnard as Minister for Defence with thirteen other portfolios. As Whitlam told Parliament the following year: 'The First Whitlam Ministry was the smallest ministry with jurisdiction over Australia since the Duke of Wellington formed a ministry with two other Ministers 138 years previously.'

At his first press conference as Prime Minister, Whitlam was asked why the Ministry had been limited to two; in particular, why the Senate Leader, Senator Murphy, and the Senate Deputy Leader, Senator Don Willesee, had been excluded. Whitlam's public explanation was that there were members of the House of Representatives of greater seniority than Murphy or Willesee who would undoubtedly be elected as Ministers by Caucus. He said, 'It would be invidious to presume to pick out which of my House of Representatives colleagues would be senior in service, significance, and portfolio, and accordingly it was most appropriate to choose two persons alone.' Some journalists believed that Murphy's exclusion was intended as a personal slight. It was not. It was, however, a conscious attempt to place the position of the Senate in Whitlam's own perspective. It was not only a matter of seniority. To have included the Senate leaders would have acknowledged the equality of the Senate and implied that the Senate was of equal relevance in forming a government. The Labor Party itself had given the Senate a kind of equality by appointing the two Senate leaders *ex officio* to its Federal Executive in 1967. Whitlam refused to accept this as a precedent in forming a Labor government. He frequently rebuked staff, colleagues or journalists who used the shorthand, 'the four leaders'. 'There are not four leaders; there is a leader, deputy leader and four parliamentary office-bearers.'

That a principle, not personal animus, was involved is shown in the announcement of the first decision of the first Whitlam Ministry, and, for many Labor supporters, its most emotionally-charged decision. Press Release No. 1 stated: 'Immediately upon the return of the Labor Government, Senator Lionel Murphy, Q.C., set out to arrange for the release of seven young men serving eighteen-month gaol sentences under the National Service Act. Last night, I signed letters to the Governor-General recommending remission of the sentences.'

This procedure of involving Ministers-designate in decisions was followed throughout the thirteen days of the duumvirate. The flurry of announcements invariably invoked the name of the appropriate 'shadow minister'. Press Release No. 2: 'The Prime Minister this afternoon authorized a written application to the Industrial Registrar asking to have the recent hearing on equal pay before the Full Bench of the Arbitration Commission reopened. The application was made after detailed consultations with Labor's Minister (sic) for Industrial Relations, Mr C. R. Cameron.' Press Release No. 4: 'Takeover Bid Freeze' ... The Prime Minister did this after discussion with Mr Frank Crean, MP, and the Treasury.' Press Release No. 5: 'The Prime Minister today forwarded two references to the Tariff Board (on colour television and electrical goods). The decision was made after consultation with Labor's nominated Minister for Trade, Dr Cairns.' Press Release No. 13: 'The interim committee for the Australian Schools Commission will hold its first meeting in Canberra next Thursday, 21 December. The Prime Minister decided to call the meeting after talks with Mr Kim Beazley today.' Throughout the fortnight of the duumvirate, the former 'shadow ministers' visited their designated departments, gave instructions, made statements and, in general, behaved as commissioned ministers. The two-man Ministry was never a two-man band, though the Labor Government was probably better orchestrated in its first two weeks than at any time thereafter.

The legal basis for all this activity, including quasi-Ministerial actions by men whose constitutional authority was no higher than that of any other private member of Parliament, was the fact that Whitlam and Barnard held the Governor-General's Commission for all portfolios. The moral basis was the mandate. Whitlam's most significant, most characteristic and most controversial use of the dummvirate was to develop his theory of the mandate. Whitlam's concept of the mandate is fundamental to any understanding of what the Labor Government did, why it behaved in the way it did, why it succeeded, why it failed and, ultimately, why it fell.

The most conservative interpretation of the meaning of the mandate is that it is no more than a general mandate to govern for a prescribed period. Whitlam pushed the doctrine to its limits by asserting that his Government had not only a general mandate to govern but a specific mandate to implement each and every undertaking of the policy speech, line by line. Specifically, the hard-line interpretation of the meaning of the mandate was a reply to Senate obstruction. The more obstructive the Senate became, the more Whitlam insisted on the principle that the mandate was, indeed, mandatory.

Whitlam also used the mandate as a means of asserting parliamentary authority over the Labor Party. He insisted that the mandate given to the Parliamentary Party and the expression of the mandate contained in the policy speech had absolute primacy over any decision taken by the Party at any level, including Federal Conference itself. In March 1973, he headed off moves originating in the Victorian branch to revise Party policy on the presence of American bases by asserting that such a change would be a breach of the mandate. No other Labor leader had dared such an audacious assertion of the prerogatives of the Parliamentary Party and it was only the exceptional situation existing in early 1973 which enabled Whitlam to argue this line successfully without serious challenge to its correctness or logic. Whitlam's doctrine of the mandate was a necessary part of an even higher doctrine—the primacy of the House of Representatives.

Late in the life of his Government, in his second Chifley Memorial Lecture at Melbourne University in August 1975,

Whitlam analysed his own doctrine of the mandate:

> The meaning of the 'mandate' in a Parliamentary system has been subject to critical analysis by parliamentarians and academics for more than a century. In Australia, the meaning has been especially scrutinized in the past two and a half years because of the very great emphasis my own Government has always placed on the fulfilment of the programme set out in the policy speech of 1972 and confirmed in the policy speech of 1974. Another reason why the meaning of the 'mandate' has become significant in the current Australian political debate is because of the existence of a hostile majority in the Senate and the use to which that hostile majority has been put. So the debate about the meaning of the mandate has centred on the question of whether in 1972 and again in 1974 the Australian Labor Party was given only a general mandate to govern or a specific mandate to implement each part of its programme.
>
> Is the mandate merely general or is it specific? Is it a grant of permission to preside or a command to perform? Our opponents naturally interpret it in the weakest sense as a general and highly-qualified mandate to govern — on their terms and indeed by their grace and favour. I interpret the mandate as being both general and specific — a general mandate to govern for the term for which we were elected and a specific mandate to implement the undertakings we made, within that term. But even when I speak of a general mandate I cannot accept the conservative definition of a mere mandate to govern, a permit to preside over the administration of government and, hopefully, to administer the existing system in a sufficiently acceptable way to give reasonable prospects of re-election — for a further renewal of the mere mandate to preside. The mandate as I interpret it is to move by specific programmes toward the general goals and the general objectives accepted by the people at the elections.

The mandate was invoked in his first administrative act and his first official communication with the Governor-General. In his letters to Sir Paul Hasluck recommending the release of the seven youths imprisoned for refusing to be conscripted, he said:

> In the recent elections, it was part of my party's election programme that this action would be taken and that a recommendation would be made that Your Excellency exercise

the prerogative [of mercy] to bring about the release from custody of those offenders against the National Service Act who are still in custody. It may be assumed that this was endorsed by the electors and that Your Excellency's Government has a mandate to take these steps.

Forty decisions were announced by the First Whitlam Ministry between 5 December and 17 December. Listed baldly, they appear unco-ordinated and unrelated. In fact, they were the logical extension of Whitlam's doctrine of the mandate. For thirteen days of the duumvirate, Australia was governed not so much by two men, but by the Whitlam doctrine of the mandate. The duumvirate was the apotheosis of the mandate.

For Labor supporters, the decisions on Indo-China, conscription, and China were the most important achievements of the duumvirate. The Foreign Affairs Department had moved more quickly than any other department to show its adhesion to the new regime. The Ministerial duumvirate was supported by the public service triumvirate of Bunting, Waller and Cooley. The Foreign Affairs Department had the advantage of Whitlam's powerful interest. When Whitlam made it clear that he wanted Australia's voting on the southern African resolutions in the United Nations General Assembly reversed immediately, and to express Australia's positive support for a zone of peace in the Indian Ocean, Foreign Affairs subsided into acquiescence. Moreover, it provided proof in writing that it understood the spirit as well as the letter of Labor policy. At his first conference as Prime Minister on 5 December, Whitlam stated, in words suggested by Richard Woolcott, who settled smoothly into the role of Foreign Affairs Department spokesman: 'The change of government provides a new opportunity for us to reassess a whole range of Australian foreign policies and attitudes . . . the general direction of my thinking is towards a more independent Australian stance in international affairs, an Australia which will be less militarily-oriented and not open to suggestions of racism; an Australia which will enjoy a growing standing as a distinctive, tolerant, co-operative and well-regarded nation not only in the Asian and Pacific region, but in the world at large.'

It was a fair enough summary. But it was intended as a signal of goodwill from the Foreign Affairs Department to the Government as much as a signal of goodwill from the Government to the world. Both sides gladly accepted it as such. Waller also made a personal gesture of conciliation — 'reconstruction' as Whitlam defined it — when Whitlam instructed that the Australian journalist Wilfred Burchett be issued with an Australian passport, denied him by the former government on the grounds that his Korean war reports had been pro-communist. Waller asked that he be allowed to write the letter to Burchett.

By the time of the 1972 policy speech, twenty years of argument, blunders, hatred and hostility could be resolved in the simple formula: 'We shall transfer Australia's China Embassy from Taipei to Peking'. On his first day as Prime Minister, Whitlam cabled the Australian Ambassador in Paris, Alan Renouf, to open negotiations with his Chinese counterpart. In the months before the election, Renouf had made futile attempts to carry out McMahon's instructions to keep the Chinese door open. Such efforts were a waste of time as long as the Australian Government remained locked into the two-China policy. The renewed Paris negotiations were swiftly successful; indeed, they were hardly more than a formality to allow Whitlam to announce (on 22 December 1972) that the two Ambassadors in Paris had signed a communiqué establishing diplomatic relations between China and Australia. Whitlam said:

> We have recognized the Government of the People's Republic of China as the sole legal government of China. It follows that we can give no recognition to others claiming to be the Government of China ... Our diplomatic relations with Taiwan came to an end with the signing of the communiqué in Paris.

Renouf's role in the consummation of Whitlam's twenty-year-old policy on China was rewarded a few months later by his appointment to succeed Sir Keith Waller as Secretary of the Department of Foreign Affairs.

The Department of Defence, under Sir Arthur Tange, the toughest and most determined of all the permanent heads, also moved quickly to carry out Labor's specific policy commitments. Tange had studied the policy better than any other Secretary and was one of the few to have drawn up a specific plan of action ahead of the elections. However, even Tange was surprised at the speed with which Whitlam and Barnard wished to move on the abolition of conscription. The Defence Department's contingency plan had assumed that legislation would be needed and that the repeal of the National Service Act would be delayed until the first meeting of Parliament. Clem Lloyd, Barnard's press secretary, pointed out that the operation of the Act could be suspended by regulation. On that basis, Barnard was able to announce on 5 December: 'Under Section 31 (1) of the National Service Act the Minister has power to defer liability of such classes of persons as he determines where it appears to him to be necessary or desirable to do so in the public interest.' Under this cover-all, Barnard ended liability for all further and future national service. There were to be no further call-ups and those currently serving were given the choice of completing their service or ending it forthwith.

If Tange resented Lloyd's successful intervention he was soon to have his revenge. When the British Defence Minister, Lord Carrington, visited his Australian counterpart in January, Tange insisted that Lloyd, then Barnard's private secretary, should not attend the meetings. Barnard agreed, scarcely taking in the implications. Lloyd had framed Labor's defence policy, including the massive reorganization and integration of the five defence departments which Barnard and Tange were to carry through with conspicuous success. To a man of Lloyd's pride and profound patriotism, the exclusion was an unforgivable affront. His resignation was a deep blow to Barnard and the whole government.

Of equal symbolic importance was Barnard's announcement on 11 December that the Australian Army Assistance Group would be returned to Australia within three weeks. He also announced that Australia's participation in the Cambodian Army training programme, involving twenty-six instructors, had finished. Thus ended Australia's military intervention in Indo-China, begun ten years before with the decision to send thirty advisers. It was the end of the military chapter, but by no means the end of the political story of Australia's Vietnam adventure.

Another chapter of the distortion of Australian politics caused by Vietnam remained to be written in 1975.

Was the duumvirate a mistake? In their perceptive book analysing the work of the Labor Government between 1972 and May 1974, Clem Lloyd and Professor Gordon Reid wrote:

> The solid and enduring achievements of the interim Government have to be sifted out from a mass of plebeian decisions. One might wonder looking back from the perspective of a year or two later what all the fuss was about.
>
> Textual analysis of this sort does not do justice to the psychological achievement of this daemonic onslaught of government. As a means of getting into government quickly and showing that the Labor Party could manipulate the machinery of government effectively, although 23 years had elapsed since it last had the levers in its hands, the first Whitlam Ministry was supremely successful. It would not have been possible to symbolize the regeneration of an infirm political party in a more impressive way.

Yet for all its excitement, the duumvirate had its drawbacks. The regeneration of Labor as an effective political force raised the enthusiasm of Labor supporters to a high pitch. Yet it was precisely this that first raised the doubts of the conservative Establishment about their wisdom in giving Labor too easy a ride into office. The peculiarly passionate nature of the response to the Whitlam Government, unique in the history of Australian Governments, the hopes and enthusiasm it raised, the fears and hostility it aroused, existed right from the first days. It was in the first fortnight of the duumvirate that those opposing passions took seed. At the end, the most common self-criticism among Labor supporters was, 'We did too much too soon'. It was less common to hear criticism of the timing of any specific decision. So with the duumvirate: approval of each specific decision, examined one by one, added up to vague uneasiness.

The style of government established by the duumvirate set a pace which could not be sustained and a pattern which should not have been sustained. The flurry of decisions seemed to follow no discernible order of priorities. In fact, the priorities were decided by what could or could not be done without parliamentary sanction; but, in the rush of announcements, it was easy to depict this as a government which drew no distinction in importance between the withdrawal of the ban on advertising contraceptives in the Australian Capital Territory and the withdrawal of military aid to Vietnam. Though the Ministers-designate were involved much more than the strictest convention allowed, the whole operation looked very much a one-man show. Though Whitlam fairly claimed the mandate of the electorate for each specific decision, the total effect could easily be made to appear high-handed, peremptory and arbitrary. It was easy for Whitlam's opponents to depict the duumvirate as an impatient and greedy grab at power after twenty-three years in Opposition. More seriously, the ease and effectiveness of the operation masked the long-term problems of decision-making. If so historic a decision as the abolition of conscription could be accomplished literally by the stroke of a pen, it was tempting to believe that other large matters could be as easily settled. The duumvirate encouraged in the new Ministry a spirit of emulation akin to rivalry. The normal self-assertiveness and independence of Ministers like Cameron, Connor and Murphy received a boost and a justification from the example of the duumvirate. It affected their attitude to Whitlam, to Cabinet, and to their departments. Whitlam was to say: 'I don't care how many prima donnas there are in the Labor Party, as long as I'm the prima donna assoluta.' There is no doubt, however, that the personal *tour de force* of the duumvirate encouraged some of the chorus to become prima donnas.

Yet as far as these things can be gauged, it seemed that the nation as a whole approved of the duumvirate and its decisions. The fairest opponents accepted Labor's right to act and even its right to enjoy it. Such overt criticism as there was tended to be couched in terms of lofty amusement. Killen rejected Whitlam's Duke of Wellington analogy and preferred the precedent of 395 A.D. when the Roman Empire was divided between the Emperors Arcadius and Honorius.

What went unnoticed in the excitement of the time was the first stirrings of uneasiness in business circles. There is no greater myth than that the Labor Government came to power on a tide of goodwill from all sections of the community, and only through Labor's own folly, was this goodwill lost. Among businessmen, there was very little positive goodwill and at best a degree of tolerance. A Labor Government was to be tolerated on condition that it would make no real change. The hyperactivity of the duumvirate aroused uneasy suspicions. When Whitlam spoke of change, he might well be serious.

Further, two State Premiers, Sir Robert Askin and Johannes Bjelke-Petersen, immediately invoked the doctrine of State sovereignty against the doctrine of the national mandate. Askin refused a request to close down the Rhodesian Information Centre in Sydney, which acted as a virtual embassy for the illegal Smith regime in Rhodesia. Bjelke-Petersen refused point-blank to consider any changes in Queensland's border with Papua-New Guinea, which runs within 500 yards of parts of Papua's shore. Thus, the two factors which contributed so much to the destruction of John Gorton — conservative unease about change and the hostility of State Premiers — set in against the Whitlam Government from the very beginning.

The duumvirate was, of course, a massive assertion of Whitlam's personal ascendancy. Not until the Government's last weeks, nearly three years later, would there be a comparable assertion. The very tight limits the Labor Party imposes on the authority of a Labor Prime Minister reasserted themselves as soon as the Parliamentary Party met to elect the full Ministry on 18 December.

Since 1905, it has been impregnably established in the rules of the Australian Labor Party that Caucus shall elect the Ministry. Throughout 1972, Caucus debated the rules and standing orders which would determine the shape of a Labor Cabinet after the election. As the certainty of a Labor Government grew, the debate became more urgent, more relevant and, at times, more bitter. At no time, however, was there any suggestion that a Labor Prime Minister should appoint his Ministers. The right of the Caucus to choose ministers points to the heart of Labor's unresolved problem in defining the nature and purpose of its leadership. There is a continuing tension between its desire for strong leadership and its suspicion of those it calls upon to lead.

The Australian Labor Party is, at once, deeply distrustful of its leaders and deeply loyal towards them.

Whitlam himself traces this contradiction to the treachery of William Morris Hughes who 'undermined the Labor Party's trust in the very concept of leadership, so much so that a quarter of a century after Hughes' expulsion, Curtin himself could be charged in the Caucus room with hankering after a wish to lead a government from the other side'. The ambivalence is implicit in the nature of the Party, with all the rich diversity, confusion and illogic of its structures, membership, aims, methods and motives. In refusing to grant the Parliamentary Leader the right to choose his Ministerial colleagues, the rule expresses the anti-élitist strand in Labor's philosophy. The contradiction is that the rule asserts the equality of an élite within the Party — the élite possessing the power to vote for Ministers. Possession of this power makes Caucus itself an élite. The contradiction was apparent in moves made in Caucus in 1972 to establish the rights of Caucus over the leadership. The moves foreshadowed many of the deepest difficulties of the future Labor Government.

J. C. Watson, in 1904, was the only Labor Prime Minister who had been allowed to choose his Cabinet colleagues. He was able to bring Henry Bourne Higgins into the Cabinet as Attorney-General. Higgins was not a member of the Labor Party but later became the architect of the Australian system of wage-fixing. By 1909, Caucus had claimed the right to 'recommend' ministers to the leader and thereafter this power of recommendation became absolute. At the height of his authority in war-time, Curtin demanded and was given the right to dismiss an errant Minister without reference to Caucus. This was a remarkable concession to the leadership principle; not surprisingly, the right died with Curtin. The Prime Minister's right to allocate portfolios remained. A move was made within Caucus in 1972 to curtail even this right by making the Leader's appointments subject to Caucus approval and veto. The move failed. Nonetheless, the freedom of a Labor Prime Minister to remove or even to change Ministers, once appointed, remained profoundly circumscribed. This was to be the cause of untold trouble. To change Ministers became almost as difficult as the outright removal of a Minister.

The feeling grew up that a Minister, once appointed, had an inalienable right to that post. It became as difficult for Whitlam to change a Minister as for a Liberal Prime Minister to sack one. Whitlam's changes were used as evidence of an unstable government led by an unstable, not to say cruel and arrogant, Prime Minister. Few things were more damaging to the Government and to Whitlam's personal reputation than the ministerial changes he chose and sometimes was compelled to make. The nature of the system ensured that such changes would be made only with the greatest difficulty, untidiness and heartburn. Each change created a crisis. This problem, inherent in Labor's system, was used to build up an impression of the Labor Government as a crisis-ridden government.

The principle of equality demanded equal status for all members of the Ministry—that is, there should be no division between an inner Ministry and an outer Ministry, between Cabinet and the rest. Menzies had instituted such a system to cope with the expansion of Federal responsibilities and functions. By 1972, the McMahon Government had been expanded to twenty-seven Ministers, twelve of them members of the Cabinet. Caucus, in 1972, narrowly rejected a move to follow this sensible system and insisted that the full Ministry should comprise the Cabinet. This was done in the name of equality; it meant in practice that the élite would number twenty-seven instead of twelve. Far from cutting down the tall poppies, it merely ensured there would be more of them. From whatever motive, and by a margin of only one vote, Caucus thus ensured that a Labor Cabinet would be as unwieldy as possible. That unwieldiness did not diminish the power of the Prime Minister. On the contrary, it almost certainly guaranteed his ascendancy. The Caucus decision of 1972 doubled the likelihood of division within Cabinet. It diluted the voice and the vote of each individual Minister, excepting the Prime Minister, and thus reduced the strength of the powerful members to parity with the weakest. In seeking to curtail Whitlam's authority, Caucus had in fact enhanced it. There is no end to paradox in politics.

During the glittering days of the duumvirate, there was intense activity among the Parliamentary members to secure election to the first Labor Ministry for twenty-three years. Whitlam's scrupulous practice is not to run a 'ticket'. His refusal to do so whenever Caucus positions are open to ballot has been the basis of charges of indifference to the fate of friends and disloyalty to loyal supporters. The truth is that he holds fast to the principle that Caucus must take responsibility for its own decisions. To run a ticket would invite the charge of dictatorship. In any case, the independence, even the waywardness, of Caucus is such that a leader's intervention would be as likely to ensure failure as success for his candidates. Curtin and Chifley were never able to guarantee success for their protégés or exclude those of whom they disapproved. Curtin had tried hard to keep Ward and Calwell out; the most he succeeded in doing was to delay their entry and he paid a high political and emotional price for it. Of course, Whitlam's preferences and dislikes were well enough known; the openness with which he declares them is one of the criticisms made against him.

In the event, the outcome of the Caucus election on 18 December was almost entirely predictable as far as the cumbersome preferential system of voting, based on elaborate rules reflecting the complexities of the Federal system itself, allows. There were no big surprises. Whitlam, Barnard, Murphy and Willesee were elected unopposed. The other members of the old 'shadow Cabinet' were confirmed. For the rest, the only significant outcome was that members of the House of Representatives enjoyed a clear preference over senators of equal or greater ability. In the positions which, under the Caucus rules, were equally open to members and senators, the House secured nine and the Senate none.

For the Prime Minister, there were personal disappointments in the result—notably the failure of Manfred Cross (Brisbane) who, if successful, would have been Minister for Aboriginal Affairs, and J. M. Berinson (Swan), one of the most able of the 1969 contingent and one of the five best speakers in Parliament. He had to wait until 1975 to enter the Ministry. On the other hand, Everingham, who would not have been on any Whitlam list, was to prove, as Minister for Health, one of the most effective of all appointments.

The general predictability and acceptability of the result is shown by the speed with which Whitlam was able to announce the new Ministry and to slot the chosen Ministers into the portfolios created by his new administrative structure. Three hours after the Caucus election, the Ministry was announced: Prime Minister and Minister for Foreign Affairs: Edward Gough Whitlam; Deputy Prime Minister, Minister for Defence, Minister for the Navy, Minister for the Army, Minister for Air and Minister for Supply: Lance Herbert Barnard; Minister for Overseas Trade and Minister for Secondary Industry: James Ford Cairns; Minister for Social Security: William George Hayden; Treasurer: Frank Crean; Attorney-General, Minister for Customs and Excise and Leader of the Government in the Senate: Senator Lionel Keith Murphy; Special Minister of State, Vice-President of the Executive Council, Minister assisting the Prime Minister and Minister assisting the Minister for Foreign Affairs: Senator Donald Robert Willesee; Minister for the Media: Senator Douglas McClelland; Minister for Northern Development: Rex Alan Patterson; Minister for Repatriation and Minister assisting the Minister for Defence: Senator Reginald Bishop; Minister for Services and Property and Leader of the House: Frederick Michael Daly; Minister for Labour: Clyde Robert Cameron; Minister for Urban and Regional Development: Thomas Uren; Minister for Transport and Minister for Civil Aviation: Charles Keith Jones; Minister for Education: Kim Edward Beazley; Minister for Tourism and Recreation and Minister assisting the Treasurer: Francis Eugene Stewart; Minister for Works: Senator James Luke Cavanagh; Minister for Primary Industry: Senator Kenneth Shaw Wriedt; Minister for Aboriginal Affairs: Gordon Munro Bryant; Minister for Minerals and Energy: Reginald Francis Connor; Minister for Immigration: Albert Jamie Grassby; Minister for Housing: Leslie Royston Johnson; Minister for the Capital Territory and Minister for the Northern Territory: Keppel Earl Enderby; Postmaster-General: Lionel Frost Bowen; Minister for Health: Douglas Nixon Everingham; Minister for Environment and Conservation: Moses Henry Cass; Minister for Science and Minister for External Territories: William Lawrence Morrison.

In retaining Foreign Affairs for himself, Whitlam did what he said he would do in 1967. Barnard, Cairns, Hayden, Crean, Murphy, Cameron, Uren, Jones, Beazley and Stewart each received the portfolios matching their designated field as 'shadow ministers'. Douglas McClelland, Patterson, Daly, Connor, Johnson, Grassby, Enderby and Morrison (at least in his External Territories role) received the portfolios Whitlam had foreshadowed for them in Opposition. Bryant, Everingham and Cass were given portfolios for which their qualifications or known interests rendered them entirely appropriate. Given the inescapable difficulties of Cabinet-making, the vagaries of Caucus selection, the contradictions between personal choices, ambitions and qualifications, the relative inflexibility of administrative structures, there can have been few more predictable sets of Cabinet appointments than this Whitlam Ministry. The order of seniority in the list of Ministers followed the voting in Caucus, with the sole variation that Cairns, Hayden and Crean preceded the two office-bearers in the Senate. There was nothing in Caucus standing orders to impose this upon Whitlam, but it was in the spirit of the Caucus decision. In constructing his Ministry, Whitlam went to the limit in fulfilling long-standing public undertakings, in implementing the will of Caucus, in indulging the wishes of his colleagues, and keeping his own prejudices and preferences out.

Few of his new Cabinet colleagues were as frankly delighted as Fred Daly, despite his decidedly unglamorous title of Minister for Services and Property: 'I don't care what he's given me. I'd take anything, as long as I'm there.' None, on the other hand, felt as Hasluck records his reaction to Menzies' offer of a Ministry in 1951: 'Resistance stirred in me. I was on the point of saying that I would have to consider . . . there was no thrill about it. I felt rather cross.' Perhaps the most difficult decision was filling the Primary Industry post. Among those elected, only Patterson, Grassby and Everingham represented rural areas. Only Patterson felt himself positively qualified for the portfolio. Whitlam believed, however, that a Minister in the House of Representatives representing a marginal seat would be too vulnerable to industry lobbying and parochial pressures. He compromised with Patterson by including responsibility for the sugar industry within the Northern Development portfolio. Wriedt, a former merchant seaman, confessed that he could not tell the difference between a Corriedale and a Merino. In the event, he became one of the most successful Ministers, personally popular with farmers. Initially, however, his appointment was derided and used as evidence of Labor's lack of concern for agriculture. Wriedt's genuine achievements, including the establishment of an Australian Wool Commission, failed to offset this suspicion which was to deepen into fierce hostility and even hatred, directed with special venom against Whitlam.

The new Ministry involved the abolition of six departments inherited from the McMahon Government, the remaining of two departments and the creation of sixteen new ones. Thus the number of departments was increased from twenty-seven to thirty-seven, although the number of Ministers stayed pegged at twenty-seven. The new structure was designed to give a practical expression to the priorities set out in the election programme. The basic plan had been devised in the weeks before the election by Dr Peter Wilenski, who was appointed Whitlam's Principal Private Secretary. He had two problems to solve: to devise a structure reflecting the priorities of the new Government and to recommend a procedure for change within the legal framework of the Public Service Act. The Public Service Board itself had worked on proposals for the kind of departmental structure an incoming Labor Government might need, but this was not ready when the new Government was so swiftly installed. Armed with the Wilenski report, Whitlam was in a strong position to shape his new Administration. A Machinery of Government Committee, comprising the four Labor leaders, Whitlam, Barnard, Murphy and Willesee, Peter Wilenski and Jim Spigelman from Whitlam's staff, and representatives of the Public Service Board, Prime Minister's Department, Attorney-General's Department and the Treasury, worked largely from the Wilenski report. Its real importance was that it enabled the politicians, inexperienced as they were in administration, to establish an immediate ascendancy over the public servants.

*

'The Platform is the Old Testament. The Policy Speech is the New Testament.' The Minister for Education, Kim Beazley, summed it up for his colleagues at a Cabinet meeting during the settling-in period. For the new Ministers and for their Departments, the 1972 policy speech did indeed become Holy Writ. The policy speech became the bridge of initial co-operation between new Ministers and their Departments. It was the first point of contact between Ministers, advisers and the public service. When, in April 1904, the six members of the first Labor Government suddenly found themselves to their amazement commissioned to form a government, Hughes asked, 'What the blazes do we do now?' Watson suggested, 'Well, we can take off our coats for a start.' In 1972, the answer would have been: 'Read the policy speech.'

The apotheosis of the policy speech came with the Governor-General's speech opening the Parliament on 27 February. This speech set out the Government's programme for the new Parliament. It was basically the policy speech stripped of its campaign rhetoric. The 'Speech from the Throne' (Whitlam's preferred description) was the instrument by which the policy speech received Vice-Regal utterance and benediction; Labor's legitimacy could go no higher!

The author was given the task of co-ordinating the speech — the same task, in a very different context, performed with the policy speech itself three months before. The proprieties were scrupulously observed; great care was taken to protect the Governor-General's feelings. The first time Sir Paul Hasluck had performed this task was the 100-word speech in November 1969; Gorton had given him the courtesy of reading it the night before delivery. The Labor Government presented him with two drafts, one a week before delivery. Whitlam invited him to make suggestions, confident that Sir Paul knew the correct relations between Government and Governor-General. Sir Paul demurred on no matter of substance and in his covering letters to the Prime Minister made it clear he knew he could not. Whatever his feelings, he accepted without comment Labor's abolition of God. The traditional invocation had been: 'Relying on the blessings of Almighty God, I leave you to carry out your high and important duties.' This became: 'With the utmost confidence that you will fulfil to the utmost of your abilities the deep responsibility the Australian people have placed upon you, I leave you to carry out your high and important duties.'*

Thus, even on the highest matters, Sir Paul Hasluck accepted the advice of the elected Government. Sir Paul made eighteen suggestions, most of style or grammar, all accepted. One phrase in the draft referred to 'a range of new initiatives, both national and regional'. Hasluck pleaded in a marginal note: 'This could be set to music by Sullivan. Must I really be made to sound like the very model of a modern major-general?'

The transition from the frustrations of Opposition to the responsibilities of power was not as easy as the breezy confidence expressed in the Governor-General's speech. Ministers, the handful of staff who had been part of the old Opposition and the new staff recruits found the adjustment sometimes painful. The enthusiasm of staff engendered resentment, though not yet resistance, from the permanent public servants. Ministers who as Opposition private members had felt free, even duty-bound, to range over all issues in their public statements now found it difficult to confine themselves strictly to their portfolios. The first and greatest test came, appropriately, over Vietnam. Cairns, Cameron and Uren found it impossible to remain silent in the face of what they believed to be a great evil for the sake of some doctrine of Cabinet responsibility. Their fierce public condemnations of President Nixon and the Christmas bombing of Hanoi produced a minor crisis in relations between Australia and the United States.

Between 18 and 30 December, 'Flying in groups of three, the B-52's dropped their bombs at the same time, with the payload falling in a rectangular pattern about a mile long and half a mile wide. Each plane could carry about two dozen 500-pound bombs and more than forty 750-pound bombs. About 100 of the big strategic bombers and 500 smaller Air Force, Navy and Marine fighter-bombers flew round-the-clock raids against what the

---

* It is interesting to note that the former invocation was used in Sir John Kerr's Speech opening the first Fraser Parliament in March 1976.

Pentagon described as "the most heavily defended anti-aircraft area in the world."¹ President Nixon's landslide victory in November 1972 had been heralded by Secretary of State Kissinger's proclamation: 'Peace is at hand'. The Christmas bombing of 1972 was the new Labor Government's introduction to American–Australian relations and its baptism in dealing with Nixon and Kissinger. Nixon's justification for the Christmas bombing was that it was the last turn of the screw to get agreement on the accords to end the war and to save American lives—in other words, the same justification in December 1972 which had been used to justify every escalation of the war since 1964. Cairns and Uren could see no reason why they should accept this any more silently now that they were Ministers than they had accepted the argument when they were lonely, powerless voices against Vietnam. Newspaper editorials urged Whitlam to repudiate his outspoken and irresponsible Ministers. The question was not the morality of the American bombing but the propriety of Australian Ministers.

At the urging of his staff, notably Peter Wilenski, Richard Hall and Gordon Bilney (the Foreign Affairs Department liaison officer within the Prime Minister's office), Whitlam sent a letter of protest to Nixon. The letter, drafted by Waller, was circumspect and courteous. It suggested that the bombing would prove counter-productive; it would, as all previous escalations of the level of violence in the war had done, increase Hanoi's intransigence. Nixon and Kissinger reacted angrily. The American Ambassador in Canberra, Walter Rice, was instructed to remonstrate with the Australian Prime Minister. Rice, a likeable, ineffectual businessman, more at home in the back-slapping world of the Elk and Rotary than in high diplomacy, came for his interview ill-briefed and ill-prepared for Whitlam's anger. Whitlam told Rice flatly that of all the wrong things the United States had done in Vietnam, the Christmas bombing was the least defensible. Rice reported back to Washington that this new Labor Government was proving more difficult than he had thought, and the United States Embassy made this known around the Canberra Press Gallery. This episode was the genesis of the propaganda that the Australian Labor Government was unreliable. The American Embassy (which had predicted a McMahon victory in its cables to Washington) was largely ignorant of the nature of the Labor Party, its purposes and its internal workings. The CIA was wholly ignorant relying as it did upon its Australian Intelligence partner, the Australian Security Intelligence Organization.

There had been a 'security' problem even before the Government had been sworn in. It demonstrated the thorniness of the problem for a Labor Government. It started on Sunday 3 December. After his first meeting with Bunting, Tange, Cooley and Waller, Whitlam introduced them to his new staff. Indicating Richard Hall, Jim Spigelman, Eric Walsh, David White and the author, he said, 'These appointments are not subject to the security check. These men are not to be harassed by ASIO.' It was a generous and imprudent gesture. He meant to indicate that he accepted personal responsibility for our loyalty. Beyond that he was signalling his strict definition of the role of staff and that 'the need to know' principle would be strictly applied in his office; people engaged in the task designated for us would not see highly-classified documents. And that principle was, in fact, strictly applied throughout the period of the Whitlam Government, as far as his own office was concerned. For example, the author never saw a Cabinet submission in three years in Government. Soon after, the author was accosted in King's Hall by a former gallery journalist, then working for the Joint Intelligence Organization in the Defence Department, who harrumphed: 'No security check eh? We'll see about that.' Soon after, stories, and an editorial in the *Canberra Times*, appeared in the press asserting that the Americans were concerned at security risks within the Prime Minister's office. The staff, new and old, then volunteered compliance, but the damage had been done.

A significant point is that at no time did any officials demur or suggest to Whitlam that his course was unwise. Some of them chose the indirect method. This was the sort of thing that, from the beginning, sowed the seeds of mutual mistrust. It was in this atmosphere that the Government addressed itself to the question of the future of the Australian Security Intelligence Organization.

The Australian Security Intelligence Organization was a by-product of the Cold War. It had been set up by Evatt in 1949 in the atmosphere of fear created by the Stalinist domination of Eastern Europe and the Chinese revolution, a period in which Menzies warned that Australia should prepare for World War III within three years. Because of its role in the defection of the Soviet diplomat Vladimir Petrov, with its shattering aftermath, ASIO came to occupy a special place in Labor demonology. Labor attitudes ranged from derision for the bumbling incompetence of its operatives to fear of its potential as an instrument of repression. At the Launceston Federal Conference in 1971, the Labor Party by the barest margin rejected a move to adopt abolition of ASIO as Party policy; the vote was equal so the motion was lost. The move failed because Murphy opposed it. He argued that a Labor Government would require a security organization to combat right-wing extremism, particularly among Croatian migrant groups, which he believed were being organized on para-military lines for the purpose of subversion and assassination in Yugoslavia. Throughout 1972, Murphy conducted a running duel with the Liberal Attorney-General, Senator Ivor Greenwood, who denied the existence of any such organizations.

Against his better judgement, Whitlam was persuaded by Murphy that the policy laid down at Launceston meant that ASIO should be responsible to the Attorney-General, not the Prime Minister. Within hours of being sworn in as Attorney-General, Murphy summoned the Director-General of Security, Peter Barbour, to discuss the future of ASIO. Two days later, he issued a statement which might variously be interpreted as a truce or a declaration of war: 'The Attorney-General is anxious to find means by which he can satisfy himself, the public and Parliament that ASIO is adhering strictly to its charter and Mr Barbour is understood to be drafting for him an explanation of the regulations and procedures under which his officers perform the functions prescribed by the ASIO Act. Mr Barbour is understood to welcome these measures as a means of rebutting criticisms and allegations which have been directed against ASIO in the past.'

The test of this 'understanding' came with the projected visit of the Prime Minister of Yugoslavia, Dzemal Bijedic, in March. Because of information from the Commonwealth Police, Whitlam and Murphy believed they had reason for a very real fear that Croatian terrorist groups would attempt violence and even assassination. The political significance of the warnings from Commonwealth Police sources was that their information conflicted with the official ASIO line, repeatedly put by Senator Greenwood, that there were no Croatian terrorist groups or organizations in Australia. Murphy became convinced that ASIO was withholding essential information from him. Late on the night of 15 March, less than a week before Premier Bijedic's arrival, a member of Murphy's staff, Kerry Milte, himself a former Commonwealth Police officer, showed him an ASIO report of an inter-departmental meeting between officers of the Attorney-General's, Prime Minister's and Foreign Affairs Departments. The report suggested that the officers present had agreed that any public explanation of the security measures taken for the Bijedic visit should exclude any admission of terrorist activities in Australia and should not conflict with statements by the previous Attorney-General, Greenwood. The document confirmed all Murphy's worst suspicions. Neither the time, the occasion nor the subject were conducive to clear thinking. In a rage, Murphy called on the regional director of ASIO in Canberra for a 'please explain'. He failed to gain a satisfactory explanation of the document and thereupon determined to visit ASIO headquarters in Melbourne early next morning.

At that point, Murphy fell victim to the mutual ill-will between the Commonwealth Police Force and ASIO. He expressed concern to Milte that files might be removed, destroyed or withheld if ASIO had advance warning of his visit. Milte interpreted this as an instruction for a full-scale operation by the Commonwealth Police in Melbourne to seal off ASIO headquarters. At about 8 a.m., police cordoned the building in St Kilda Road, entered the premises and sealed safes and files. The Melbourne *Herald* had received notice from a source in the Commonwealth Police, ensuring that Murphy's arrival and the presence of Commonwealth Police received spectacular and graphic attention. Thus was Murphy's visit transformed into a 'raid'.

In a bland statement that evening, Murphy said:

> At ASIO's Melbourne headquarters I conferred with Mr Barbour and addressed the staff. I was accompanied by Commonwealth Police Officers involved in investigations of Croatian Terrorist activities in this country. I inspected certain files ... The most stringent security measures are necessary for Prime Minister Bijedic's safety because of the existence in our midst of Croatian revolutionary terrorist organizations. These were tolerated by the previous Government, which even denied their existence.

The ASIO 'raid' was to have consequences out of all proportion to its intrinsic importance. It marked the end of the honeymoon period normally enjoyed between a new government and the press. The break came on a matter in which the element of farce predominated in such a way as to cast doubts on the Government's executive grasp. It was one thing to accept, perhaps admire, the purposeful style of the duumvirate; but the ASIO visit provided a focus for the vague uneasiness already stirring. For a public service accustomed to quiet enjoyment of their prerogatives, Murphy's use of one arm of the service to discipline another was deeply disturbing. It was disturbing, too, even to Labor supporters who shared Murphy's suspicion of ASIO. The ASIO visit smacked of that authoritarian tendency which had been the most unattractive characteristic of the Chifley Labor Government. When, in May, both State and Commonwealth police staged a dawn raid on a number of Sydney residences, purportedly in search of weapons, this impression deepened. The ASIO visit was the first serious blow to the self-confidence of the Labor Government and the first serious check to the unqualified enthusiasm of its supporters.

In the longer term, however, the two most serious consequences were the damage to Murphy's reputation and the effect on the mood of the Opposition.

Murphy had already shown a determination to achieve the programme of reform equalled only by Whitlam himself. Hastily and badly advised, he had attempted to achieve vast divorce law reform by promulgating badly-drafted Orders-in-Council; although these had to be withdrawn, they were the forerunner of his monumental Family Law Act of 1974. More skilfully and successfully, he carried Australia's opposition to French nuclear tests in the South Pacific to the World Court in The Hague. By March, he had already set well in train the Government's proposals for legal aid, aboriginal land rights, Freedom of Information legislation, a Federal Ombudsman, censorship reform, the protection of endangered species, the prohibition of kangaroo product exports. He had even announced the end of restrictions on home-brewing. It was ironic and unfair that a Minister whose thrust was so strongly libertarian should be cast in an authoritarian light. On top of his personal achievements, he had, as Attorney-General, overall responsibility for the vast mass of legislation being prepared to implement the new Government's programme. All this substance was vitiated by his ASIO intervention. From this he never fully recovered, despite a record of achievement as a reformist Attorney-General not matched since Deakin. In the debates and a Senate Inquiry following his ASIO visit, the Opposition failed to establish any impropriety in his conduct; yet he was unable to overcome fully the impression that there was something sinister about it. And it was this impression, still with a life and force of its own two years later, which enabled Labor's opponents to use his elevation to the High Court to set in train the events which destroyed the Labor Government on 11 November 1975.

The ASIO intervention brought the Opposition out of its post-election lethargy and pointed them towards basic strategy to undermine the new Government by impeding and then denying its legitimacy. For the first time, in the debates on the ASIO affair, the Opposition developed the vocabulary of denigration which was to become the litany of 1974 and 1975 – scandal, impropriety, illegality, corruption. It was Murphy's ASIO intervention which provided the first damaging opportunity to use the strategy.

A bizarre footnote to the ASIO visit was to be written in June 1977. A former senior CIA officer, James Angleton, claimed in an interview given to the ABC in New York that the CIA had considered breaking off intelligence relations with ASIO in retaliation for Murphy's action, on the grounds that it endangered the integrity of CIA operations. If this claim is true, it permits only one of two interpretations: either the CIA grossly overreacted, in ignorance of the Australian system, to the Attorney-General's lawful prerogatives and Murphy's actual conduct; or ASIO misrepresented the nature, purpose and results of Murphy's visit. Either way, overreaction and misrepresentation characterized the whole episode.

Yet, the magic still held. Summing up 'the first hundred days', Robert Drewe wrote in the *Australian*:

> If you want to put a label on it, the New Nationalism does as well as any. Call it that, or a greater spirit of national identity, or an increased sense of Australian purpose, or whatever, but the chances are that unless you're a 67-year-old mining magnate who's a member of the League of Empire Loyalists you're aware of a certain rare feeling of national self-respect these days.
>
> It's not as if we're suddenly a bigshot country—more wise, sensitive or creative than we were a year ago. But, God knows, to this 30-year-old child of Robert Gordon Menzies out of World War II, it almost feels like it. Nor has Australia suddenly accomplished any remarkable national catharsis. So we had a change of Government and the slack was taken up! Big deal, your European businessman or British diplomat might think.
>
> But the fact is that Labor restored some dignity to the conduct of our national affairs at a time when we had all come more or less to expect nothing but ill from political action.
>
> Without precedent in the history of British-style governments, it set out to make up for lost time by immediately implementing its campaign promises. Australians blinked as within weeks we recognized China, ended conscription, abolished race as a criterion of our immigration policy, began reform of the health service, supported equal pay for women, banned racially selected sporting teams, abolished Federal British honors, increased arts subsidies, put contraceptives on the medical benefits list, took the tax off Australian wine, moved to stop the slaughter of kangaroos and crocodiles and searched for a new national anthem.
>
> Along the way, the Government attempted to make our

> relationship with America (negotiated by Labor, conservatives tend to forget, in the face of scandalized tsk-tsking from their side of the fence 30 years ago) a bit less one-sided. The End of The Ice Age, is how Russel Ward describes the new era in a current Meanjin article . . .
>
> It seems, too, even after a few more than 100 days in power (and despite the fracas over the Asio affair) that Gough Whitlam has personally had an important effect on Australia. He has been responsible for reviving the frustrated and almost impotent Australian nationalism merely by bringing panache to his office and by using (and being seen to use) the idea of the Australian Government, as he prefers to call it, as a direct and intelligent instrument for the general good . . .
>
> At the moment his personal image, right across the political-class board, is about as good as a realistic politician could hope for . . .
>
> Whitlam does have the confidence of one assured of his own sense of excellence and Australians being what they are, they seem to find the feeling contagious.

Yet it was precisely this idea of 'contagion', in its unhealthy sense, which fed the hopes of Labor's opponents. On 8 March 1973, the Opposition Leader in the Senate, Reginald Withers, told the Senate: 'Because of the temporary electoral insanity of the two most populous Australian States, the Senate may well be called upon to protect the national interest by exercising its undoubted constitutional rights and powers.'

Thus, from the outset, the Opposition established the four lines that were to govern its attitude throughout the ensuing three years: that the election of a Labor Government was an aberration; that the Labor Government lacked legitimacy; that there was a national will above and beyond that expressed by House of Representatives elections; and that the Senate was the repository of this higher national will.

17
Opposition

'We embarked on a course some twelve months ago to force an election for the House of Representatives.' Thus, the Opposition Leader in the Senate, Reginald Withers, explained the tactics of the Opposition leading to the Double Dissolution of 1974. It was a simple statement of something that had been long obvious. By Withers' own admission, almost from the beginning it was never intended that the Labor Government would be allowed to run its normal three-year term. Yet, the Senate, which proposed to force the House of Representatives to another election, had not itself been elected in December 1972. It comprised senators who had been elected in 1967 and 1970. During the campaign, the role of the Senate had not been raised. A few commentators had mentioned in passing that an incoming Labor Government might have some difficulty with its legislative programme, but in general, Whitlam and his colleagues discounted the problems arising from their inheritance of a hostile majority in the Senate. As Whitlam was to say in his third and last speech as Prime Minister to the Melbourne Press Club on 10 November, the day before his dismissal:

> It's like Churchill explaining in his war memoirs why he failed to make any inquiry at all about whether the great guns at the Singapore Naval Base could shoot north as well as south—landward as well as seaward. He didn't raise the question because it never occurred to him. As he wrote: 'It no more occurred to me to ask that question than to ask if a battleship were launched without a bottom.'

He meant that in December 1972 neither he nor, for that matter, the Australian electors, seriously contemplated that the Senate would attempt to use the ultimate weapon of refusal of Supply to bring down a government with a majority in the House of Representatives.

*

There was a curious joylessness about the election of Billy Mackie Snedden as Leader of the Liberal Party on 20 December 1972. It is, after all, a high and honourable estate to be Leader of the Opposition and Leader of a great political Party. Yet, after choosing him his colleagues disappeared. The press showed little interest in the new leader; a perfunctory interview on the steps of Parliament House and that was it. He and the newly-elected Deputy, Phillip Lynch, returned to inspect his new office—the office of the Leader of the Opposition—which for the first fortnight Whitlam had continued to use as Prime Minister.

Snedden was an excellent example of the young Liberals of the Menzies generation, an archetype of thousands of young Australians who, in the late forties, responded enthusiastically and idealistically to Menzies' call on behalf of free enterprise and individual freedom. In the atmosphere of post-war austerity, the attractions of Menzies' philosophy were powerful. In his book, *Labor's Role in Modern Society*, Calwell put it this way:

> Towards the end of World War II there arose in one of the most politically astute minds that Australian politics had produced an idea that was to prove one of the most successful political gambits known to Australian history. It was then that Robert Gordon Menzies established the Liberal Party and unfurled the flag of 'free enterprise'. The new philosophy—for it appeared to many as new, and he made it sound like a philosophy—struck a chord in the hearts and minds of a diverse and large section of the community. It was a doctrine which, when expressed with glib arguments and in eloquent tones, appealed in many ways to some of man's best instincts; it appealed also, despite the moving eloquence that called us to higher things, to the worst side of human nature. It made an appeal on the grounds of 'liberty', 'freedom from restraint', 'individual effort', 'just rewards', 'the end of austerity' . . . It was tempting for Sir Robert to argue . . .

that Labor was addicted to control for the sake of control . . . He did not resist the temptation. It was in that atmosphere that the glorious doctrine of latter-day liberalism was born and nurtured . . . The spell was woven and the magic worked. And Sir Robert Menzies came to power.

Snedden was a natural recruit under this banner. By character and force of circumstance he was an achiever. As a boy in Perth, he sold newspapers to help support a deserted mother. Although he had the advantage of being able to attend Australia's only free university, he graduated in law through his own efforts and self-sacrifice. He gravitated to the Liberal Party as naturally and inevitably as Whitlam to the Labor Party. But, as with Whitlam, his class, background and experience deeply influenced his attitude towards his Party. After unsuccessful attempts to enter Federal Parliament from his home State, he transferred to Melbourne, headquarters of the Liberal Party and home of the Establishment. In that milieu, he was an outsider. The Establishment will freely give patronage to a useful outsider; it will not give acceptance. The key to Snedden's political character is that he was an outsider who wanted acceptance. He had a pressing need to prove himself, often painful for those who liked him, to watch. In particular, he seemed to feel the need to prove that he was tough. As a leader, he confused a show of toughness with strength. By trying to prove his toughness, he failed to exploit his real virtues and genuine gifts which may be quite as serviceable for a leader — his basic honesty, a willingness to learn and take advice, a good intelligence and a fundamental decency. In the sixties, he had sometimes made himself ridiculous by his transparent mimicry of Menzies, but his true exemplar was Harold Holt with whom he shared this quality of an uncomplicated decency. His real qualities, properly applied, might have won him the thing he most needed and which he never had — the loyalty of his colleagues. Instead, he sought to win their support by proving his toughness. As he was not genuinely tough, much less ruthless, the attempt was bound to fail. In their final devastating confrontation in Parliament on 4 March 1975, Whitlam would say: 'The Leader of the Opposition suffers from the abiding defect of those who feel insecure or inadequate — the desperate need to prove their toughness ... the boasting and bravado of the impotent.' If it seemed cruel, it was only because so many of Snedden's nominal followers—the men he had recently claimed would 'walk across hot coals' for him—felt it to be true.

For the deputy leadership, the Liberals chose Phillip Lynch, Minister for Labour and National Service in the McMahon Government. If Snedden represented the typical young Liberal of the early post-war period, Lynch was quintessentially the young Liberal of the high Liberal hegemony. Before 1949, indeed until 1955 when the Split guaranteed indefinite Liberal rule, there was a genuine element of crusade for a cause within the Liberal Party. By the sixties, this spirit had disappeared to be replaced by a preoccupation with the details and mechanics of government. The rhetoric of free enterprise and individualism gave way to jargon about 'monetary and fiscal policy within the parameters of a mixed economy'. Lynch mastered this language. If he had not been a Catholic and had gone to a Great Public School (it was a sign of changing times that he achieved promotion in the Liberal Party despite these disadvantages) he would have been called a swot. He was the perfect organization man.

There had been no effort by the Liberal Party, either at the Parliamentary or organization level, to develop a comprehensive programme since 1949. While Menzies ruled there was little need and less opportunity for it. The turmoil under Holt, Gorton and McMahon prevented it altogether. Such policy initiatives as there were came largely through the public service or in response to immediate political exigency. Snedden accurately identified his Party's chief need—a period of consolidation, in which to develop new, relevant and comprehensive policies. Realizing how much, and in many cases, how completely, the old ministers had depended on their departments, he saw the need for outside advice and a more effective Party Secretariat. He sought to involve the Party organization and its membership in a way no previous Liberal leader had dreamt or deigned. He wrote letters to 2,000 branches of the Liberal Party throughout Australia inviting responses to policy questions. He sought advice from the

British Conservative Central Office. He established good lines of communication with the State branches. He realized that the preoccupations of the next three years were likely to be economic. Perhaps he understood this better than most, because he knew better than most the economic consequences of his last Budget and his last year as Treasurer.

Although the approach (and the nature of his problems) was very different from that of Whitlam as Opposition Leader, the aim was the same — to prepare a comprehensive programme, the broad themes of which would be understood by the public well before the next elections. This was a three-year programme and a three-year task. Snedden's instruction to the newly-formed committees and research units called for interim reports within six months, to be 'deepened and extended during the next two years'.

Just as the Government needed at least three years to implement its programme, so did Snedden and the Opposition need at least three years to draw up and explain a new programme. Both Government and Opposition were starting from scratch. Both were working in strange territory. The Opposition was as inexperienced in working without the public service as the Government was inexperienced in working with it. It was a three-year exercise or it was nothing. And in the end it all came to nothing. When Snedden, in the pursuit of toughness, allowed himself to be side-tracked from his three-year goal, he destroyed his own work and ultimately destroyed his own leadership. As it turned out, the hasty production of half-finished documents on the eve of the 1974 election transformed the hard and serious work of 1973 into a national joke. Of the policy work done by the Liberal Party throughout 1973, almost nothing remains. Snedden worked harder and more constructively than any Liberal leader, not excluding Menzies; he worked harder than any political leader of his time, except Whitlam. He threw it all away — the work, the leadership, the certainty of being Prime Minister of Australia.

Snedden believed that the Liberal Party had to recapture the middle ground, particularly the younger, urban middle class whose support tipped the scales in Labor's favour in 1972. As well as trying to develop suitable policies, he made overtures to the founder of the Australia Party, Gordon Barton. Barton had no illusions about what was involved. After a private dinner in one of Sydney's most expensive restaurants in February 1974, he commented, 'What Bill Snedden really wants is for me to join the Liberal Party so that he will be able to set me off against Malcolm Fraser ... With the Liberal Party at present, with people like Malcolm Fraser going further to the right, it will polarize the position and leave the middle ground to Labor or to emerging forces like the Australia Party.'

Little, sometimes ludicrous things pointed to trouble in store for Snedden's leadership. The Liberal senators insisted on meeting separately to elect their leaders; the first indication that the senators saw a role for themselves independent of the House of Representatives. The Country Party abandoned the practice of joint party meetings with the Liberal Party and appointed a separate 'shadow ministry', although its leader, Anthony, was resentful when Snedden retaliated by appointing his own spokesman for rural industries. With the cheerfully mischievous connivance of Speaker Cope, Anthony moved peremptorily into the office reserved for the Deputy Leader of the Opposition before the Liberals had met to elect their leaders, consigning Lynch to an office renovated for Arthur Calwell on the former site of a members' lavatory. As Killen put it, 'The spirit of the squattocracy lives on.' The subsequent wrangle over titles and salaries resulted in a compromise formula, devised by Snedden, that Lynch was indeed Deputy Leader of the Opposition but that Anthony was 'the second most important person' in the Opposition. As the Country Party moved to distance itself from the Liberal Party, it moved closer to the Democratic Labor Party. In Queensland and Western Australia, a formal merger took place. Whitlam, recalling that Sir Henry Bolte had once described the Country Party as the whore of Australian politics, said, 'The old whore churched at last'. Snedden mistook the reference and when the DLP later agreed to support a bill for a referendum on Prices and Incomes, jibed, 'Who's churching the old whore now?' The DLP suspected some dark reference to the whore of Babylon. Nor was the DLP overly impressed by

Snedden's boast that during his talk with Chou En-lai in Peking in June 1973, he had the Australian Ambassador, Stephen Fitzgerald, 'sitting on the edge of his chair'.

Senator Gair said he thought there was an all too probable reason for the Ambassador's nervousness. This is just the small change of politics; yet Snedden's reponse to his difficulties with these three groups—the Liberal senators, the Country Party and the DLP—was to lead him into a disastrous miscalculation in April 1974.

In the day-to-day Parliamentary duel, Whitlam established an easy ascendancy over Snedden. Nonetheless, Snedden could demonstrate his substance when he chose his occasion and advice with care. His speech on the Budget in August was particularly effective. As Opposition Leader, Whitlam had always used this occasion to set out Labor's programme. Snedden's speech conformed to a more traditional concept of a speech on the Budget; he made a speech on the economy. He perceived that inflation was already the dominant issue. He proposed a ninety-day freeze on wages and prices to act as a 'circuit breaker to cut the wage-price spiral'. Nearly four years later, this proposal was revived by Malcolm Fraser, lived again briefly and died ingloriously.

A by-election for Parramatta (N.S.W.), caused by the departure of Nigel Bowen to the New South Wales Bench, provided Snedden with his first electoral test as leader. The circumstances were all in his favour. Whitlam, acknowledging the down-turn in public favour, made no psychological commitment to Parramatta as Holt had done in Capricornia in 1967. He also refused to make concessions to local opinion; he refused to qualify or modify a Cabinet decision for a feasibility study for Sydney's second international airport at Galston, north-west of Sydney, which it was claimed would put part of Parramatta in the flight-path. The need for such a hard-and-fast decision may be judged by the fact that by 1977 Sydney—and Galston—was no nearer obtaining a second airport than in 1973. But the announcement had been made, and the Government was stuck with it; so, as Whitlam curtly told the electors of Parramatta, was Galston. The Liberals retained the seat comfortably.

The Parramatta result encouraged the hard-line senators in taking any risk to bring down the Labor Government. In their over-confidence they temporarily over-reached themselves. The most important and the most popular part of the programme for 1973 was the establishment of the Schools Commission and the acceptance of the interim committee's recommendations for a trebling of Federal funds for all schools on the 'needs' formula laid down by the Government. As Minister for Education, Beazley had skilfully won support for the principle—and the cash—from the Church schools organization, although the Bishops still expressed preference for Malcolm Fraser's per capita grants formula. Fraser urged the Opposition to push their rejection of the Labor plan to the limit and risk an election. Beazley was temporarily incapacitated by overwork. The tactical struggle was placed in the hands of Lionel Bowen, the Postmaster-General. He warned the State Governments and the independent schools that no funds would be made available at all for the 1974 school year. Anthony saw the folly of intransigence. McManus, now Leader of the DLP, warned that rejection was as sensible as killing Santa Claus at Christmas time. Bowen reached an arrangement with Anthony, by-passing the Liberals. The arrangement phased out per capita grants for the wealthier schools instead of abolishing them immediately. The Schools legislation, ending an issue which had profoundly divided Australian society for a century, received the Governor-General's assent on 19 December 1973.

The new ALP National Secretary, David Combe, succeeding his fellow South Australian, Mick Young, who had been appointed to Whitlam's personal staff, made a report on the Parramatta by-election. He identified four reasons for Labor's loss of support: the Galston controversy (described by Hawke as 'political imbecility'); the Government's failure to explain its programme (the so-called 'communications gap'); the Liberals' skilful exploitation of inflation; and a decision in the last week of the campaign to raise interest rates. If this judgement, based on surveys commissioned by Combe, was correct, it meant that the Labor Government had been rebuked both for its style and its handling of the economy. Faults of style could be easily corrected. But from Spring 1974 the problems of economic management threatened to engulf the Labor Government.

18
The Economy v. The Programme

When Australia joined the Organization of Economic Co-operation and Development – the association of the industrially developed non-communist nations – it was almost a signal that Australia was now mature enough to share the big troubles of advanced capitalist societies. Fittingly, the first OECD survey of the Australian economy was published in December 1972. It said: 'In the past two years, after almost a decade of brisk and relatively smooth sailing, the Australian economy is running into troubled waters: domestically, as in most other OECD countries, wage-price inflation has sharply accelerated, while economic activity has slipped down markedly and unemployment has risen.' The West was now beginning to pay the price for Vietnam.

Professor Fred Gruen, a consultant to the Prime Minister from 1973 to 1975 and Professor of Economics at the Australian National University, wrote in 'The Australian Quarterly' in December 1976:

> The present world-wide inflationary wave had its origin in the economic policies pursued by the United States government during the Vietnam war, and the associated mismanagement of aggregate demand after 1965. The spread to the rest of the world operated through a variety of transmission channels. Of particular importance were the liquidity effects of the very large increase of world international reserves produced both by record U.S. government deficits and record U.S. balance of payments deficits. The increase in the prices of internationally traded foodstuffs, raw materials and manufactured products, the permissive climate created by these developments generally and by the removal of the

threat of balance of payments crises were supplementary channels of transmission of inflation.

Most far-reaching of all, Vietnam contributed significantly to the oil crisis of 1973–74, with all its immense economic consequences for the West. America's preoccupation with Indo-China where her essential interests did not lie, led to a corresponding loss of influence in the Middle East, where the West's vital interests do lie. In the Vietnam years, the Soviet Union established ascendancy in the Arab world. When the Yom Kippur war broke out on 6 October 1973, American prestige and influence was at its nadir. The Arab oil embargo prepared the way for the massive oil price increase by the oil-producers. All this was done contemptuously and flauntingly in the face of a weakened and humiliated United States, led by an embattled President caught in the web of Watergate — itself the ultimate distortion and corruption of American politics by Vietnam. Nixon claimed (in May 1977) that he ordered the American forces around the world on full alert (including the Australian bases, without the knowledge of the Australian Government) because of an 'ominous' message from Moscow calling for a joint Soviet-American military force to keep peace in the Middle East. Perhaps most of the world, including Whitlam, believed at the time that the alert was a move by a desperate President needing an escape route from Watergate. Such was the state of trust to which Vietnam had brought the United States and her allies. In any case, because of Vietnam, Israel's enemies knew that there was one thing they did not have to fear — American military intervention, alone or under the flag of the United Nations. If Moscow was proposing a joint force, it did so in the knowledge that the United States was in no position, politically, to provide part of it.

When the Labor Government came to office, registered unemployment stood at 100,000, about 2 per cent of the workforce, the highest Australia had experienced since the credit squeeze of 1960–61. This had a significant psychological effect on the Ministers and on Caucus. Everyone, from the Prime Minister down, was anxious to get on with the task of implementing the programme, fulfilling the mandate, getting on with the job. The high level of unemployment did nothing to encourage restraint, patience or caution either in Cabinet or Caucus. Although Labor inherited an inflation rate of 7 per cent, the highest since 1951, there could be no doubt where the economic priority lay. For the first six months at least, the view, shared generally by the community, was that the first priority was to get unemployment down. The relatively high level of unemployment reinforced Labor's sense of mission and its commitment to the rapid implementation of its programme.

The defeated Government's economic policies in its dying months strengthened the new surge of inflation which was to undermine the Labor programme and the Labor Government. Its last Budget was expansionary—in Treasurer Snedden's phrase: 'Taxes down, pensions up, growth decisively strengthened.' It was understandable for a government confronted both by relatively high unemployment and a difficult election, but it was a stimulus to a situation in which domestic consumer demand was already recovering strongly. Both Governments—the 'old' and the 'new'—were victims of a break down in the political-economic cycle.

Less excusably, the Australian dollar was grossly undervalued. Overseas capital flooded in and Australian reserves had risen to $5,000 million. With an excess of frankness and lack of caution, Whitlam had foreshadowed revaluation during an ABC interview on Budget night in August 1972. 'I am quite convinced that the Australian dollar ought to be appreciated in value,' he said. In the same interview, he said that Australian protective tariffs were too high. Whitlam's statement on revaluation brought a strong reaction from the Leader of the Country Party, Douglas Anthony, who excoriated Whitlam for his 'irresponsibility' and unsuccessfully urged McMahon to call the election immediately, even before the Budget had passed. Whitlam's television comments were scarcely more than asides in a lengthy interview about the Budget, but he had identified two of the most critical decisions of his future government—revaluation and tariff cuts.

At their first post-election meeting on 3 December 1972, Sir

Frederick Wheeler had warned Whitlam of the dangers of inflation already at work. Contrary to the myth later propagated by Treasury, Whitlam did not brush aside these warnings. Despite the priority given to the unemployment problem, most of the specifically economic decisions of the Whitlam Government in its first year were aimed against inflation.

The first of these was revaluation of the dollar. Whitlam was determined not to repeat the disaster over which McMahon had presided in December 1971. Exactly a year after the McMahon Government had come to the brink of breaking up over the question, Whitlam announced, on 23 December, a 7.05 per cent overall upward change of the value of the Australian dollar. He also announced measures to control the volume of capital inflow, including the introduction of a variable deposit scheme requiring borrowers to lodge a proportion of their overseas borrowings with the Reserve Bank. If the thing were to be done cleanly and quickly, there could be no question of full Cabinet consultation. The 1972 revaluation was decided by a Cabinet committee consisting of Whitlam, Crean, Barnard and Cairns. While the decision was generally applauded — not least for the quick, clean way in which it had been done — it caused the first stirrings of uneasiness among rural and mining interests which were to rise to a tempest over the next year.

The proposal to establish a Prices Justification Tribunal was also seen as part of the Government's armoury against inflation. This long-standing proposal had originated in a suggestion by John Menadue after he had left Whitlam's staff to join Rupert Murdoch's publishing organization, News Limited. He had gained some insights into the character of Australian enterprise and suggested that the persuasive power of such a body, with the attendant publicity of an open inquiry in which a corporation had to justify its price increases would be more effective than any other form of control. In January 1973, Australia's largest corporation, Broken Hill Proprietary Company Limited, signalled its desire to co-operate with the new Government by asking for an inquiry into its case for steel price increases. Pending the legislation to set up the Tribunal, Whitlam obliged by appointing Mr Justice Moore, then Acting President of the

Conciliation and Arbitration Commission, to conduct the inquiry. The Act establishing the Prices Justification Tribunal was given assent on 1 June. It provided that a company with an annual turnover of more than $20 million must notify the Tribunal of any proposed price increase. Price increases were prohibited until the Tribunal had determined whether or not it would inquire into the proposed increase. The legislation did not provide powers or penalties to enforce the Tribunal's recommendation. In practice, however, its moral authority proved persuasive and companies abided by its decisions.

On 3 April, Whitlam announced the appointment of a 'task force', under the leadership of Dr H. C. Coombs, to review government spending and to 'undertake a close scrutiny of the continuing policies of the previous government'. This followed a Cabinet decision that 'action be set in train to apply a close scrutiny to continuing policies of the previous Government so that room may be found for our own higher priority programmes'. In his letter of appointment to Dr Coombs, Whitlam wrote: 'I regard the importance of such actions as self-evident . . . a thorough-going review of the policies of our predecessors seems to me to be highly desirable in its own right . . . we have a lot to do ourselves and it will be important that we should have the widest possible options available to us in instituting our own programmes.' Whitlam asked Coombs to have his report ready well before the Budget in August; their report was in the Prime Minister's hands by the first week of May.

The Coombs Task Force set out some 140 programmes which might provide scope for cut-backs in government spending. Under the terms of its appointment, it made no recommendations, set no order of priorities and made no assessment of the desirability of dropping or keeping any of the programmes it mentioned. That was wholly Cabinet's responsibility. The Coombs Report was a useful document — a kind of shopping-list in reverse. It was also a political time-bomb. From the time parts of the report were published prematurely before the Budget, every beneficiary of the concessions, subsidies and bounties mentioned felt threatened. The Budget incorporated some of the Report's suggestions, but the remaining beneficiaries felt no gratitude for their reprieve, but rather added fear that their day was coming closer. A large number of the items in the Coombs Report involved rural industries, reflecting the proliferation of special arrangements for farmers during the years of Country Party ascendancy. The very existence of the Coombs Report was felt as an affront to agricultural interests. It became a symbol of Labor's alleged hostility to agriculture; the name Coombs became a sort of incantation against the Labor Government. 'This very dangerous document ... attacking and ruining primary industry ...': the Country Party member for the Northern Territory, Samuel Calder, spoke for farmers around Australia. For many, the Coombs Report epitomized everything they found distasteful in the style and decisions of the Whitlam Government: a group of bureaucrats and advisers — the 'longhairs' and 'intellectuals' — remote from reality, sitting in judgement on the fate of the wealth-producers of Australia. It became a menace to a thousand accumulated privileges, anomalies and perquisites and has become the classic warning to governments that the political damage done by an existing benefit removed can never be offset by a new benefit conferred. Thus, when the Government decided in February 1974 to abolish the superphosphate subsidy, the actual political impact of the decision was magnified ten-fold by six months of resentment and uncertainty. As in so much of its history, the Whitlam Government suffered not so much from what it did but what particular groups feared or were made to fear what it might do.

The new Government's tendency to look for different advice was the beginning of the mutal suspicion between the Labor Government and the Treasury, which grew into hostility on the Government side and disloyalty within the Treasury which powerfully, perhaps decisively, would contribute to the destruction of the Whitlam Government. Treasury weakened its own influence with the Labor Government not because its advice was often unpalatable but because it was always monolithic. Treasury refused to set out the economic options available to the politicians. It presented its advice on an all-or-nothing, take-it-or-leave-it basis. If its first advice was rejected, Treasury volunteered no other. If Cabinet persisted in rejecting the original advice, some Treasury officials then used their contacts with other departments, the press, business, and Opposition spokesmen, to build support for its line. In 1976, Fraser, as Prime Minister, attempted to break the monolith of Treasury advice by dividing the Treasury into two Departments. In 1973, the most the new Labor Government felt able to do was to appoint a handful of economic consultants to provide an alternative source of advice.

Long after, Whitlam acknowledged that 'our greatest economic mistake was that the credit squeeze was too late and lasted too long'. By that he meant that the Government had waited until September 1973 to introduce it and waited until September 1974 to remove it. The Treasury and Reserve Bank did not advise a credit squeeze earlier because they believed that the excessive liquidity springing from the 1972 Snedden Budget would have been reduced by the revaluation and foreign investment restriction to a greater degree than actually happened. Nor did either the Treasury or the Reserve Bank fully warn the Government of the impact of the credit squeeze. As late as April 1974, on the eve of the election, Whitlam specifically questioned Sir Frederick Wheeler and Sir John Phillips, Governor of the Reserve Bank, on public allegations by Russell Prowse, Assistant-General Manager of the Bank of New South Wales, that the actual squeeze was far in excess of the Government's specifically announced measures, such as the increase in interest rates. This interview took place at Kirribilli House, after Jim Spigelman had expressed concern to Whitlam that the Government was not being fully informed. Whitlam accepted their reassurances. In any case, for the next two months, the Government was effectively paralysed by the election challenge and its aftermath. In August 1974, Sir John Phillips admitted to Cairns that 'a mistake was made' about the timing and severity of the credit squeeze.

The high-point of the anti-inflation decisions of the Whitlam Government in its first year was the across-the-board tariff cuts on 18 July. It remains the most controversial and hotly-contested of any single economic decision of the Whitlam Government, inside and outside the Labor Party. Yet, at the time of its making, few decisions of the Whitlam Government were more favourably received. The paradox is simply explained: in July 1973, it seemed so obviously correct in general economic terms; by July 1975, it was the one obvious decision that could be blamed for unemployment in the manufacturing sector. Within the Labor Party itself, it was so much more palatable to blame the tariff cuts and the people associated with the decision — the 'egg-heads' and academic advisers surrounding Whitlam, according to disgruntled former fervent allies like Sir John Egerton — for the difficulties in the textile industry than to acknowledge how much of these difficulties were due to equal pay for women or the wage explosion of 1974.

Essentially, the decision to reduce tariffs was part of the same strategy as the Coombs Task Force — to prepare the way for Labor's first Budget for twenty-four years. By June, Treasury was urging an austerity budget as the appropriate strategy against rising inflation. The alternative was a large increase in taxation. Although the formula in the policy speech, 'there should be no need to raise taxation', had been devised to provide a semantic loophole, Whitlam argued that tax increases would go against the spirit of his undertakings and against the spirit of the mandate. The mandate was becoming a strait-jacket. In November, Dr Coombs, not inhibited by his role as an economic adviser, told the National Press Club in Canberra: 'If I had to criticize the Government, I believe that they were unwise to allow themselves to be persuaded that their pre-election policy precluded their facing the community with an income-tax increase.' Hawke made the same criticism of the Budget, but incorrectly said that the policy speech undertaking had been made without consultation. Hawke had apparently forgotten the debate during the Federal Executive meeting in September 1972 at which the matter was thoroughly canvassed. The latter part of Hawke's statement roused Whitlam to describe Hawke's statement as 'petulant'; Hawke, not surprisingly, suggested such a description was 'a bit rich' coming from Whitlam.

It was at this point and partly on this issue of taxation that both Whitlam and Hawke made a serious and consequential error of judgement. For the first time, they adopted adversary positions in public and allowed their public stances to flow over into their private relations. Hawke felt he was not being sufficiently consulted on matters affecting the trade union movement, by which he meant the whole of economic policy. Whitlam could accept no such obligation to consult. Hawke's mistake was to put the entitlement to consultation too high. Whitlam's mistake was to put the political advantages of consultation too low. So instead of consultation either informally or through the formal agency of the Commonwealth Labor Advisory Council, which the Labor Party had established for the express purpose of achieving consultation between the political Party and the industrial wing, these two formidable leaders with a long history of genuine mutual respect and mutual co-operation had a short period of sharp spats. At a critical point in the life of the Labor Government, co-operation between the political leader and the industrial leader temporarily broke down.

That blessed word 'consultation'. It became the standard complaint of everyone who felt aggrieved by a particular decision that they had not been consulted. At one time or other, almost everybody made the complaint — Ministers, Caucus members, departments, staff, Party officials, State Premiers and State leaders, trade unions, industry organizations, women's groups and welfare organizations. To a very large extent, the complaint confused consultation with decision-making. Nonetheless, Whitlam underestimated the value of consultation, or the appearance of it, as a public relations device. There is the example of the annual pre-Budget discussions with leaders of the employer, union and farm organizations. These had been instituted in 1962 by Menzies and McEwen as a gesture to the principal pressure groups after their near defeat in 1961. Thereafter, every July, spokesmen from the invited groups made the pilgrimage to Canberra to put their views on the economy and the forthcoming Budget to senior ministers. There is no evidence that these solemn proceedings made the slightest difference to any Budget; but it remained a useful exercise in political massage, soothing to the aggrieved and the self-important. In 1973, the more than usually perfunctory nature of the exercise, with Whitlam rarely attending the meetings, meant that the Government failed to make the most of a convenient, inexpensive chance to preserve the illusion of 'consultation'. Partly because of this lost opportunity there arose another standard complaint—the so-called 'communications gap', identified by the National Secretary, David Combe, as one of the causes of the poor result in the Parramatta by-election.

Within the Labor Party, this complaint was an echo of the more general complaint, 'too much too soon'. In the first year, at least, the generalization was much more often and easily made than any specific criticism of particular decisions. Whitlam had, in fact, embarked on a very elaborate, time-consuming process of both consultation and communication—the method of using invitations from business organizations to make formal statements of Government policy. Whitlam had used such forums successfully in Opposition. In Government, the method proved generally unrewarding and sometimes disastrous. Organized functions of this kind, known in the United States as the 'rubber chicken dinner circuit' are not designed to elicit information; they are organized to entrench the views of the promoters. Thus, when a Labor Minister accepted an invitation to address the Melbourne Chamber of Commerce, he did not possess a captive audience to propound his views; he was the captive on exhibit. In September and October 1973, Whitlam, on advice, accepted a series of invitations to such functions with the object of explaining Labor's policy. It was worse than a waste of time. Whitlam's declarations of good intent were received with scepticism, his call for a more vigorous competitive system of private enterprise with suspicion; his assertions that business was prosperous with hostile derision. At a time when the annual reports of most corporations were showing record profits, Whitlam's assertion at the Adelaide Chamber of Commerce and Industry in October that the 'stock exchange was being masochistic' brought howls of rage; four years later there were businessmen around Australia who have never forgotten or forgiven that statement.

Perhaps the most serious failure of communications was with the Labor Government's strongest supporters—the trade unions. The basis of Whitlam's 1972 programme was that government services—health, education, welfare, improvement of the physical environment—were as important in determining a family's standard of living as the pay packet and more important in reducing inequality than redistribution of personal income. This part of the Labor programme and this expression of the Whitlam philosophy was never accepted by the trade unions and not accepted by half the Cabinet or Caucus. Union opposition to the referendums to give the Federal Government power over prices and incomes guaranteed their defeat in December 1973. That opposition was partly an expression of union scepticism towards the thesis that the provision of services was a substitute for higher cash incomes. From the second half of 1973, successful wage claims, accompanied by a high level of industrial disputes, led to a 35 per cent increase in wages and a squeeze on profits—in effect, a considerable redistribution of income in favour of employees.

The so-called 'communications gap' was produced not so much by a failure in public relations as by a gap between expectations and their tangible fulfilment. Impressive as the record of achievement for 1973 was in terms of honouring the undertakings of 1972, its benefits were not always immediately apparent. By November 1973, Cabinet had made 1,675 decisions, 254 bills had been presented to Parliament and 223 had been passed, and 39 reports by the 94 inquiries, task forces and commissions had been tabled. The problem was not that so much activity was indigestible or incomprehensible; it was that so much of it, particularly in the fields of education, urban affairs, the environment, aboriginal welfare, foreign ownership, the status of women and law reform, could produce tangible benefits only in the distant future. Committed as it was to such a massive programme of reform, no government needed its full three-year term more than the Whitlam Government. It was to be denied the thing it needed most.

On 13 December 1973, Whitlam delivered a first year report to Parliament:

> I sometimes hear it said that the Government is doing too much, we ought to slow down. This is an argument that comes easily and naturally from conservatives. I reject it. I reject it because the task

before us is still a great and formidable task, because much that could have been achieved last year still remains to be achieved because a nation that has waited twenty-three years to see essential and overdue social reforms cannot be expected to wait any longer.

The Government is proud of its record. We have abolished unemployment and restored a healthy rate of economic growth. Australia has a new strength and confidence at home and a new respect abroad. We are no longer a cipher or a satellite in world affairs. We are no longer stamped with the taint of racism. We are no longer a colonial power. We are no longer out of step with the world's progressive and enlightened movements towards freedom, disarmament and co-operation. We are no longer in thrall to bogies and obsessions in our relations with China or the great powers. We have strengthened old friendships and established new ones . . .

After all the carping and niggling, backing and filling by the Opposition, the changes we made will remain — like all great Labor legislation — permanent landmarks in our history. I am confident that the Australian people — when next they are asked to choose between the alternatives before them — will endorse our record and renew our mandate.

The test was to come sooner than Whitlam expected.

19
Double Dissolution

Early in 1974, the ALP National Secretary, David Combe, wrote a report expressing concern at apparent apathy within the Labor Party about the forthcoming election for half the Senate, scheduled for 18 May. 'It is', he wrote, 'of much greater importance than any Senate election since 1961.' Combe pointed out that a vote comparable with the 1972 House of Representatives vote could give the Government equality of numbers in the Senate and the possibility of ending Senate obstruction. Other Party officials, like the New South Wales General Secretary, Geoffrey Cahill, complained that the Prime Minister's determination to put four referendum proposals to the electors on the same day as the Senate elections might cause confusion and diminish Labor's Senate chances. They need not have worried. For Whitlam was about to attempt a master-stroke which would transform what was at stake on 18 May. He had decided to make Senator Vincent Gair Australia's Ambassador to Ireland.

The motives for the Gair appointment were simple. Its purpose was to give Labor a chance to control the Senate. Its execution and its aftermath were as complicated as the mechanics of the Federal system, the contradictions of the Australian Constitution and the anomalies of the proportional voting system by which the Senate is elected.

The Senate in 1974 was composed of sixty senators, ten from each State, elected in 1967 and 1970. Of the thirty senators whose term would expire in mid-1974, twelve were Labor senators, leaving fourteen Labor senators elected in 1970 with only half of their six-year term completed. Combe's scenario in his report of February 1974 envisaged the election of three Labor senators out of five in New South Wales, Victoria, South Australia and Tasmania, and two out of five in Queensland and Western Australia — a difficult task, but not impossible. With the fourteen 1970 senators, this result would give the Government thirty senators; with the general support of an independent senator for Western Australia, the Government could then expect to pass most of its legislative programme. If, however, another vacancy could be created in one of the States where Labor could hope to win only two of the five seats — Queensland or Western Australia — the proportional representation voting method by which the Senate is chosen would guarantee another Labor seat — three out of six. This was the arithmetic of the Gair appointment.

The other side of the equation was the record of obstruction by the Senate. By April 1974, the Senate had twice rejected ten bills, and rejected nine other bills. The twice-rejected bills included four referendum bills for the alteration of the Constitution (among them a proposal for simultaneous elections of the House of Representatives opposed by the coalition in 1974, but sponsored by the Fraser Government in 1977, and once again defeated by being lost in the three smaller States despite a large national majority). The most important of the other six twice-rejected bills was the legislative basis of the national health insurance scheme (Medibank). A number of other bills, including a stronger Trade Practices Bill and a bill to strengthen the Australian Industry Development Corporation, were shelved or stalled by various devices. Beyond the number and substance of rejected measures, there was a distinct change of mood in the Senate Opposition, raising the spectre of the previously unthinkable — Senate refusal of Supply. Thus, control of the Senate assumed an importance and urgency never previously contemplated.

By 1974, Vincent Clair Gair was a bitter, lonely, frustrated man. Discarded by his four DLP colleagues as their leader in October 1973, he could look forward to three gloomy, barren years in the Senate, and then political oblivion. In late night, maudlin conversations with the Tasmanian Labor Senator, Justin O'Byrne, he expressed his resentment of his colleagues. O'Byrne told him, 'You know, they have kicked you into the back-benches. They have given you a rough trot. Why don't you go for a holiday?' O'Byrne was thus established as a sympathetic channel to Gair's old enemies in the Labor Party. On 5 March, Gair put a breath-taking proposition to O'Byrne: he would be interested in accepting a diplomatic post from the Labor Government. O'Byrne relayed the message to Murphy, who carried it as swiftly to Whitlam.

Whitlam reacted with glee and despatch. Control of the Senate was worth a deal with Gair. The prospect of such a glittering coup was irresistible. Against it, there was only one argument: it involved a deal with a bitter political enemy — and worse, a Labor renegade, a former Labor Premier who had led the DLP. If this argument determined the morality of politics, the most important political alliances in Australian history — all of them alliances against Labor, from Deakin through to Hughes, Lyons, Menzies, McMahon and Fraser — could never have taken place. Conservative politics is the history of alliances and deals between hitherto irreconcilable enemies. The Gair appointment was audacious, opportunist and cynical; it was only when the manoeuvre was circumvented, when Whitlam was bested by a superior tactician, that the question of its 'immorality' was raised. The first response from the Labor Party, the Opposition and the press, was astonishment, either admiring or hostile, at Whitlam's audacity.

If the prospect of gaining control of the Senate was irresistible to Whitlam, the offer he made was equally irresistible to Gair. On 14 March, the Governor-General, Sir Paul Hasluck, approved the appointment of this 73-year-old Irish-Australian Catholic as Australian Ambassador to the Republic of Ireland. A week later, the Irish Government agreed to the appointment. Whitlam handled the arrangements personally. The secrecy of the operation was ensured by the fortuitous circumstance that he was acting as Foreign Affairs Minister in the absence overseas of Willesee, who opposed the whole idea. Whitlam's success in preserving secrecy was to prove his undoing. It meant that there was no consultation or examination of the gaps in the plan or the means by which the plan could be thwarted. Because of this, Whitlam and Murphy made two astonishing omissions: they failed to get Gair's resignation in writing and they failed to foresee the way in which a determined opponent could counter them. Whitlam was playing in the wrong league. It was an exercise which required a certain deviousness; Whitlam has no natural talent in that area. The essence of his style and technique is openness; in the game of back-room intrigue, he is not a good politician at all.

Secrecy was maintained until the eve of the day Whitlam had intended to make the public announcement in Parliament on 2 April. That morning, the story appeared in the Melbourne *Sun News-Pictorial*, written by Oakes, from a tip-off received by his junior colleague, John Lombard. The significance of the premature revelation was that it allowed a vital few hours for the Opposition's astonishment, indignation and disbelief to be digested and for cooler minds to find a way to circumvent Whitlam. There were two steps in this process: to make sure that Gair did not resign and to get the Queensland Government to issue as soon as possible the writs for the 18 May Senate election. Various people have since claimed credit for devising this counter-manoeuvre; there is no need to go beyond its executor, the Premier of Queensland, Johannes Bjelke-Petersen, obsessive in his hatred of the Whitlam Government. On the night of 2 April, the Queensland Government Printer worked overtime to produce a special single-sheet edition of the *Government Gazette*, proclaiming that the State Governor, Sir Colin Hannah, acting under Section 12 of the Constitution, had issued writs for the election of the five Queensland senators whose terms expired on 30 June 1974. It meant simply that on 18 May, Queensland voters would elect five senators instead of six; of these it was almost certain that Labor would win only two. As a means of securing an election for six Queensland senators, three of them Labor, the Gair appointment had become irrelevant. And, within twenty-four hours, arguments about the propriety or legality of Bjelke-Petersen's action, focusing on whether or when Gair had actually resigned, became equally irrelevant. The Opposition had decided to force an election for the whole Parliament.

More than a year after the events of April–May 1974, when

Snedden had been deposed as the Liberal leader, the author had a conversation with him in King's Hall (though unlike many alleged conversations in King's Hall, this one actually took place). 'I have never really understood', I said, 'why you forced that election. The probabilities are that except for it, you would still be leader, we would be heading up to an election, you would have won and you would be Prime Minister.' Snedden replied, 'The pressure was on me from Anthony. We thought you had a chance of getting control of the Senate at the half-Senate election or at least enough to get a redistribution through. With a gerrymander, you'd be in forever.' This does not altogether square with the declaration by the Liberal Leader in the Senate, Senator Withers, that since April 1973 they 'had embarked on a course to bring about a House of Representatives election'. Yet it helps explain Snedden's and Anthony's fury at the Gair appointment; it sprang not from moral principle but from political fear. Snedden's explanation also provides an insight into the coalition's attitude towards the Labor Government: Labor was simply not entitled to preside over an electoral redistribution; it could not be trusted. The last time a Labor Government had presided over a redistribution was in 1948. That redistribution was so principled or so inept that it secured for Labor defeat in 1949, 1951 and 1954, despite the fact that in 1954 Labor secured 50.2 per cent of the national vote. The Gair appointment was an attempt to manipulate the electoral laws to the Government's advantage. It was lawful, clever, expedient — and unsuccessful. The supreme irony is that the State Premier who thwarted the attempt maintained his power by the most remarkable gerrymander Australia has ever known.

The Gair appointment ended the Opposition's year-long search for an issue on which to force a House of Representatives election by using its numbers in the Senate to reject Supply. It provided an issue, which in the judgement of Snedden and Anthony, would make the refusal of Supply, never attempted by the Senate since Federation, more acceptable by alleging that the Gair appointment represented another even worse breach. To create the proper climate, they alleged 'corruption'. According to Snedden, it was 'the most shameful act by any Government in

Australia's history . . . worse than any Tammany Hall effort ever made in the United States.' Anthony said it was an 'act of corruption to buy off a Senator', although his gloating question, 'Can the Prime Minister say whether he has ever been taken for a ride by the Premier of Queensland?' indicated that he accepted the manoeuvres and counter-manoeuvres as a valid political exercise in which Whitlam had been outsmarted. On 4 April, Snedden announced that the joint Party meeting had decided to oppose the Appropriation Bills in the Senate. Whitlam responded immediately: if the Senate voted against Supply, he would advise the Governor-General to dissolve both Houses. Events were to show that Snedden and Anthony had miscalculated. But their miscalculation was minor compared with that of Gair's former DLP colleague, Senator McManus. Snedden lost an election; McManus lost a party.

Francis Patrick Vincent McManus was one of the founders of the DLP, having been Victorian assistant-secretary of the Australian Labor Party at the time of the 1955 Split. He had dedicated his life to making and maintaining the DLP as a key political force. His features and physique reflected his views: narrow, but straight. He had all the melancholy of the Celt with none of the ebullience. Bitterness was his companion; the bitterness of a man who felt he had been betrayed — betrayed by Evatt, betrayed by his fellow Victorian Catholics, Calwell and Kennelly. Now he had been betrayed by Gair, the man for whose sake he had for years foregone his own claims to the leadership of the DLP. He arrived in Canberra by train — he hated flying — to read Oakes' story that Gair was to go to Dublin. The political judgement with which he had steered the DLP through survival to a position of great influence deserted him. Betrayed again! He made two disastrous miscalculations. He calculated that in a double dissolution, with ten senators to be elected from each State, the DLP could muster one of the necessary quota of votes to elect one senator in Victoria, New South Wales and Queensland. Moreover, he believed he had an undertaking from Snedden that the non-Labor parties would run a joint Senate ticket in a great national crusade against Labor, socialism and centralism, and that in the major States at least, DLP senators would be included on the ticket. If Snedden had entered any such agreement, it was an undertaking he could not possibly fulfil. The State machines of the Liberal Party rejected it out of hand. But, in fact, McManus had betrayed himself and his whole career. He boasted of his tough education in the old school of Victorian Labor politics. Two of the maxims of that school were 'Get the numbers' and 'Get it in writing'. In his blind indignation at Gair's defection, he forgot these basic precepts. By supporting the coalition in voting against Supply, he signed his own political death warrant and destroyed the Party he had done so much to make. One of the few positive results of the 1974 Double Dissolution was the parliamentary annihilation of the DLP, exactly nineteen years after its creation. Frank McManus may have felt he was fated to be betrayed, but in the end he betrayed himself.

McManus records in his book, *The Tumult and the Shouting*, that as early as mid-1973 he had discussions with Snedden about bringing about a double dissolution in order to end the Labor Government after less than a year. He was entitled to believe he had an agreement. By failing to enforce that agreement in April 1974, Snedden robbed the DLP of its traditional role, indeed its *raison d'être* — to keep Labor out. In his book, McManus correctly points out that the breach of the understanding meant that the DLP was unable to do what it had done in every Australian election since 1955 — throw its preferences and the full power of its organization and propaganda behind the Liberal Party. In this sense, Snedden and the Liberal Party were responsible for the destruction of the DLP, which had guaranteed the Liberal ascendancy for a generation.

The end of the Twenty-eighth Parliament came quickly. Although the principle at stake was the rights of the House of Representatives versus the power of the Senate — the same issue of the great Constitutional crisis of 1975 — the content was not pushed to the extreme in the way it was to be in 1975. There were several reasons for this. The election machinery was already in operation, preparing for the half-Senate election scheduled for 18 May. The proposition that the Senate should purport to refuse Supply was so novel that neither side had thought through or were prepared to face the consequences of a prolonged dispute. Most importantly, the Government knew that a double dissolution in May would be as good a time for an election as any in the next eighteen months. For these reasons, Whitlam did not push a great constitutional question to the edge of crisis.

On 10 April, Withers moved that 'because of its maladministration, the Government should not be granted funds until it agrees to submit itself to the people'. After a short debate, Murphy, responding for the Government, said, 'The Government . . . has kept the position that if the Senate Opposition were foolish enough (to refuse Supply) . . . by having some kind of slick motion passed which will say that they are not really refusing Supply but are asking that it be deferred conditionally upon the Prime Minister agreeing to submit to a dissolution, it will not be threatened . . . They sit back there and think about how they can salvage their consciences . . . in some way that will enable them to pretend that they did not really refuse Supply. They have come up with this absurd amendment'. Unwittingly, Murphy had described the scenario of October–November 1975. But the matter was not to be put to the test in 1974, for Murphy then said, acting on an arrangement with Whitlam:

> Mr President, to put an end to this matter, I will tell you that I intend to move 'That the question be now put': if that motion is defeated, the Government will treat that as a denial of Supply. If that motion is carried and this absurd amendment is carried, the Government will treat that also as a denial of Supply. In either event, the Prime Minister, who is conversant with the absurd proposition which has been put here, will call forthwith upon his Excellency the Governor-General . . .

And so it transpired. At 8.30 p.m., Whitlam curtly told the House of Representatives that he had 'advised the Governor-General that a situation had arisen under Section 57 of the Constitution which would entitle him to dissolve both Houses' and that his advice for an immediate double dissolution had been accepted. So, without ceremony, the Twenty-eighth Parliament, a House of Representatives that in little more than a year had passed more legislation than any since Federation, and a Senate which had rejected or deferred more than any other since Federation, passed into history.

*

The 1974 campaign was the most difficult Whitlam has ever fought. He was under a serious restraint. The cost-of-living figures for the March quarter appeared to show a slowing-down in the inflation rate. This masked temporarily the deepening economic difficulties ahead. The head of the Treasury, Sir Frederick Wheeler, warned that it was a 'false dawn'. Whitlam told his colleagues that the election was being held at a more favourable time than any likely to occur in the next eighteen months. It was a knowledge he could hardly share with the public. This accounts for the air of uneasiness which settled over Labor's campaign and explains Whitlam's unaccustomed defensiveness.

The policy speech included only one major proposal for new spending — $130 million for child care services. He asked for a renewal of the interrupted mandate. Labor's campaign strategy rested almost entirely on the appeal for a 'fair go'.

In a way not intended, the campaign began a day before the scheduled official opening in Blacktown Civic Centre. Whitlam had committed himself to officiating at the foundation-stone ceremonies on the site of the proposed national headquarters of the Australian Labor Party in Canberra, now Curtin House.

It began in the rain, a cold and grey Canberra autumn day. As Whitlam began to speak a fitful sun came out:

> For all the lustre of his achievements, for all the grandeur of his leadership of this nation in its most perilous years, it cannot be disguised that the story of John Curtin was essentially tragic. It was a story of noble failure.
>
> If ever a man was born to lead this nation in times of peace and in the paths of peace, he was John Curtin. If ever a man was born to apply his vision of what Australia at peace could be, his vision of what Australia at peace should become in his time, he was John Curtin. And yet his place in history is as our war-time leader.
>
> He did not choose his time. He was enslaved by the times; and for him, time was a cruel master. He was slave to his own destiny.

He was a pacifist who had to lead his country in war. He was called upon to take decisions which sent men to die. He was an anti-conscriptionist, imprisoned for his conscientious objection in 1917, who had to impose Australia's first conscription.

There was above all that terrible time in 1942. More than half our army, our only battle-tried and battle-trained, our best and bravest, were in unescorted transports on the Indian Ocean — then a deadly ocean — coming home from the Middle-East to form our only defence in the Pacific War. It was his decision, his responsibility. With all the weight of this fearful knowledge upon him, unable to tell his colleagues, even those of them who were mates, he was alone. He was the loneliest man in Australia, whose Prime Minister he was...

Even so, he never forgot his real aims and ideals. He never lost his vision for Australia. As soon as the moment of highest danger passed, he turned to the task of building a new Australia, an Australia as he wanted it to be after the war. He wanted to turn the prodigious energies the Australian people had shown in war to the task of rebuilding, remodelling and reforming this country, in peace.

He expressed those ideals and that vision in the great referendum campaign thirty years ago. So far as the dry and confining language of constitutional and legal statements can convey, John Curtin's vision is expressed in the 1944 referendum proposals. He won the war; but he lost that battle. I invite you to reflect how different, how better, this nation would now be if those proposals had carried. They were proposals for unity, for equality, for human dignity.

The referendum was lost and thirty years have been lost, thirty lost years, in building a better Australia, in bringing Australia to the full promise our resources, our skills, our energies, our national spirit would allow. The struggle continues. For John Curtin, it was one of his last political battles, just months before his death. It was for me, in a very small way, publicly insignificant, privately important, my first. I have never forgotten it and I have never forgotten John Curtin's efforts to modernize and revitalize Australia's Constitution and thereby remould and reinvigorate the Australian nation.

As he came to this passage, Whitlam's voice faltered, and with difficulty he moved on to make what Laurie Oakes in his book

Grab for Power — Election 1974 aptly describes as a 'call to arms':

> Not by our choice, this ceremony falls on the eve of another campaign, another hard struggle for our party, another landmark in our nation's history. The challenge we now face is, of course, not so perilous as any of those John Curtin faced, but it is momentous nonetheless. For we are all involved in decisions which for good or ill will turn the history of this nation as crucially as any decision taken in his time, as important as any decision the Australian people took in his time. They are fundamental decisions about the future of Australia, about the direction in which Australia should go for the rest of this century and beyond.
>
> In the times of his deepest difficulty, his most awesome responsibility, John Curtin drew strength from his deep belief in the intelligence and idealism of the Australian people and their sense of fair play. They are, I believe, things upon which any Australian leader can always rely with confidence.

The return of Whitlam to Blacktown for the policy speech was meant to symbolize continuity. As Whitlam said: 'The Government you elected for three years has been interrupted in mid-career. Our programme has been brought to a halt in midstream.' The audience was as enthusiastic as in 1972 but the mood had changed: the joyful exuberance of 1972 had given way to a kind of fierceness. Whitlam concentrated on the achievements of 1973.

> For the first time Australia has a government determined to promote Australian ownership and control of Australian industries and resources.
>
> For the first time for a generation Australia has a government dedicated to equal opportunity for all its citizens. We have more than doubled spending on schools. We have abolished fees at universities, colleges of advanced education and — going one better than our pledge — at technical schools.
>
> For the first time Australia has a government determined to make the conditions of life more equal for all Australians, wherever they live in Australia.
>
> For the first time Australia has a government seriously concerned to give equality of opportunity to women.
>
> For the first time Australia has a national government involving itself directly in the affairs of our cities.

For the first time Australia has a government ready to give local government direct access to the national finances.

For the first time Australia has a national government prepared to co-operate in renewing our decaying urban transport systems.

For the first time Australia has a national government determined to fulfil its constitutional obligation towards the aboriginals.

For the first time Australia has a government determined to preserve, protect and enhance Australia's national estate — our natural and historical inheritance, what we keep from our past, what we transmit to the future.

For the first time Australia has a national government which recognizes the significance of the arts and artists in our society. Our support for the arts has released an unparalleled burst of creativity in this nation.

These achievements are just some of the fruits of programmes based on expert advice. We sought and obtained the co-operation of the most highly qualified Australian men and women to inquire into and to report upon basic requirements in Australia's social and economic structures. These reports are public. The work of these inquiries and commissions is the basis of a continuing, coherent, comprehensive programme. The new initiatives I announce tonight are the new dimensions and expanded fruits of this work.

The new initiatives included abolition of death duties on the matrimonial home, a Bureau of Consumer Protection, the appointment of a Royal Commission into the Australian Public Service headed by Dr Coombs, a Schools Curriculum Development Centre, a National Employment and Training Programme (NEAT), expansion of the Australian Assistance Plan Shelters for homeless men and women, increased assistance for nursing homes, and a new child care programme. He undertook to implement the report of the National Rehabilitation and Compensation Committee under Mr Justice Woodhouse of the New Zealand Court of Appeal, due in June, and to establish an Australian Government Insurance Office. These two promises began a costly and unrelenting campaign against the Labor Government, mounted by the Australian insurance industry. The emphasis of the speech, however, was on the continuation and completion of the 1972 programme and for a 'fair go, not just a fair go for an elected Government, but a fair go for our programme to work'.

Snedden led the best Liberal campaign since 1966. His strategy was simple and correct — to concentrate on economic issues. Advice from the Liberal Party Secretariat led him into a disastrous tactical error: he committed himself to giving a daily press conference for the entire campaign. Three things resulted: he lost control of the issues, his own lack of economic solutions was exposed, he fuzzed his chosen issue by allowing himself to be side-tracked by persistent and antipathetic pressure. The most damaging example was a dispute over oil price policy, with Anthony advocating world price parity for Australian oil.

Mightily assisted by the pressmen, who were tougher on Snedden than they had been even on McMahon or Gorton, Whitlam was easily able to exploit this, not only as a policy split in the coalition leadership, but as a nullification of Snedden's anti-inflation line. To the extent that the election campaign itself decided the election result, Snedden lost because he accepted the advice which locked him into an inflexible agenda of daily press conferences. One other decision designed to help the Liberal Party rebounded. The New South Wales Premier, Sir Robert Askin, sponsored a series of television commercials of an extraordinarily virulent and unsubtle kind. One in the series became notoriously known as 'the Estonian woman', in which a woman from Estonia expressed her disillusion with Australia as a land of freedom now that it had become 'a disguised communist' under Labor. In 1974, this was unacceptable and it discredited the entire Liberal Party advertising campaign.

Fundamentally, the Labor Government won the 1974 election because Whitlam was able to hold together the urban coalition he had patched together in 1969 and consolidated in 1972. The coalition held in the cities, notably in Sydney and Melbourne, and in Perth and Canberra where new seats offset Labor's losses elsewhere. In extraordinarily difficult circumstances, Whitlam had won an election at a time when governments around the world facing elections were being defeated, notably the Heath Conservative Government in Britain. He became the first Labor

Prime Minister to win a second successive election.

For a Government in difficulty, the actual result was remarkably good. Labor's 49.3 per cent national vote in the House of Representatives was only 0.3 per cent below its 1972 level. The 4.2 per cent rise in the coalition's vote was almost entirely at the expense of the DLP, which fell by 3.9 per cent to a catastrophic 1.4 per cent, dropping below the Australia Party's 2.3 per cent. Labor held at the centre, increasing its vote in Victoria by 0.5 per cent, in New South Wales by 0.8 per cent and in the Australian Capital Territory by 3.4 per cent. Labor also increased its vote in Western Australia by 0.2 per cent and in the Northern Territory by 5.3 per cent. It declined in Queensland (3.2 per cent), South Australia (0.2 per cent) and in Tasmania (3.8 per cent). In urban areas, its vote increased by 1 per cent. The result in terms of seats was less satisfactory. In a House of 127 (with an additional seat created in the Australian Capital Territory and Western Australia, both of which Labor won), Labor finished with 66 against the coalition's 61. Three seats were lost in New South Wales and two in Queensland. Two were won in Victoria, both in Melbourne. Four of the five seats lost were in country areas, Al Grassby losing in Riverina after a virulent racist campaign.

The Government increased its strength in the Senate by three to twenty-nine. By the barest margin, it failed to win a sixth place in New South Wales and a fifth in Queensland. The failure in New South Wales was entirely due to the huge number of informal votes—more than 10 per cent—because of the large number of candidates on the ballot paper. Even so, the Government secured 296,000 more votes than the coalition, and 6,000 more than the coalition and the DLP combined. So close, and yet so far.

20

A Question of Legitimacy

'We were not defeated. We just didn't win enough seats to form a government.' When the Leader of the Opposition gave this post-election assessment, the press and most of the nation were disposed to jeer. But it was no joke; it was declaration of renewed war. It was to be continuation of war by the same political means as before — Labor's lack of a Senate majority. Snedden was serving notice that the victory of May 1974 no more guaranteed Labor the three-year term which it believed it had fairly won than had the victory of December 1972.

From the outset, the delay in the final returns sapped the authority of the re-elected Government. On Sunday 19 May, the Chief Electoral Officer, Frank Ley, told Whitlam, 'You may win by seven and cannot win by less than five.' His assessment was accurate. Whitlam's staff urged him to make a national broadcast immediately to establish the Government's legitimacy and its renewed mandate. He refused. 'It would be improper.' During the days of delay, the psychological force of a genuine if patchy win bled away. Whitlam's decision to wait was proper and mistaken. He had achieved something no other Australian Labor Prime Minister had ever achieved — a second successive victory; he had achieved equality of numbers with the coalition in the Senate and had come close to achieving a majority there. But the delay in completing the count seemed to cast a doubt on the outcome. The very fact that the Government came so close to achieving a majority in the Senate, but in the final, complicated count, failed to do so, itself became a cause to question its authority and the validity of its new mandate. As the votes were tallied in the days after 18 May, it seemed possible that Labor would win at least thirty seats in the Senate; all interest turned on the Senate, as if a majority in the Senate had been the sole object of the Double Dissolution; the failure to achieve that impossible objective then was deemed a real defeat. Thus, the consequence of the Double Dissolution of 1974 was not the reaffirmation of the right of the House of Representatives to elect the government; its ironic consequence was to strengthen the power of the Senate.

The Opposition quickly showed that the election verdict on 18 May had changed nothing and settled nothing. The Constitution provides that if bills which have been the subject of a double dissolution are still not passed by the Senate in the new Parliament, they shall go to a joint sitting of both Houses. The Opposition determined to go the long road. At the first Joint Sitting of the House of Parliament in its history, the bills which had been the subject of the Double Dissolution passed into law. (The legislation to establish the Petroleum and Minerals Authority was later challenged in the High Court and was ruled invalid on the technical ground that it had not properly constituted a cause for the Double Dissolution. Of all the challenges against the validity of Labor's legislation, this was the only one to succeed.) The most important was the Health Insurance Bill, the legislative foundation of Medibank. The electoral bills, which provided for more equal electorates, remained a dead letter, because the Senate continued to refuse to pass a new distribution of electorates.

From May 1974, the Labor Government was under siege. At no time thereafter could it plan confidently on more than six months of existence at a time. It lived in the expectation that each November and each May it would face another fight for Supply in the Senate. The two Independent senators, Steele Hall from South Australia and Michael Townley from Tasmania, were both former Liberals. Townley quickly signalled his adhesion to his old party. Labor's only comfort was that Hall, a former Liberal Premier of South Australia, had stated he did not believe the Senate should reject Supply; he had a record for high integrity, particularly on constitutional matters. He had brought about the fall of his State Government and split from his Party in the cause of a more just electoral system. If the Senate as elected by the people in May 1974 remained intact and if Senator Hall stuck to his declared principle, then Labor could survive. In the event, Labor's hopes about the first condition for survival were wrecked. The Senate elected in May 1974 was to be radically changed, not by the electors, but by State Premiers. But Labor's confidence in Steele Hall's integrity proved completely justified. But from the beginning, it did not inspire the Labor Government with self-confidence or a sense of stability to know that so much hung by the thread of a former Liberal Premier's decency, that everything depended on whether Steele Hall would keep his word.

The Labor Caucus itself reacted in such a way as to suggest that it regarded 18 May as an interruption rather than as a fresh start. In the election for the new Cabinet, Caucus confirmed all existing Ministers. The newcomer was the Western Australian senator, John Wheeldon, to fill the vacancy left by the defeat of Grassby. Yet, under the surface of the status quo, Caucus had made two fundamental changes of great portent for its future as the governing Party. It elected Cairns over Barnard as deputy leader and gave its heaviest vote for the House of Representatives Ministers to Connor. Thus the duumvirate was broken, not just the fleeting duumvirate of 5–18 December 1972 but also the Whitlam-Barnard partnership which had steered the Labor Party to power through seven tumultuous years. So while the overall result seemed to confirm the old team, Caucus had decisively changed the balance of personal relations within the Government.

In the wake of alarming new figures showing a 4.1 per cent cost of living increase for the June quarter, Treasury submitted an economic package including tax increases on middle and higher incomes, a 10 cent increase in petrol excise, increases in postal and telephone charges and in the tobacco and spirits excise. Whitlam found himself almost alone in putting the deflationary view. Cabinet rejected most of the Treasury package and Caucus, unable to stomach a policy deliberately designed to create unemployment, threatened to reject most of what was left. Caucus found a spokesman for their doubts in Jim Cairns, the man they had just elected Deputy Prime Minister. In the weeks leading up to the presentation of the Budget, Cairns became the spokesman for the Caucus view against the Treasury view. In Cabinet, the Treasurer, Frank Crean, failed to argue strongly or effectively for his department's view. He accepted it; he did not fight for it. So the Treasury line fell by default, thereby ending the days of the Treasury's economic ascendency. It was at this time that Whitlam told Cairns and Crean that since Cairns was making the running on economic matters, he should take the Ministerial responsibility for them.

As the economy deteriorated, Caucus became increasingly alarmed at its government's apparent lack of control. The May elections had robbed the Government of three months normal administration. In the crucial weeks between early April and the end of May, no decisions could be taken at all. The Budget itself even had to be delayed. It could not be brought down until September instead of the normal August. For a brief period, Whitlam was inclined to accept the Treasury line of 'the short, sharp shock'. He did so in the pleasing belief that his victory in the May elections had gained him three years of office in which to get the economy into balance. So, for his speech at the Premiers' Conference, he accepted a Treasury brief which foreshadowed tough, draconian policies to come. The other Treasury brief, espoused by Crean, predicted company losses and bankruptcies, particularly in the land and building development field. In September, the prediction came true with the collapse of the financial firm of Cambridge Credit Corporation Limited (Receiver appointed) and Mainline Corporation Limited (In liquidation). The Government had to bail out the Leyland Motor Corporation of Australia Limited. Unemployment in the September quarter rose to 123,000 and retail prices rose 5.4 per cent. A devaluation of 12 per cent in September was generally well-received, although it negated the tariff-cutting efforts of the Industries Assistance Commission. It also rather contradicted Whitlam's criticism of those of his colleagues, notably Barnard, who were uneasy about tariff cuts. Whitlam dismissed such critics as 'nervous nellies'. The pre-Budget weeks represented a low point in Whitlam's relations with his colleagues. They were some of the unhappiest weeks in the life of the Labor Government.

Crean remained Treasurer to present the 1974 Budget. Against the gloomy background of rising prices, rising unemployment and collapsing companies, the Budget was, Crean said, 'designed to make the best of things as they are in the world today.' It contained tax cuts for those on incomes of less than $202 weekly. It reduced the maximum allowable tax deduction for education from $400 to $150 — a move as deeply unpopular with the higher income earners with children at private schools as the removal of the superphosphate bounty had been with farmers. It abolished television and radio licence fees and provided for the partial implementation of Whitlam's 1974 pledge of $130 million for child care services. Treasury had mounted strong opposition to this proposal. Largely through the lobbying of Whitlam's adviser on women's affairs, Elizabeth Reid, Cabinet ruled against Treasury. The largest spending increase was in the field of urban affairs, an increase of 175 per cent over 1973. While Treasury's deflationary line had been rejected, it was not a wildly expansionary Budget. Many commentators, notably the Melbourne University Institute of Applied Economic and Social Research argued that it was not expansionary enough and at best neutral.

With the safe passage of the Budget, Whitlam decided to formalize Cairns' role as the Government's principal economic spokesman by placing him at the Treasury. The announcement was complicated by Cairns' professed reluctance and his absence overseas in November. The impending change was an open secret, since it had been in the air since June. Whitlam's embarrassment and Crean's discomfiture was made acute by the continuing press speculation, avidly fuelled by the Opposition in questions. Whitlam felt he could not make the announcement until he had a firm assurance from Cairns that he would take the job and he was not willing to do this over the international telephone. Whitlam had no desire to humiliate Crean. He had offered Cairns the portfolio of his choice in June, after he had been elected Deputy Prime Minister. Whitlam urged Treasury as the most appropriate, citing the example of Holt under Menzies. Cairns had declined. He said he preferred to keep his existing portfolio of Overseas Trade, augmented with responsibility for the operations of the Australian Industry Development Corporation. However, he had asked for the right to comment on economic matters generally. Whitlam had agreed.

Crean was depressed and disheartened after his treatment by Cabinet and Caucus over the 1974 Budget. He sounded Whitlam out on the possibility of his appointment to the chairmanship of the Commonwealth Bank Corporation, soon to fall vacant. Whitlam was agreeable. Hawke got wind of this proposal, presumably from Cairns, who was the only other Minister privy to it. He objected strongly on the grounds that the Government could not face a by-election in 1974. This was the more surprising because Hawke himself would have been the most likely candidate in Crean's seat of Melbourne Ports. The proposal, however, was dropped. When the decision to make Cairns Treasurer was finally accomplished, the straight exchange of portfolios minimized Crean's embarrassment. Yet it was an unhappy time for Whitlam and Crean. For the Treasury too, it was proving traumatic.

*

The eclipse of Treasury in the last months of 1974 was to have a profound effect on the fate of the Labor Government. The first reaction of some of the senior officials to the rejection of their Budget advice was sullen resentment. They restricted their involvement in the preparation of the Budget to the minimum needed to carry out Cabinet's decisions. They refused to take part in the 'selling' of the Budget, the process of public relations, particularly with the finance writers in the Press Gallery, to secure better understanding of the purpose of the Budget and hopefully, favourable press reaction to it. The Budget Speech is a key part of this process. The speech presented to Crean was curt in its explanation of the Budget to the point of being derisory. A last-minute attempt to provide the barest dressing-up was made in the Prime Minister's Office. Crean's replacement by Cairns transformed Treasury sulkiness to hostility. It was then that certain Treasury officials, known by the code-name of 'Mr Williams', began clandestine contact with some members of the Opposition. For the next six months, the Opposition was closely informed of the Government's intentions, particularly its overseas loan raising activities.

This defection was the ultimate expression of the demise of legitimacy to the Labor Government. Some of the officials in the principal department of state were prepared to use their position and their knowledge to damage the elected Government — and they did so.

This was conduct of an entirely different order from the common practice under all Governments of leaking to the press by public servants in order to push a policy rejected or ignored by a Minister. It cannot be fully explained by the Government's rejection of Treasury views in 1974. That explains the hostility; it cannot explain the treachery. It is impossible to avoid the conclusion that Treasury reacted against the Labor Government in exactly the same way and for exactly the same motives as other power groups — the coalition, the Senate, the State Premiers, the mining corporations. Some people in the Treasury acted as they did because they saw the Labor Government as a threat to their own power, privileges and interests. Cabinet and Caucus rejection of its 1974 Budget strategy was only a minor brush in what parts of Treasury saw as a concerted drive to diminish its power and authority. Whitlam's appointment of his former private secretary, John Menadue, as head of the Prime Minister's Department made this threat vivid. Menadue wished to make the Department an alternative source of economic advice as a counter to the Treasury monolith. In November, Whitlam — not the Treasurer — announced a new economic package devised entirely in the Prime Minister's Department. The ambitions of the Department of Minerals and Energy under Connor, culminating in a proposal to raise $4,000 million overseas, posed a similar threat. Like the other power groups who believed Labor threatened their interests, some Treasury officials justified their conduct by positing a higher national interest and a higher loyalty than that of the elected Government. In 1975, Malcolm Fraser and Phillip Lynch used this justification when admitting that they were receiving information from Treasury. It was, in fact, all part of the same pattern: the denial that a Labor Government represented the legitimate national interest.

The challenge to Labor's legitimacy extended to the House of Representatives itself. There the Opposition pursued a deliberate strategy of disorderly conduct to reduce the prestige of the House and thus the prestige of the Government whose authority derived solely from its majority in that House. In this strategy, they were helped by Labor's choice as Speaker, James Cope, first elected in December 1972, who had survived a challenge from Joseph Berinson in the Caucus elections of June 1974.

The Speakership of the House of Representatives was a glittering prize for James Cope, proud to be known as the boy from Redfern, once Sydney's toughest, roughest inner suburb. He had two qualifications for the post: for years he had trained himself for it by close study of parliamentary procedures and standing orders; and he had a gift for the pointed interjection which punctured pomposity, cooled tempers and defused tense situations both in Parliament and in Caucus. As Speaker, however, he commanded affection but not respect. Handicapped by a weak voice rising to a raucous shrillness under stress, his incessant plea for 'Order, order', against a background of shouts and jeers conveyed to radio listeners a damaging impression of a House of chaos, beyond the control of an impotent Speaker. Throughout 1974, the Opposition pursued a deliberate strategy of disruption. A stronger Speaker may have fared better (and Cope's successor, Gordon Scholes, did restore a degree of control) but Cope was denied any semblance of a fair go. By February 1975, the frustrations on all sides, Cope's own sense of frustration, the Government's frustration at its lack of control over the House, had created a flash-point situation. When the inevitable explosion came, it was over a triviality. Yet it was to have shattering results — for Cope, for Whitlam, but most of all, for Snedden, Fraser and the Liberal Party.

*

By February 1975, the Whitlam Government seemed to be on the verge of a recovery. The Queensland State elections of December 1974 seemed to have marked a low point from which the way could only be upwards. The disastrous result with Labor retaining only eleven seats, one third of its pre-election strength, was the more damaging because Whitlam had thrown himself vigorously and visibly into the campaign. Once he had been the most popular national politician in Queensland. It had all turned to ashes under the unrelenting attacks of Premier Bjelke-Petersen. Whitlam's personal prestige fell further when he went on a lengthy overseas trip, interrupted by his return to Darwin after its devastation by Cyclone Tracy. At The Hague, he made ill-considered remarks about the cause of an accident in which a tanker had destroyed the Tasman Bridge over the River Derwent at Hobart. For the first time in his career, he came perilously close to becoming a figure of ridicule.

Whitlam recovered rapidly from the low-point of January 1975. The national Conference of the Labor Party in Terrigal in early February went off smoothly. When Parliament re-assembled, Whitlam reasserted his ascendancy. Snedden, already under threat from Malcolm Fraser, failed lamentably either to prevent or to justify the move by the New South Wales Premier, Tom Lewis, to appoint a non-Labor senator to replace Lionel Murphy, transferred to the Bench of the High Court. Snedden conceded that this was an unwarrantable breach of a convention established by Menzies in 1950 and followed by all State Governments, Liberal and Labor, since then. This unbroken convention was that a deceased or retiring senator should be replaced by a person of the same party. At a rally at Randwick racecourse in Sydney, designed to assert his leadership with the Liberal rank and file against the looming threat of Fraser's challenge, Snedden declared, with that front of toughness which exposed the weakness of his position:

> I, Billy Mackie Snedden, have no power to determine who that successor (to Murphy) will be. I have no power under the Liberal Party constitution. I have no power under the Australian Constitution. I have stated my view clearly, frankly, without fear and without seeking favour. That is my view and I maintain it.

Whitlam used this quotation in Parliament with devastating effect, provoking Snedden into one of the most curious, damaging and embarrassing interjections in memory. Hansard records:

Mr Whitlam: The first casual vacancy to arise after proportional representation was introduced in the Senate in December 1951. It occurred in Western Australia. A Labor Senator died. The Liberal Premier . . .

Mr Snedden: Come on. Woof, woof! . . .

Mr Whitlam: The right honourable gentleman seems to be more than usually hysterical. I have never known him to giggle so much. He is going ga-ga.

The fatal thing for Snedden was that his nominal followers observed his embarrassment with either resignation or contempt. He had allowed the pressure for an early election to build up, yet daily demonstrated in the eyes of his own Party his inability to provide leadership to match Whitlam in such an election. The Parliamentary in-fighting, intrinsically a side-show, took centre-stage.

On 27 February, in a bitter exchange, a former Liberal Minister, James Forbes, accused his fellow South Australian, Clyde Cameron, of telling 'a monstrous lie'. Speaker Cope dithered about securing a withdrawal, provoking Cameron to say, 'Look, I don't give a damn what you say, I . . .' There was uproar and disorder on all sides. Three times Cope asked Cameron to apologize: 'Is the Minister going to apologize?' Whitlam answered defiantly for him, 'No.' The Speaker then 'named' Cameron, the step which leads automatically to a vote to suspend the offending member. During the division, Whitlam told Cope behind the Speaker's chair, 'If you lose this division, you should resign.' With the Government voting against its own Speaker, the division was lost and Cope left the Chair for the last time. Snedden was saying the truth when he said, 'We have witnessed a matter the like of which has never occurred in the life of this Parliament.' Yet his own conduct in permitting his followers to create an explosive situation in Parliament had contributed directly to Cope's resignation.

It was Snedden who was to pay the highest penalty for Cope's dismissal. Given a superb opportunity for a great parliamentary attack, he failed utterly. His speech on 4 March to his own no-confidence motion was badly-prepared, delivered in the worst Snedden style, tearing a synthetic passion to tatters. Whitlam turned the attack back upon Snedden, asserting that the whole incident was 'the logical end to the pattern of disruption established by the Opposition'. Claiming that Snedden's own performance had precipitated the events leading to Cope's humiliation, Whitlam said, 'He did it for the same reason that he does everything else: this embattled pygmy has to show his failing followers that he is a big boy after all.'

The ominous thing for Snedden was passive acceptance of Whitlam's onslaught by his own followers. As the demolition of their leader went on, his friends sank into deeper gloom; his enemies did not bother to hide their satisfaction at their leader's discomfiture. In King's Hall that day, Anthony Staley, Fraser's lieutenant in an abortive putsch at the end of 1974, gleefully and openly accosted anyone who cared to listen—journalists, Labor members and Ministerial staff (including the author) to deride Snedden's performance. It was the end for Snedden. The Fraser group decided on that day to organize their second move against Snedden. For the rest of the week, they campaigned through the Press Gallery to foment a leadership crisis. The line they took was that Snedden had been mortally wounded by the Cope debate.

Tony Staley had organized an abortive coup against Snedden in November 1974. Ill-planned and ill-timed, a snap challenge in the Party room was defeated. Fraser claimed he had no knowledge of the move on his behalf. In a brief press statement on 27 November, he said: 'I took no part in what occurred this morning in the Party room. I asked no one for support, nor would I do so. What occurred was initiated by other people without my knowledge or encouragement. I support the parliamentary leader of the Liberal Party.' In the *Nation Review*, Mungo MacCallum predicted on 27 December: 'They hate him and distrust him; but sooner or later they'll elect him.' On 16 January 1975, Fraser felt compelled to state: 'In view of rumours and reports which never at any stage were inspired by me of a challenge to Mr Snedden, let me say that I made a statement after the Party room discussion on 27 November. In that statement I made it plain I supported the Party room decision.' On 6 February, Fraser was reported in the *Financial Review* as saying: 'Bill Snedden has my full support. I repeat, as I have said on numerous occasions, that I supported the Party room decision of last November and that I support the elected leader. Despite that, there has been continued widespread speculation that I or colleagues of mine on my behalf are promoting a challenge to Bill Snedden. This is not so. There is no contest. The issue was decided in November.'

Whitlam's parliamentary triumph over Snedden sent Fraser's supporters on the trail again. Their leader was wounded and they smelt blood. His deputy, Phillip Lynch, reached an agreement with the Fraser group: Snedden must go; Lynch would stay. On 14 March, Andrew Peacock issued a statement saying that it was totally unreal 'to ignore the leadership rumours'. 'Mr Snedden should call a meeting and ask for a vote of confidence so that speculation can be ended.' From the day of this statement by Snedden's closest public ally, the pressure on Snedden became irresistible. On 17 March, after a meeting with Fraser and Lynch at which Fraser still refused to state openly that he was a challenger for the leadership, Snedden announced a Party meeting for 21 March. At last Fraser declared his candidacy and gave as the reason: 'The high hopes in February were for an election now or in May to get rid of this government. For some time those high hopes have disappeared.' On 21 March, Fraser defeated Snedden by ten votes for the Liberal leadership.

After Snedden's defeat the question naturally arose whether Whitlam had not made a great political error in using overkill against Snedden and opening the way for Fraser, a far more formidable opponent. On this point, four things should be said. Firstly, Whitlam badly needed a parliamentary triumph to counter the damage done by his conduct over Cope. Secondly, parliamentary debate, based on the adversary system, does not readily allow for subtle adjustments to meet a longer-term political need. The object of a parliamentary debate is to win if you can; there is no known example of a party leader consciously losing a debate to save a lesser opponent; there is sometimes chivalry, but seldom selective mercy. Thirdly, nobody could have foreseen how badly Snedden would muff his chance in the Cope debate. Fourthly, and most importantly, Fraser was in any case determined to make his challenge. A good performance by

Snedden in the Cope debate might have delayed the Fraser challenge; it could not have stopped it. It is the supreme irony of Whitlam's career that two of his greatest parliamentary triumphs led to disaster. In February–March 1975 and again in October–November 1975, he achieved by a sustained and single-handed effort unparalleled parliamentary ascendancy. The result of the first triumph was the emergence of Malcolm Fraser as leader of the coalition. The result of the second was the dismissal of his Government.

*

For the Government, the immediate effect of Fraser's victory was that it relaxed. The threat of an election forced by the Senate receded. At his first press conference, Fraser said:

> The question of Supply — let me deal with it this way. I generally believe if a Government is elected to power in the lower House and has the numbers and can maintain the numbers in the lower House, it is entitled to expect that it will govern for the three-year term unless quite extraordinary events intervene.
>
> I want to get talks about elections out of the air so the government can get on with the job of governing and make quite certain that it is not unduly distracted in these particular matters.
>
> Having said that, let me also say that if there are questions that when an election might be held or when someone might want to do something about it, I would not be wanting to answer that particular question or trailing my coat about what our tactics or approach would be, because if we do make up our minds at some stage that the Government is so reprehensible that all opposition must use whatever power is available to it, then I'd want to find a situation in which we made that decision and Mr Whitlam woke up one morning finding the decision had been made and finding that he had been caught with his pants well and truly down.

After his abortive bid for the Liberal leadership in November 1974, Malcolm Fraser invited a journalist from the *National Times* to interview him at his country property at Nareen, in the Western district of Victoria. It was thoroughly unusual for Fraser to do such a thing; he was sending signals both to Snedden and Whitlam. Neither of them really saw the signals. On 23

December, the *National Times* reported:

> Fraser believes that the Opposition's best chance for overthrowing the government would be if it could force an election in the new year but admits that he sees no opportunity of doing this. Fraser believes that a Federal election should be forced by the Opposition in 1975, but he believes that the timing must be immaculate. He is aware that the Government can delay a Supply until late June and thus stave off possible moves to force an election until then. He is also aware that if this happens the Labor Party stands a better than even chance of controlling the Senate even if it loses control of the House of Representatives. As he says: 'It is not inevitable that there will be an election in 1975 even if it is desirable. Queensland will not be the yardstick. The issues which will count will be those around in mid-year, not those that will be six months old by June. The technical opportunities for the Opposition to force an election are somewhat limited. Supply can be delayed until late June. If it is, Whitlam could dissolve the Reps and only half the Senate and he would have to win a majority in the Senate in only two States to gain control of the Senate. We would have to win a majority in all States to keep control.'

Thus, within six months of the 1974 election, it was plain that all Liberal strategy was once again turned towards one main point — the forcing of a House of Representatives election by the Senate. Whether Snedden or Fraser was leader was immaterial; it was never a question of what was to be done, but only a question of how it was to be done. In the same month, in another interview, Fraser said: 'The way things are done are as important as what is done. The timing must be immaculate.' In December 1974, he had, in fact, served notice on both Snedden and Whitlam. The events of 1975, right up to 11 November, unfolded according to the scenario Fraser had painted in December 1974.

Whitlam and his colleagues made exactly the same mistake as Snedden: they took Fraser's words at their face value, without seeing the escape clause. In Snedden's case, all that Fraser had to do to remain true to his word was to create the conditions in which a leadership contest became inevitable. In the matter of Supply, all he had to do was to create a climate in which almost any failure on the part of the Government could be depicted as 'an extraordinary and reprehensible circumstance'. This he proceeded to do with great skill, powerfully assisted, as he had been in the case of Gorton and Snedden, by his opponents.

21
'Scandal'

Throughout 1975, the Labor Government was subject to an unrelenting attack on its morality. There was a skilful and extraordinarily successful effort to portray it as shoddy, tainted, even corrupt. Almost every important matter that arose was treated as a moral question. The Labor Government was put to trial by standards of personal morality as well as political morality. The whole object of the exercise was to undermine the moral authority of the Government and the moral reputation of Ministers, above all, of the Prime Minister. By this tactic, the word 'affair' became synonymous with 'scandal', and the word 'scandal' was used in Parliament and in the press loosely to describe almost any action of the Government which might be a matter for dispute or criticism. Thus: the Morosi 'affair' was easily elided to 'scandal'; a controversial High Court appointment became the 'Murphy scandal'; these were followed by the loans scandal, the overseas trips scandal, the jobs for the boys scandal, the Cairns scandal, the Vietnamese refugee scandal and, again and again, the 'loans scandal'.

There were sound reasons for this strategy. In the final analysis, the strongest ground on which the Labor Party and the Labor Government stood was its claim to idealism, the assertion by its leaders and the assumption by its supporters that it stood for a set of ideals of service and justice. To take that away is to undermine it at its strongest point. That was the strategy. The campaign on morality stepped up when Fraser became leader of the Opposition.

The politics of 1975 were a continuing effort to find the one 'extraordinary and reprehensible circumstance' which would stick and create a climate in which a minor incident could justify extreme measures. The desired result was achieved not so much by the particular incident which led to the refusal of Supply, but by a cumulative process in which almost any incident could be used as the clinching evidence of the Labor Government's moral, if not actual, corruption.

Fraser's performance in 1975 was one of the most concentrated, single-minded and effective exercises in political destruction ever undertaken in Australian history. His great success was that, in the end, he had the majority of Australians believing not merely what he believed but believing that there *must* be much more in it all than he was willing to say. In the end, for every one who would dare say it, there were ten who believed the Labor Government was corrupt. The constant repetition of the word 'scandal' and its indiscriminate application to any controversial action of the government worked.

It would be possible to write a relevant account of the last year of the Whitlam Government solely in terms of the political consequences of certain official, judicial and diplomatic appointments made in 1974 and 1975. None of these appointments was improper, unprecedented or unconventional. The political consequences were not foreseen and could not reasonably have been foreseen. Yet, collectively, their impact on the fortunes and reputation of the Labor Government was devastating; their consequences for the leadership of both parties were immense.

These crucial appointments were: John Menadue as Secretary of the Department of the Prime Minister and Cabinet; Dr Peter Wilenski as Secretary of the Department of Labour and Immigration; James Spigelman as Secretary of the Media Department; Miss Junie Morosi (Mrs David Ditchburn) as Jim Cairns' Office Co-ordinator; Senator Lionel Murphy to the High Court; and Lance Barnard as Australian Ambassador in Stockholm. None of these diverse appointments was ever seriously criticized on grounds of competence (except retrospectively in the Morosi case); not one was exceptional in terms of precedent; yet all were used as evidence of a unique propensity of the Labor Government for jobbery and impropriety. In the appointments of Menadue, Wilenski and Spigelman, the fact that each at one time had been principal private secretary to Whitlam was urged as a disqualification rather than an outstanding qualification. The Morosi appointment undermined Cairns' standing in Caucus, alienated his closest colleagues, disrupted his office, imposed crushing personal strains on his private life at the peak of his public career, distracted him when his public duties required undivided concentration, broke his self-confidence, transformed his personality, ruined his reputation as an idealist, provoked gross breaches of loyalty within the Treasury, and led, directly and indirectly, to his political destruction.

The Murphy appointment or, rather, the reaction to it, caused the first in a series of breaches of constitutional practice which led, step by step, to the dismissal of the Whitlam Government; it also led indirectly to the overthrow of Snedden as the Liberal leader. The Barnard appointment led to a damaging Cabinet reshuffle and caused a by-election for his seat of Bass, the result of which enormously increased the pressure on Malcolm Fraser to use the Senate to force an election. Above all, these appointments and their aftermath created an atmosphere in which the dismissal of the Government could be seen as an act of moral necessity.

*

Junie Morosi was born in Shanghai in 1931. Her family fled to Manila in the face of the Japanese invasion. She came up through the toughest society in Asia, through its toughest years, the years of the Japanese occupation and its aftermath. In Australia, she made a career for herself in one of the toughest and most competitive fields, the travel business. Through her business association, she met Senator Lionel Murphy's wife. In 1974, she secured a position in the office of the Special Consultant on Community Relations, Al Grassby, former Minister for Immigration, who had lost his seat in the 1974 election. Late in 1974, she was introduced to Cairns who, on 2 December, announced her appointment as his Office Co-ordinator. There is no evidence that before 1974 she had held any special commitment to Labor. Beyond all her undoubted drive, energy, ambition and physical magnetism there was one special quality which held particular attraction for Cairns, as indeed it did for almost anybody who met her. She was a deeply sympathetic listener. Cairns, in his new, lonely and immensely difficult post, felt he needed just that.

As soon as the appointment was announced, controversy erupted. On 4 December, a New South Wales Liberal backbencher alleged on the adjournment that Miss Morosi's business activities were under investigation by the New South Wales Corporate Affairs Commission. Whitlam immediately sent an inquiry to the New South Wales Premier, Sir Robert Askin, asking for information. Askin replied that there was no evidence of impropriety on the part of Miss Morosi or her husband. Another Liberal back-bencher, John McLeay (S.A.), held a press conference to allege that he had information which cast 'grave doubts' on the fitness of the appointment. An anonymous dossier on Miss Morosi's alleged activities was circulated around the Press Gallery. Meanwhile, a document calculated to embarrass Murphy was leaked to the press. It was a letter, obviously tongue-in-cheek, to Gordon Bryant, then Minister for the Australian Capital Territory. In facetious terms, Murphy asked Bryant to provide a Government flat for Miss Morosi, 'a most engaging employee of the Commonwealth', and suggested that Bryant should 'at one stroke exercise your ministerial responsibility'. Murphy had previously asked Bryant if it was possible for Miss Morosi to obtain a Government flat; Bryant said he needed a request in writing — 'but don't write me one of those dull letters full of officialese'. In the event, no flat was made available to Morosi. On 6 December, Morosi announced that she would not take the job with Cairns 'because of the barrage of innuendo'. Cairns refused to accept this on the grounds that it would be a surrender to the press and an admission of guilt. Junie Morosi took up her new duties in Cairns' office in January.

In its 9 December issue, the *National Times* tried to put the matter in perspective:

> The Opposition and the media wanted the public last week to read between the lines and detect a sexual innuendo in the reports of the Morosi affair. Which the public duly did.
>
> Dr Cairns referred to this innuendo in Parliament and Senator

Murphy said correctly that if Miss Morosi was a father of three the fuss would have been less.

It is a convention of the reporting of public affairs that public officials' private lives are left unreported unless they impinge on the public arena. But nevertheless newspapers, television and radio will seize an opportunity, if it is there, to ventilate scandal or gossip. And that, after all, is part of their role. In terms of political tactics, what the Opposition has successfully pursued is a linking of Miss Morosi's name with Dr Cairns and Senator Murphy.

Dr Cairns and Senator Murphy now undoubtedly believe they have been smeared by the media and in a sense they are right.

But they are also rightly on the defensive for other more concrete reasons: for this latest event is part of a pattern of Labor politics since the Whitlam Government took power.

The Government is rightly sensitive about trying to find a cheap flat for Miss Morosi because this is not the first time that Labor appointees have benefited ahead of more deserving cases from government perks. Even when the Whitlam Government has looked at its best, a recurring flaw has spoilt the image. This flaw has been the atmosphere of petty venality, minor nepotism and bureaucratic extravagance ... a strong element in the Labor Government's thinking is 'The others did it' ... But if the Government wishes to indulge itself as previous Liberal-Country Party governments did, then it must pay the consequences. Mr Holt, Mr Gorton and Mr McMahon had the disasters of the VIP aircraft affair, the destructive row over Ainsley Gotto and the allegations associated with Jet Air.

The Labor Government is acquiring a similar reputation.

One of the significant things about this editorial is its date—9 December 1974. The use of the Morosi appointment to denigrate Cairns and, through him, the whole Labor Government, began as soon as the appointment was made. The compression of events in the public memory, and even the memory of some of those who played a part in these events, has led to a rationalization that it was Morosi's actual use of the appointment, and not the appointment itself which caused the subsequent criticism of it. In December, there was no real questioning of her professional competence for the job, if only because nobody had the slightest idea what being 'office co-ordinator' involved. It was only months later when, far from co-ordinating Cairns' office or the relations between his office and the Treasury, she broke up his office and broke down the lines of communication with both the Treasury and his Cabinet and Caucus colleagues, that her unsuitability became manifest.

The greater part of what has been written about Junie Morosi and her association with Jim Cairns is presumably true, because she wrote or said most of it herself. Press harassment and press hypocrisy there certainly was, but for people who complained so bitterly about invasion of privacy, Morosi and Cairns courted publicity with extraordinary indiscretion. So interesting a pair and so stylish a woman could scarcely have escaped attention. But whenever public interest, whether real or press-created, seemed likely to fade, a new interview or magazine article restored them to centre-stage. Most notorious and damaging of all was Cairns' remarkable avowal of 'a kind of love' for Junie, made in a prearranged interview with the most sensation-hungry and anti-Labor of all the Australian newspapers at that time, Murdoch's Sydney *Daily Mirror*. This interview was given during the biennial Conference of the Australian Labor Party at Terrigal, New South Wales. Tremendous efforts had been made by the National Secretary, David Combe, the President, Bob Hawke, Whitlam and Cairns himself to ensure the success of this crucial conference and to make a showing of stability, responsibility and seriousness. The Morosi affair seemed to be dying. Cairns himself had gained considerable prestige from his conduct as Acting Prime Minister (especially in the aftermath of Cyclone Tracy, which had devastated Darwin on Christmas Day) during Whitlam's long, unpopular European tour. All this was undone by four little words. For all its touching naïvety and engaging frankness, Cairns' interview was perfectly calculated to restore the Morosi affair to the front pages in the most embarrassing and ludicrous way. It was this kind of conduct which enraged some of Cairns' firmest defenders and made Morosi the object of unforgiving hatred in the Labor Party—not least among its women supporters.

By the end of 1974, Cairns was passing through a time of deep personal change, a period of self-doubt and self-questioning. He found himself caught up in massive self-contradictions. He was going through a phase of profound personal indecision, yet sought responsibilities which called almost daily for important decisions. He had reached the peaks of power as Deputy Leader of the Labor Party, Deputy Prime Minister and then Treasurer at the very time when he had become deeply disillusioned with the exercise of Executive power. By seeking and accepting these great offices, he committed himself to responsibility for the system at a time when he had never been so alienated from it. He claimed a dominant role in solving the immediate, concrete problems of inflation and unemployment at the very time he was coming to believe in the impossibility of any solution. He found himself at the head of Treasury, the most cynical, arrogant and hard-nosed of departments, at a time when he was developing a theory of economics based unashamedly on the language of love. On 3 September 1974, within weeks of becoming Treasurer, he stated as the Chifley Memorial Lecturer at Melbourne University: 'I have come to a position in life in which I value most as a source, or as a confirmation, the words of men who are simple and real. I have come also to a rejection, and even a fear of men who make basic philosophy complex and intellectual, or who base it upon an untested or abstract assumption. We have learned little from the halls of fame, or from intellectual towers, or even from the campuses that helps to understand how to live or to love ...' The hope of any meeting of minds, much less hearts, between the Treasury officials who were then in a state of virtual mutiny because the Labor Government had refused to accept their recipe of 'a short, sharp shock' to cure inflation and a Treasurer who said 'to me the values of co-operation and even love, become paramount ... What is needed is not just a way to control inflation but a way to reform society so that it may avoid inflation' was slim indeed. Morosi did not cause Cairns' isolation from Treasury; she merely confirmed it. Once he had received the assurances from Premier Askin, Whitlam's refusal to intervene further in the Morosi matter rested on two grounds: he insisted that Ministers were personally responsible for their staffs; and no allegation had been proved against Morosi. The principles were admirable, but the consequences of his neutrality were deeply damaging to his Government. Snedden, unwilling to follow his back-benchers, McLeay and Wentworth, who in the New Year became the hyper-critics of Morosi, said, 'I'm not prepared to say that Miss Morosi should go. That's a matter for Mr Whitlam. Has he got the leadership or hasn't he. Is he prepared to say to a member of his own Ministry: "I want this matter solved?"'

When everything has been said on the Morosi controversy, the inescapable conclusion remains: in the final analysis, it was sexist. She was that most disturbing thing — a woman with influence.

*

The first suggestion that Lionel Murphy should go to the High Court came from Cairns in December 1974. The death of Mr Justice Menzies had created a vacancy. Cairns acted as Murphy's emissary. Whitlam told Cairns that while hitherto he had not considered Murphy's appointment, he now felt that Murphy's work as Attorney-General had been largely accomplished. In particular, he had shown great skill in steering through the Senate the complex and controversial Trade Practices legislation, and through his work on the Family Law Bill he had secured a lasting social and legal reform. The appointment would have been announced in the first or second week of December except for allegations made in the Senate of his involvement with Junie Morosi. Whitlam immediately saw that to announce the appointment in such an atmosphere would only fuel suspicions of impropriety. The appointment was not announced until February. This delay was the first fruit of the Morosi controversy. It was to have enormous consequences. It is very probable that had the announcement been made in December in a calmer atmosphere it would have been made before the disastrous Queensland election and the Premier of New South Wales, Tom Lewis, would not have been able to take the extraordinary action of refusing to appoint a Labor senator to replace Murphy. It was this first breach of convention which led step by step to the dismissal of the Whitlam Government.

Partly because of Snedden's decent reluctance to exploit the

Morosi controversy, it remained a sideshow in the parliamentary circus. Nor did Fraser, after he deposed Snedden, find it particularly suitable for his ends. So for all its political ramifications, it was left largely to the press, or a section of it, to keep the pot boiling.

*

In the attitude of the press in all these matters, from Morosi to the loans affair, there was a special factor at work. These were the months after Watergate and the example of the *Washington Post* and the achievements of Bob Woodward and Carl Bernstein in 'investigative reporting' became the fashion. Australian journalism has virtually no tradition of investigative journalism. The press proprietors have never been prepared to pay the price for it in law suits, in unpopularity and in the time and money needed to allow their reporters to spend weeks or months in hunting down an important story which may never be printed. But the events of 1975 provided opportunities for the appearance of investigative reporting without any of its substance. There was no true investigation; there were only leaks. There was no financial risk in the work of the reporters; there were only a few pay-outs for documents. For instance, in the whole of the so-called Khemlani affair, the only Australian money exchanged was that paid by newspapers to get hold of Khemlani's only real asset — his documents. In the post-Watergate atmosphere, a leaked document took on the appearance of a true investigation. All sorts of documents were freely available and there was a rush to print without much investigation into their authenticity or relevance. As Whitlam put it: 'The *Age* will not query them; it will just condemn.'

There was often in this period a wide gap between the news columns of the papers and their editorials. It is easy to dismiss newspaper editorials as impotent and irrelevant. The fact remains, however, that when a great newspaper — and the *Sydney Morning Herald* and the *Age* are great newspapers by any standard — decides to exert itself greatly it must have an influence, if not a power. For whether its readers are affected or not by its editorials, the people to whom those editorials are first directed will certainly be influenced by them. And they are first directed to politicians and public servants, who are the very people who do read them, whether the generality of readers do or do not. The key to personal political influence is access to those with power. Editorial writers of five Australian newspapers, the *Financial Review*, the *Age*, the *Sydney Morning Herald*, the *Canberra Times* and the *Australian*, in that order of importance, enjoy proxy access daily. The editorial writers also enjoy a source of influence the political commentators of the same papers do not — the mystique of anonymity. There is an ineffable wisdom about anonymity. The most hardened politicians, the most experienced public servants are not altogether immune from this influence. The effect on television journalists is even greater: without the self-confidence which comes with a newspaper training, they fall completely under the spell of what they read in the newspapers.

The technique of using leaked documents to pass for genuine investigative journalism received its first test on a matter even more far reaching than the Morosi controversy, the Murphy appointment or the loans affair. It was first tested — by Fraser as well as the press — on the greatest matter of our times. The first test came on Vietnam. And it was the conduct of the Whitlam Government in the last days of the Thirty Years War in Vietnam which was to earn the condemnation of Fraser, using the words of the *Sydney Morning Herald*, as 'the greatest scandal in the history of Federation'.

22
The Last Casualty

It may seem grotesque to link such matters as the controversy over Morosi with the cataclysm of Indo-China. Yet it is part of the tradition of grotesqueness which marked Australia's Vietnam adventure and the Australian debate on it from beginning to end. And it was part of the pattern to destroy the moral authority of the Labor Government. For one last time, the sufferings of the people of Vietnam were used as the small change of the Australian domestic political brawl. Vietnam degraded everyone who touched it. With obscene irony, its last casualty in Australia was the Labor Party which, however ambiguously, however ambivalently, had opposed intervention on the grounds of its immorality and its military futility. When the inevitable happened, when the military failure in Vietnam was complete, it was to be a Labor Government which found itself under attack for the immorality of its conduct in Vietnam.

By March 1975, North Vietnam was poised for final victory to bring about unification — the result obtainable by elections in 1955, if they had been held in accordance with the Geneva Accords of 1954. Both sides had consistently breached the Paris Accords of January 1973. On neither side was there any will to observe them. The only result of the Paris Accords of January 1973 was their intended purpose — to allow a decent interval between American withdrawal and the collapse of the Thieu Government; a decent interval for Nixon and Kissinger to justify their policy. As fighting in the Northern and Central Provinces of Vietnam intensified, the President, General Nguyen van Thieu, hastened his own end by ordering his forces to withdraw south to the defence perimeter around Saigon. In the wake of their headlong flight streamed thousands of refugees in the path of the unopposed and uniting forces of North Vietnam and the National Liberation Front. A Central Intelligence Agency briefing made to the Australian Embassy in Washington summed up the situation at the beginning of April in these terms:

> The analyst said that there is now uniform agreement in CIA that the communist side will shortly achieve complete victory in South Vietnam. Differences remained on the question of how long it would take them -- estimates ranged from a week upwards to three months -- and as to the exact circumstances of the final defeat of the Government of South Vietnam (GVN). The analyst's personal feeling was that Saigon was likely to be in communist hands by the end of the month.
>
> The CIA analyst represented the GVN's defeat in terms of psychological collapse, stemming from Thieu's decision to undertake withdrawals in The Highlands and the North without any advance preparation, which had shattered the confidence of the Army of the Republic of Vietnam (ARVN). What had followed was not a victorious communist military sweep down the coast but the progressive collapse of the GVN and ARVN. There had been little heavy fighting along the way, except for ARVN's stand for a few days at Nha Trang. The North Vietnam Army (NVA) tactic has been to follow up quickly and to give the developing situation 'a military nudge' where necessary to sustain its momentum.
>
> The current balance of forces around Saigon was not entirely unfavourable to ARVN but this fact was of secondary importance to the disintegration in ARVN's morale. The analyst thought that, in the Saigon region, it would take about two weeks for the collapse of morale to reach the point where it sapped the will to resist further. The analyst left open the question of whether the final defeat would come about as a result of a major communist military thrust or through political collapse in Saigon.
>
> As to the political scene in Saigon, the analyst said that Thieu was completely discredited in the eyes of his senior generals but it would now be hard to find a candidate to replace him. Most current political activity was on the part of the traditional leaders in the National Assembly and among the Buddhists but they had little to offer in present circumstances. The political realities of the hour were beginning to sink in. While the analyst agreed that

there was considerable potential for anti-American and anti-foreign feeling in Saigon, such feeling has not so far been evident. The predominant reaction was one of despair.

The analyst thought that the GVN would not undertake further operations to evacuate refugees as there was nowhere to send them other than Saigon and they were not wanted in the capital.

It was at this stage that the Ford Administration began to mount a refugee-rescue operation on the grounds that a communist victory would inevitably mean a 'blood-bath' — that same massacre of Catholics and others which Whitlam had been told in 1954 would cause a split in the Australian Labor Party.

As late as mid-March, before Thieu's self-destructive panic, the speed of total collapse was not foreseen in Canberra or Washington, and probably not even in Hanoi. Indeed, the Provisional Revolutionary Government, representing the pro-communist groups in South Vietnam had renewed on 1 April its offer to join in any Saigon government which excluded Thieu. In March, therefore, an attempt to secure a political settlement, however temporary, as an alternative to a total military solution still seemed within the range of possibility. On 13 March, Whitlam wrote two letters, one to President Thieu and one to the Foreign Affairs Minister of the Democratic Republic of Vietnam, Nguyen Duy Trinh. His letter to Thieu said:

> Your Excellency, I have the honour to reply to Your Excellency's letter of 25 January 1975 concerning the situation in Viet-Nam, and the implementation of the Paris Agreement.
>
> Although the Australian Government was not one of the signatories to the Paris Agreement, we have consistently supported its implementation and have looked to all of the parties to the Paris Agreement to abide by its terms and to ensure that another opportunity to achieve an enduring peace in Viet-Nam is not lost.
>
> Unfortunately, Australia's hopes have not been realized. There has been continued lack of progress in the implementation of both the political and military provisions of the Paris Agreement, now more than two years old, and continued fighting in South Viet-Nam.
>
> The responsibility for this situation cannot be placed on North

Viet-Nam alone. I trust that your Government will so act as to enable the violence which is causing so much suffering to be reduced, and as soon as possible brought to an end. In particular, I trust that it will for its part take early and positive steps towards the establishment of the National Council of National Reconciliation and Concord as specified in Article 12 of the Paris Agreement, and will observe Article 7 of the Paris Agreement concerning the piece-for-piece replacement of armaments, munitions, and war materials.

The Australian Government confirmed to your Embassy in Canberra on 11 November 1974 that it supports your Government's call for the resumption of the peace talks both on the political level at La Celle-Saint Cloud and on the military level at the Two Party Joint Military Commission talks in Saigon. It is my Government's hope that it will be possible for all parties in Viet-Nam to set aside the deep-seated animosities caused by the tragic events of the last two decades and to work together — whether within the established machinery or under new arrangements — towards a peaceful and enduring settlement in Viet-Nam.

The letter to Trinh followed the Thieu letter virtually word for word until the fourth paragraph which said:

The responsibility for this situation cannot be placed on South Viet-Nam alone. I trust that your Government will so act as to enable the violence which is causing so much suffering to be reduced, and as soon as possible brought to an end. The Australian Government was disappointed at the cessation in 1974 of the peace talks both on the political level at La Celle-Saint Cloud and on the military level at the Two Party Joint Military Commission in Saigon, and supports the resumption of these talks without pre-conditions. It is my Government's hope that it will be possible for all parties in Viet-Nam to set aside the deep-seated animosities caused by the tragic events of the last two decades and work together — whether within the established machinery or under new arrangements — towards a peaceful and enduring settlement in Viet-Nam.

If the speed and finality of subsequent events made these expressions redundant and apparently naïve, it should be recalled that on 9 April France's President Giscard D'Estaing issued a declaration in similar terms. Following the despatch of Whitlam's letters, the Foreign Affairs Department discussed the need for some kind of follow-up. On 2 April, Whitlam, acting as Foreign Minister in Senator Willesee's absence overseas, approved drafts of instructions from the Department to the Australian Ambassadors in Saigon and Hanoi. This was essentially a Departmental initiative.

The genesis of what became known as the 'secret cables affair' —what the *Sydney Morning Herald* was to call 'the greatest scandal in the history of Federation'—lay in an internal submission within the Department of Foreign Affairs. In the submission, dated 27 March, the chief of the North and South Asia desk wrote to the Secretary, Alan Renouf, suggesting that: 'For Australia to go on simply calling for strict adherence to the Paris Agreements and an end to bloodshed . . . serves only Saigon's purpose (to keep Vietnam divided, which is not Hanoi's purpose and should not be Australia's) and fails to recognise that military pressure is the only way in which the status quo can be changed.' The submission went on to recommend approaches to be made to Saigon and Hanoi to follow up the Prime Minister's letters of 13 March:

> In private approaches to the DRV (Democratic Republic of Vietnam) and PRG (Hanoi and the Vietcong), Australia should say that—
> (a) we are sympathetic to the DRV/PRG aim of having in Saigon a government which will genuinely negotiate for reunification;
> (b) we appreciate that Thieu has given no indications that he is willing to do that;
> (c) we recognise that military pressure seems to be the only realistic means at present of bringing Thieu, or a successor regime, to that point;
> (d) but we continue to trust (see Mr Whitlam's letter of 13 March) that the parties to the conflict will so act as to enable the violence which is causing so much suffering to be reduced, and as soon as possible brought to an end;
> (e) we also continue to hope (Mr Whitlam's letter of 13 March) that it will be possible for the parties to the conflict to resume working together—whether within the established machinery

or under new arrangements — towards a peaceful and enduring settlement in Viet-Nam;

(f) while Australia's interests and influence in South Viet-Nam are limited, we shall be urging on Thieu the necessity of carrying out in good faith the provisions of Chapter 4 of the Paris Agreements, especially Article 12 on the NCNRC (National Council of Reconciliation) and Chapter 5 on Reunification.

In private approaches to the RVN (Republic of Vietnam), Australia should —

(a) reaffirm the hope (Mr Whitlam's letter of 13 March) that all parties in Viet-Nam will resume working together — whether within the established machinery or under new arrangements — towards a peaceful and enduring settlement in Viet-Nam;

(b) express the conviction that a prime requirement is for the RVN to carry out in good faith the provisions of Chapter 4 of the Paris Agreements, especially Article 12 on the NCNRC, and Chapter 5 on Reunification;

(c) state the view that until that requirement is met there is no chance of there being an early end, or even significant reduction, in the violence which is causing so much suffering.

In private approaches to the United States, Australia should —

(a) ask whether the USA really believes that South Vietnam will go on getting into the indefinite future the external assistance without which it cannot maintain itself as an independent country;

(b) ask whether the USA really believes that an independent South Vietnam is vital to the USA's security, or that a reunified Viet-Nam would be a disaster for the USA or even the region;

(c) ask whether it would not be better for the USA to recognise now that it cannot go on propping up Saigon for ever, and to recognise that the USA's allies, far from worrying about the USA's credibility elsewhere if it bows to the inevitable in Viet-Nam, will be relieved that the USA has got itself out of an impossible predicament and will have the time and energy and resources to devote to more manageable and more serious crises in more important parts of the world;

(d) ask whether the USA could not persuade the Saigon Government to carry out in good faith the provisions of Chapter 4 of the Paris Agreements, especially Article 12 on the NCNRC, and Chapter 5 on Reunification, and say that Australia will be urging that on Saigon as the first requirement for securing a peaceful and enduring settlement in Viet-Nam.

The messages sent by the Department on 2 April were drawn from this submission. The message to Saigon said:

> We wish you to take every opportunity in discussion with members of the RVN to reinforce the position expressed in the Prime Minister's letter of 13 March to President Thieu that:
>
> (a) All parties in Viet-Nam will resume working together — whether within the established machinery or under new arrangements — towards a peaceful and enduring settlement in Viet-Nam.
>
> (b) A prime requirement is for the RVN to carry out in good faith the provisions of Chapter 4 of the Paris Agreements, especially Article 12 on the NCNRC, and Chapter 5 on reunification, and to state the view that until that requirement is met there is no chance of there being an early end, or even significant reduction, in the violence which is causing so much suffering.
>
> 2. You should also make use of the Prime Minister's press statements to say (see separate telegram).
>
> 3. We should like you if and as circumstances permit to make this Australian position known to South Vietnamese who are not members of the Government but who might yet play an influential political role.
>
> 4. Please keep us informed of what you have been able to do.

The instructions to Hanoi were:

> We wish you to make representations as soon as possible to the DRV and PRG at a suitably senior level to reinforce the views expressed in the Prime Minister's letter of 13 March to DRV Foreign Minister, Mr Trinh, and to state the Australian attitude towards the present fighting in South Viet-Nam and the implementation of the Paris Agreements.
>
> 2. In addition to handing over a copy of Mr Whitlam's statement made today, you should say that
>
> (a) We would like to see in Saigon a Government which will genuinely negotiate for reunification as provided for in the Paris Agreements.
>
> (b) We appreciate that Thieu has given no indications that he is willing to do that, and the Australian Government understands the sense of frustration which has given rise to renewed recourse to military pressure on Thieu although it cannot condone that

recourse. In this connection, the Australian Government believes that it would have a most favourable effect if the PRG were to make it clear that recent military operations had had the aim of applying pressure to secure the observance of the ignored political provisions of the Paris Agreements and that the operations would cease when satisfactory assurances were offered by Saigon that these political provisions would be implemented.

(c) We shall be urging on the Thieu Government the necessity of carrying out in good faith the provisions of Chapter 4 of the Paris Agreements, especially Article 12 on the NCNRC, and Chapter 5 on reunification.

(d) We continue to hope (Mr Whitlam's letter to DRV Foreign Minister, Mr Trinh, of 13 March) that it will soon be possible for the parties to the conflict to resume working together — within the framework of the Paris Agreements or under new arrangements — towards a peaceful and enduring settlement in Viet-Nam, and that the parties to the conflict will so act as to enable the violence which is causing so much suffering to be reduced, and as soon as possible brought to an end.

3. You may also confirm with the DRV and PRG that the Australian Government is contributing through international organisations such as the UNHCR (The Council for National Unity and Reconciliation) (with whom the PRG has been in contact) to humanitarian assistance throughout South Viet-Nam, and that the Australian Government stands ready to contribute generously to the longer-term task of post-war reconstruction.

4. We appreciate that the foregoing approach may be interpreted by the DRV and PRG as simply an attempt to cut across an inevitable military victory. The Acting Minister nevertheless considers that the approach should be made at this point of the war and notes the continuing statements to you and publicly about the PRG desire for a negotiated settlement conforming with the political provisions of the Paris Agreements.

5. Please advise us of the DRV and PRG responses.

These Departmental instructions were what subsequently became known as the 'secret cables'. They were stamped 'confidential', one of the lowest classifications. In Australia, copies were distributed to the Prime Minister's Department, the Defence Department and the Joint Intelligence Organization. Abroad, the cables were copied for the Australian posts in

Washington, New York, London, Paris, Moscow, Vienna, Teheran, Peking, Bangkok, Jakarta, Kuala Lumpur, Singapore, Manila, Hong Kong, Rangoon, Vientiane, Tokyo and Wellington. Thus did the Whitlam Government conduct its conspiratorial and clandestine operations.

The initiative for these cables came from the Department of Foreign Affairs. Whitlam approved them but had not suggested them. There is no suggestion that the final responsibility for them rests anywhere else but with the Whitlam Government; but it is a superb irony that the last row over Vietnam started in the Department. From beginning to the end, it was so much the Department's war.

It was scarcely remarkable that these cables fell into the hands of the Opposition. The 'shadow Minister' for Foreign Affairs, Andrew Peacock, had with Whitlam's knowledge and approval kept contact with the Department. In the first week of April, Peacock visited Saigon, Bangkok and Singapore to investigate the refugee problem. On 8 April, he put this question to the Prime Minister in Parliament: 'I refer to Government statements that the Government has urged both sides in the Vietnam conflict to end the current hostilities and abide by the Paris Peace Accords of 1973. What precise representations have been made? Were, in fact, the same representations made to each side?' Whitlam's answer was: 'There have been communications certainly in the last three or four weeks, but there were also communications two years ago in the middle of 1973, and they have been to both sides. They are *not* in the same terms because while it is true that there have been gross breaches of the Paris Accords by both North and South Vietnam, and by the Provisional Revolutionary Government and perhaps by the United States of America, nevertheless the breaches have not been all of the same kind. The Government, I repeat, has on several occasions communicated, sometimes in writing, sometimes orally, with both sides.' The next day, in reply to a follow-up question from the Leader of the Opposition, Whitlam said: 'I have noticed suggestions that there was some difference in the tone of communications I sent to Hanoi and Saigon. There is no truth whatever in the allegations that the Honourable Member for Kooyong (Peacock) has made. The communications I sent to Hanoi and Saigon were substantially the same.' Whitlam then went on to specify where and why his 13 March letters had differed. In the Senate, the Foreign Affairs Minister, Senator Don Willesee, answered a similar question asserting: 'I have seen the cables . . . I had read them before and I have read them again this morning . . . there is not one atom of truth in Mr Peacock's statements.' There that particular matter seemed to rest. None of the Departmental officers involved in the drafting of the documents expressed the slightest concern that the Ministerial statements contained material for allegations of impropriety or that the Parliament might have been misled in any way. Nor was there any need for such warning – the prime responsibility of a loyal service – because on any objective reading, no such allegations could be sustained. But this was not a time or a matter marked by objectivity. The CIA prediction of Saigon's collapse by 'the end of the month' was coming true. The Australian press, particularly the *Age* and the *Sydney Morning Herald*, discovered that Vietnam was indeed a question of morality, as Cairns had pleaded ten years before. Now the question of morality turned on the fate of the refugees. Then, right at the end, was raised once more the oldest of all themes used to justify foreign intervention in Vietnam, a shrill echo across thirty years of blood and destruction – that a communist victory in the South would mean a bloodbath, the massacre of millions.

Throughout the press and Opposition treatment of the refugee question, there was an attempt to claim that Whitlam personally was obstructing efforts by the Australian Embassy in Saigon. Michael Richardson, the *Age* correspondent in Saigon, wrote an article about 'how he was ashamed to be an Australian' and implied that a compassionate Australian Embassy was being impeded by a callous Australian Government. It was throughout implied that the stringency of Australian Government requirements, imposed at Whitlam's behest, and not the difficulties of the actual situation, was placing in jeopardy the lives of Vietnamese to whom Australia had an obligation.

On 1 May, Ambassador G. J. Price sent this message to Canberra:

The majority of Vietnamese citizens who fitted into categories set by the Australian Government and who were carried out by R.A.A.F. C130 did not have valid Vietnamese travel documents and therefore could be said to have departed Vietnam 'illegally', even though they had adequate travel documents to enter Australia.

2. As you know we used Embassy cars bearing diplomatic plates to transport Vietnamese direct to waiting R.A.A.F. C130s in the belief that the chances of an Embassy car being checked by police was lessened. As it happened our gamble came off but if any Embassy car had been stopped at a police checkpoint we could have been in trouble.

3. When Australian journalists got wind of the operation (ABC, Denis Warner, Michael Richardson, Bruce Wilson) they gave us assurances that they would not mention our efforts in any way in case publicity jeopardized the chances of other Vietnamese who were eligible for uplift by Australian aircraft. No Australian journalist to our knowledge broke this agreement. Given the climate of Australian opinion we believe that the emphasis on the 'legality' of our Vietnamese departures should be lessened in case journalists are provoked into reporting that some of the refugees were not approved for travel by GVN. This was particularly the case with Vietnamese fiancés and wives.

4. If we had not acted in this way, we would have got no Vietnamese out at all, or maybe only a very few.

Meanwhile, Whitlam had given authority to his second Private Secretary, Michael Delaney, to organize flights to Australia of orphans and refugees. The end result was that Delaney, acting on Whitlam's authority, brought to Australia nearly a thousand Vietnamese, and supervised the entry of nearly as many after the fall of Saigon as were later allowed to enter Australia in the first year of the Fraser Government. Fraser called for the entry of 50,000 refugees. In April 1975, 359 evacuees arrived from Vietnam. In the months remaining to the Whitlam Government, a further 667 arrived in Australia. In 1976, the Fraser Government admitted 399.

On 29 April, ten years and one day after Menzies announced Australia's combat commitment to South Vietnam, Saigon fell. But the matter which dominated Australian attention was not the defeat of the Saigon government but the 'duplicity' of the Australian Government. After a decent interval, the cables which Andrew Peacock had seen early in April were given to the *Age*. Asserting that the 'confidential' departmental instructions were 'the Prime Minister's secret messages', the *Age* asked: 'On the basis of what vital national interest, what high principal of statecraft, can the Whitlam Government justify the apparent duplicity and callousness of its recent Vietnam policy . . . These cables . . . provide striking evidence that the Prime Minister deceived — and we would put it no higher than that at this stage — Parliament.' The *Sydney Morning Herald* had no such reservations. Its editorial of 30 April stated: 'The Government We Cannot Trust'. Then in one of the longest editorials ever to appear in Australia's oldest newspaper and the longest ever on any aspect of the war in Vietnam, using the full length of its editorial page, a treatment reserved for events of supreme importance, it stated boldly:

> The Prime Minister has lied to Parliament. He has deceived the Australian people. He has abused their trust in him . . . His duplicity has been damningly exposed by the publication (unauthorised) of secret cabled instructions sent by Mr Whitlam to our Ambassadors in Hanoi and Saigon. Their publication brings into the open the *gravest political scandal since Federation* (Author's italics). What is involved goes far deeper than questions about the correctness or rectitude of Mr Whitlam's foreign policy . . . What is involved is the trustworthiness of the Prime Minister and his Government. The text of the cables shows that Mr Whitlam cannot be trusted. The Herald believes that Mr Fraser now has a duty to withdraw his undertaking not to force the Government to an election . . . The Government should be forced by all legal means to answer at the Bar of the people . . . Throughout this entirely disgraceful episode the Prime Minister has said one thing and done the other. He has lied to prevent Parliament and the people from knowing what he has done or in what direction he was taking Australian foreign policy. His actions have been those of dissimulation and duplicity. Neither he nor the Government . . . are any longer to be trusted. A Government which cannot be trusted, which abuses its power and its command of secrecy, forfeits its right to govern. It should be brought down.

Thus did the *Sydney Morning Herald* first set the line which would doom the Labor Government—that, on moral grounds, it should be destroyed by any legal means. The constant reiteration of words like 'lie', 'duplicity', 'scandal', 'deceit'; the glossing of 'secret' into 'sinister'; but above all, the assertion of personal immorality and corruption and the consequent implication of total unfitness to govern was used to justify almost any means to destroy such men and such a government. The pattern of 11 November was set firmly in April. The search for the 'extraordinary and reprehensible circumstance' had begun in earnest.

The response of the Leader of the Opposition, who had been privy to the 'secret cables' for the best part of a month, was, by comparison, restrained. He merely said that the Prime Minister, then at the Commonwealth Heads of Government meeting in Kingston, Jamaica, should immediately return home to face these charges. He said that Parliament, then in recess, should be assembled immediately to discuss 'the deceit and duplicity of the Whitlam Government'. He said: 'He has misled the Australian Parliament and people ... he has taken sides with the Communist North Vietnamese regime in its aggression against South Vietnam ... On all the evidence available it appears that a great tradition of Parliament has been breached by an Australian Prime Minister.'

The parliamentary debate proved an anti-climax, as such pre-publicized affairs almost always do. Whitlam was easily able to dispose of Fraser's charges and little more was heard of Fraser's call for the entry of 50,000 Vietnamese. And nothing more was heard about the 'lies', the 'duplicity', the 'deceit' on Vietnam. But the damage had been done.

During his visit to Washington after the Commonwealth Heads of Government meeting at Kingston, Jamaica, Whitlam addressed the American National Press Club on 8 May:

> In the wake of the remarkable events in Indo-China, all of us—leaders in my calling, commentators in yours—are in the process of reassessing basic policies and relations. For the United States in particular, this is bound to be a difficult and perhaps a painful process. In that reappraisal the last thing the Government of the people of the United States need are sermons and homilies from

foreigners. Certainly, you don't need them from an Australian. It is true that I happen to lead a political party which strongly opposed the intervention in Indo-China; it is also true that I am the Prime Minister of a nation which for many years supported the intervention and encouraged the escalation of the war. Whatever recriminations we might have at home, it is no role for an Australian Prime Minister to lecture the United States.

It is, however, very necessary that we should prevent the creation of new myths about what went wrong. And to do that it is necessary to look at past mistakes — mistakes in which both countries shared. The great danger is that in an atmosphere of deep emotion and recrimination engendered by the suddenness of events in Indo-China, we should fasten upon explanations and self-justification and over-simplifications which would ensure a return to, a repetition of, the great mistakes of the past. We have, in particular, to resist the same sort of myths which developed after the revolution in China. Those myths, those distortions of reality, perverted our relations with China for more than a generation. They led directly to the débâcle in Indo-China.

We should have no truck with any new variation of the 'stab in the back' theory — that the war in Vietnam was lost not in Saigon but here in Washington. The truth is that the United States did not 'lose' Vietnam, anymore than she 'lost' China. Vietnam was not America's to lose. What was defeated was not the United States and her allies but a policy of foreign intervention which was bound to fail. There was no time in the past thirty years when such a policy could have succeeded. The tragedy for us all, but above all for the people of Indo-China, is that a policy so manifestly doomed from the beginning should have been carried through for so long.

For many years to come, people like us, Americans, Australians, politicians, journalists, will be examining in arguments, in articles, in speeches, in books what happened in Indo-China. This is as it should be. For so great a disaster, so great a mistake, such great suffering, cannot be easily dismissed or even forgotten. We shall all have to live with it for the rest of our lives. Yet even so, we have to go on to the future. In the two years since I last spoke to you America has undergone a vast domestic, as well as a vast international, catharsis. With those profound traumas behind you, with their bitterness and misery being purged away, what better time to profit by experience and build on the true strengths of American democracy and American idealism? Here is an opportunity—not just for America but for all of us—to end our long preoccupation with military alignments in Asia, our ideological confrontations, our cold war hang-ups, and open a new chapter in Western co-operation. Let the deeper issues of poverty, over-population and mal-distribution of the world's wealth assume their proper importance in our hearts and minds. These are the real problems of Asia. These are the real problems of the world. These, I trust, will be the real concerns of the United States. With your great tradition of moral leadership, your unexampled generosity, your vision, your energy, your sheer zest for accomplishment, you will find new inspiration in this task—a task in which Australia will be a ready and a willing partner.

The Democratic Majority Leader in the United States Senate, Senator Mansfield, who had warned Whitlam a decade before of mounting Congressional opposition to the war, had this speech read into the Congressional Record—'a speech from a great friend of America, a critic of America's war'. Thus, the epitaph on Australia's involvement in Vietnam. The final episode had a double symmetry. For the last time, Vietnam was used as part of the armoury of Australian domestic political warfare. And at its end, an Australian Prime Minister placed Australia's involvement in the context in which it had always lain—the American-Australian alliance. It had come full circle. And within that circle lay 500 Australians dead, 50,000 Americans dead and unknown millions of dead and maimed in Indo-China.

As a 'scandal', the Vietnam cables affair was easily exploded when the plain facts were revealed. Yet, however flimsy, it had done great damage. Most importantly for the Opposition, it had served as a full rehearsal for the big show—the Great Loans Affair.

23
Loans

For all its subsequent complications and distortions, the essential elements of the 'loans affair' were simple. So too were the motives of the Ministers principally involved — Rex Connor, Gough Whitlam, Lionel Murphy, Jim Cairns and Clyde Cameron. Their intention was straightforward: to raise a substantial overseas loan to finance a range of projects, some urgent, some long-term, in accordance with long-standing Labor policy of public participation in national development. One possible means was also plain: to tap the new source of international finance — the so-called petro-dollars — created by the oil price rise in 1973. In pursuit of these purposes, the Ministers involved went through every legal and official channel that the law required. One of the greatest myths surrounding the loans proposals is that it was cooked up by a handful of Ministers who kept officials in ignorance of their aims. While it is true that Treasury opposed the venture outright from the outset and set out to torpedo it, and while it is true that many of the officials involved were sceptical or uneasy, official involvement at the highest level was deep and close throughout. Yet this single proposal to raise a substantial overseas loan led to the destruction of the Labor Government.

At the centre of the loans affair there is a vacuum. The one indisputable fact is that there never was a loan. That is the real enigma — how a government could be brought down because of something that never happened. The 'affair' certainly happened; but the loan itself did not exist. It is this void at the centre which makes all accounts of the loans affair ultimately unsatisfactory, for they must deal with a series of events which themselves focus on a non-event.

There is a story of how Dr Evatt once dealt in Caucus with a demand for attention from the Irish-born member for West Sydney, Dan Minogue: '*Mr Chairman, I want to raise the matter of the pinsioner I wrote to you about.*' Impatient to close the meeting, Evatt waved him down, 'Yes, Dan . . . I know all about it, and I'm looking into it.' '*But all I wanted to say, Mr Chairman, was don't worry. The litter was niver sint.*' The loans affair dominated Australian politics throughout 1975 and led to the destruction of the Whitlam Government. It was described by the Deputy Leader of the Opposition as 'the greatest scandal in the history of Australia'. Yet the loan was '*niver sint*'.

No decision by any government has been better documented. No loan-raising effort by any government has been so thoroughly discussed. The most private and confidential communications have been blazoned in the press and tabled in Parliament. The advice of departments has been published in books. The Queensland Government sponsored an international investigation. Yet nothing has emerged to change the basic facts: between December 1974 and May 1975, the Australian Government attempted to raise a substantial loan in petro-dollars; those efforts failed. At all stages in the abortive negotiations, Australian Government officials were deeply involved; at no stage between November 1974, when the possibility of raising a loan was canvassed, to 20 May, when the Executive Council authority to raise the loan was revoked, did any Minister take any private action without the knowledge and involvement of his officials. In the end, as Whitlam said, 'not a cent was paid'. The only cost involved was the cost to the reputation of the Government. That cost was to be immense — it was government itself.

*

By November 1974, Rex Connor had emerged as the third most powerful man in the Government. Three years before, he had been an embittered man, cheerfully predicting doom for Australia, the Labor Party and, in particular, Whitlam. He had been beaten for the 'shadow ministry' after the 1969 election. He blamed Whitlam for this defeat on the ground that Whitlam had failed to use his influence with the new members. Whitlam told the press Caucus had simply chosen a younger team. Connor thereafter seemed an implacable enemy but a spent force; yet he kept a sense of his destiny alive within him. His huge frame seemed to hold down an elemental force. Perhaps more than any other member of the Labor Party, he had a vision. It was deeply patriotic; yet private and deeply personal. It was a vision of Australia Unlimited — an Australia richer and stronger than the United States, the vision of the Australians of the booming 1880s. In *The Rush That Never Ended*, Geoffrey Blainey writes: 'The old gold trail that stretched late in the last century from the Pacific to the Indian Ocean was followed again (in the 1950s), promising copper and bauxite and lead and uranium and iron. The rush to Ophir had often faltered but never ended.' Connor's vision was of the new Ophir, an Ophir in which modern technology and Australian skills would be the basis for a bright, new Australian civilization. As Minister for Minerals and Energy, Reginald Francis Xavier Connor dealt in millions and dreamt in billions. The mining companies complained of lack of communication with the Government. Connor offended them by call them 'hillbillies'. He was only echoing Winston Churchill, who said at the time of the General Strike in 1926: 'I thought that the most stupid men I had met were the coal union leaders — until I met the coal mine owners.' Yet Connor understood their language better than any politician since McEwen. In many ways, he was the Labor Government's counterpart of the Western Australian mining magnate, Lang Hancock, with the same mixture of toughness and imagination. From opposite ideological and political standpoints, they had the same vision. They both preached bigness. In 1971, during a tour of Hancock's Western empire, a member of Whitlam's staff, Lorraine Dwyer, commented on the beauty of a bank of desert flowers. Hancock rumbled, 'The only lovely flowers here are iron ones.' For both, there was no limit to what could be achieved with iron, coal, bauxite, gas, uranium. For Hancock, it was a matter of what individuals through corporations could achieve; for Connor, it was what the Australian community through governments could achieve.

Connor was to defend his handling of the loans affair in these words: 'Throughout my years as a Minister of the Crown I have stood in the path of those who would have grabbed the mineral resources of Australia. I have no apologies whatever to make for what I have done. It has been done in good faith; it has been done in honesty. I fling in the face of the little men of the Opposition the words of an old Australian poem:

> Give me men to match my mountains
> Give me men to match my plains
> Men with freedom in their vision
> And creation in their brains.'

That he was such a man, that he was *the* man to match the vision, Connor never doubted. By early 1971, he had snapped out of the sulks. With Whitlam's encouragement, he took the lead in framing Labor's minerals policy adopted by the Launceston Conference. Whitlam let him know that he was still his choice as Minister for Minerals and Energy in a Labor Government.

The reconciliation included a promise that Connor could make his own choice for the permanent head of his department. The man he wanted was Sir Lenox Hewitt, Gorton's Department chief, in eclipse since Gorton's fall. Connor had been critical of Hewitt's role in the Esso-BHP oil negotiations; but he admired the general thrust of the Gorton-Hewitt approach on Australian ownership of Australian resources. In Hewitt, Connor felt he had found a kindred spirit — both were strong nationalists, both loners, both impatient of the windy orthodoxies of 'established channels'; both saw themselves as tough-minded negotiators, both authoritarian, both secretive, both more easily able to inspire fear than affection, yet both had great charm in private; both were supremely confident in the ability of their applied intelligence to master any problem. In their qualities and flaws, they almost exactly paralleled each other. That was the trouble: they were chalk and chalk. Their most notable similarity, and in the case of the loans affair, the most damaging, was their passion for secrecy. This ran to the exclusion of their closest Cabinet colleagues and their own staff. An Australian Ambassador on his way to his post visited Hewitt's office for a policy briefing; the desk was piled with files in apparent disorder. 'Having a spring-cleaning, Len?' The answer was, 'No. These are the real files of this Department and anyone who wants to see them has to see me.' From beginning to end, secrecy was the major cause of the damage done by the loans affair. It was secrecy with the Treasury which set some Treasury officials on a course of sabotage of the loans proposal; it was secrecy which gave the appearance of a government with something to hide; it was secrecy which led Connor to withhold documents in themselves perfectly proper and innocent, which in turn led to his dismissal and ultimately to the destruction of the Labor Government.

The loans proposal was a logical extension of the philosophy Whitlam had been developing for nearly two decades—the concept of public participation instead of nationalization. In his 1961 Curtin Memorial Lecture, he had said:

> Nationalization is now the most difficult and least important aspect of socialism for an Australian government to achieve . . . A more fruitful and complete use can be made of Australia's human and natural resources through the initiation of public enterprise than the regulation of private enterprise. The Australian government is as constitutionally free as any other national government to initiate public enterprise internally or internationally. Public enterprise is not only the best but probably the only means of now staving off or counteracting private monopoly in Australia and providing continued competition where there is still competition.

In 1975, the Whitlam philosophy and the Connor dream came together in the loans proposal. Whitlam's involvement in the loans proposal was part of his determination to carry out the Labor programme.

As a result of the oil-price rise of 1973, the oil-producing countries, particularly in the Middle East, had accumulated huge surpluses. In 1974, the Australian Government was informed that $US 60 billion had accumulated in this way. In August 1974, the International Monetary Fund raised $US 3.4 billion from OPEC (Organization of Petroleum Exporting Countries) surpluses. The Governments of Japan, France and Britain had also negotiated large loans from these sources.

In October 1974, Clyde Cameron brought an Adelaide businessman, Mr G. Karidis, to see Cairns (then Acting Prime

Minister during Whitlam's visit to the United Nations) at Parliament House. Connor and Hewitt were also present. Karidis told of his business contacts who were able to tap considerable Middle-Eastern funds at low interest rates. Cairns expressed sceptical interest but suggested Karidis should continue to discuss the matter with Connor. In the following weeks, Hewitt followed up various leads suggested by Karidis. By November, Mr Tirath Khemlani, a London-based international commodities dealer, had emerged as the person most likely to be able to put the Australian Government into touch with OPEC sources — that is, to be the intermediary. Throughout his relationship with the Australian Government, Khemlani was never its agent; he was an intermediary. On 11 November, Connor and Khemlani met for the first time. He told Connor he was convinced he could assemble funds up to $4,000 million available at reasonable interest rates — 8 per cent was the figure usually mentioned. Khemlani produced as his credentials a letter of introduction from a leading firm of London bankers. Hewitt made further checks with Australia's established legal advisers in London. On 12 November, Hewitt gave Khemlani a letter stating:

> I am writing to confirm to you what was said by the Ministers at their meeting yesterday. The Australian Government is interested in overseas borrowings of the magnitude you mention, up to a total of approximately $US 4,000 million in blocks of $US 500 million, repayable at the end of twenty years. The Minister has suggested that the availability of such funds for lending should be confirmed at this stage, whereupon negotiations on amounts and other terms and conditions could be commenced immediately between the principals.

Armed with this letter, Khemlani flew off on his appointed task. On 1 December, he cabled Canberra:

> I have now arranged sufficiently for the prime banks acting for lenders from Middle East to establish the line of offer into our bank Johnson Mattey Bank Limited so that my bank can convey such message to the Reserve Bank of Australia in Canberra. At the moment, the loan is therefore to become available from 6/12/74

and therefore they would do this by Tuesday and the first batch is to amount to $US 500 and depending on my further meeting in Canberra the total $4 billion can be completed within 8 months or 4 months and thus 500 shall be granted every two weeks or every month as may be desired by your Excellencies. I therefore beg of your Excellencies to bear with me just a few more days so that I can arrange all . . . it is now only a matter of a few days.

That was to be Khemlani's continuing theme for the next six months—'only a few days'.

Ironically, few decisions of the Government owed less to the much-maligned 'outside advisers'. In no other instance were the special advisers, the private staffs, the part-time consultants so totally excluded. This is the only single instance of a major decision by the Labor Government where the exclusive involvement was between the Ministers and their full-career advisers. Staff—Whitlam's staff, Connor's staff, Cairns' staff—were involved only long after the event, only when defensive action was needed. Until the stable-door needed locking, those who were paid for their presumed professional expertise or their political experience were never consulted. 'Secrecy' required that they be kept in ignorance.

The impression still survives that the loans proposal was devised and carried through by a few Ministers without consulting officials. The truth is very different. Discussions were protracted and wide-ranging, not least on the legal aspects of the proposal. In the week before the Executive Council authority* was given, the Attorney-General's Department was involved in the following meetings: On 6 December, a meeting involving Hewitt, Mr Dennis Rose (Attorney-General's Department), and representatives of Darling and Company, merchant bankers of Sydney, and their solicitors; on 7 December, a meeting between Connor, Hewitt, Rose and Khemlani; on 8 December, a meeting

---

* Under the Constitution, the Executive Council consists, nominally, of the Governor-General and two Ministers of the Crown. It is the Australian form of the concept of the Queen-in-Council. Constitutionally, it gives validity to the concept of the Cabinet, which does not appear in the Constitution. The Executive Council gives legal authority to the political decisions of the Government of the day.

involving Connor, Hewitt, Rose and Mr Bronowski (Darling and Company); discussions between Harders and Murphy, Rose and Byers (Solicitor-General); on the morning of 9 December, a meeting involving Whitlam, Murphy, Connor, Cairns, Hewitt, Harders, Byers and Rose, Sir Frederick Wheeler, Sir John Phillips and Sir Roland Wilson; on 10 December, a meeting of officers of Treasury and Attorney-General's Department; on 13 December, a meeting between Menadue, Wheeler, Prowse and Bailey (Treasury) Byers and Rose; a meeting between Whitlam, Murphy, Wheeler, Byers, Rose and Hewitt; a meeting involving Whitlam, Connor, Cairns, Murphy, Hewitt, Wheeler, Menadue, Harders, Byers, Menzies and Rose.

It is now clear that in their enthusiasm, Connor, Hewitt — and Whitlam — underestimated the difficulties in their path. Setting aside the actual difficulties of obtaining the loan itself, there were three major obstacles: the Treasury, the States and the Senate. In their preoccupation with these obstacles, they overlooked the wider implications in seeking a formula for the authority to raise the loan which would circumvent them.

By November, the relations between the Government and its Treasury had so degenerated that in some Ministers' eyes Treasury could be bracketed with anti-Labor State Governments and the Liberal-controlled Senate as a hostile force.

It was not until 9 December that Whitlam directed that Treasury and the Reserve Bank be brought into the discussions. He did this reluctantly, at the urging of Clarence Harders, the Secretary of the Attorney-General's Department. Whitlam complained that Minerals and Energy seemed to be the only Department which could keep a secret. Treasury's opposition to the proposal was absolute. Its formal objections on economic grounds left unstated one of its main objections — the fear that the Minerals and Energy Department, once in possession of its own funds, would become a rival empire. In July 1975, Whitlam would complain: 'We are the victims of a power struggle between Wheeler and Hewitt.' He was correct at least to the extent from the time of the loans proposal, some Treasury officials openly set out to discredit the Government by a series of calculated leaks to the Opposition and the press.

From the outset, Sir Lenox Hewitt played a key role in the negotiations. His first proposition was that the borrowings should be made directly by the Petroleum and Minerals Authority, the Pipeline Authority and the Atomic Energy Commission. When the Law Officers of the Attorney-General's Department expressed doubts about the powers of such bodies to borrow, Hewitt then put as an alternative that the borrowings could be made by the Reserve Bank or the Commonwealth Bank. The Law Officers were also doubtful about this expedient. After a series of meetings between 6 December and 13 December, involving the Attorney-General's Department, the Prime Minister's Department, the Minerals and Energy Department and, after 9 December, the Treasury, it was decided that the whole loan should be borrowed by the Australian Government. On the advice of the Attorney-General, Lionel Murphy, it was decided that the borrowing be for 'temporary purposes' within the meaning of Clause 3(8) of the Financial Agreement. The Solicitor-General, Maurice Byers, expressed reservations but said that it was 'an arguable view . . .'

An Attorney-General's Department minute of 8 July 1975 states:

> On 13 December 1974, the then Attorney-General (Senator Murphy) advised the Prime Minister and other Ministers that the borrowing proposed by the Minister for Minerals and Energy would be a borrowing for *'temporary purposes'* within the meaning of Clause 3(8) of the Financial Agreement. (By virtue of Clause 3(8) borrowings (*inter alia*) for 'temporary purposes' do not have to be included in the Commonwealth or State loan programmes to be submitted to the Loan Council. Accordingly, these borrowings are not subject to the standing Loan Council resolution of 14 November 1956, which requires the terms and conditions for each proposed borrowing to be put to all the State Premiers and to have the prior concurrence of at least three Premiers.
>
> 2. The Attorney-General's reasoning began with the proposition that Australia was facing a major economic crisis. He forecast that unemployment would reach 500,000 within a few months, and referred to the serious inflation. He said that, in order to deal with these acute problems, the Australian Government needed money for the 'temporary purpose' of averting this economic disaster otherwise likely to occur by April/May 1975. He referred

to acute shortage of capital, and said that economic confidence could be restored if the Government obtained the proposed moneys on loan.

3. He recognized that the expenditures would involve long-term programmes, but said the borrowing would nevertheless be for the 'temporary purpose' of meeting the immediate crisis.

4. He also said that the borrowing would be for a fixed 20-year term. He asserted that the moneys could only be borrowed for a long-term. He also said that the 'temporary purpose' of providing capital to meet the domestic crisis would be defeated if the Australian Government had to repay the loan after (say) 6 months.

5. On 28 January 1975, Senator Murphy confirmed that his advice of 13 December 1974 still applied, in the conditions prevailing at the end of January 1975, to the proposed borrowing by the Minister for Minerals and Energy of $2 billion.

The Government's difficulties were political rather than legal. The hostility of the States precluded a formal approach to the Loan Council. The hostility of the Senate precluded an approach which would have required prior legislation. This did not mean that the Government intended to 'by-pass' Parliament or the Loan Council. As Whitlam stated in Parliament on 9 July 1975:

> There has been much deliberate confusion created about the manner of use by the Government of these overseas borrowings. Any moneys borrowed by the Government would become part of the loan fund, in the public account, requiring an appropriation by the Parliament for expenditure for the purposes of Australia. None of the moneys could have been spent except in accordance with Parliamentary appropriation.

Similar intentions applied to the Loan Council. On 20 May — the day that Connor's authority to raise a loan was revoked once and for all — Fraser asked Whitlam when the Government would seek Loan Council approval for the proposed loan. Whitlam replied: 'If and when the loan is made.' And as Whitlam said in Parliament on 9 July:

> It is of course usual and proper for loans to be sought overseas in advance of Loan Council approval. The terms and conditions of a proposed overseas borrowing are usually referred for approval to

other members of the Loan Council—the States—only when there is a firm proposition to put to them.

As early as 13 December 1974, Whitlam proposed that a Loan Council meeting would be held in January or February 1975 in order to ratify the borrowing. He believed that with the money in hand, the States would be persuaded to accept a proposal involving projects of immediate advantage to them. In the spending, as distinct from the raising of the loan, there never was and could not be any intention to 'by-pass' the States.

On 14 December, Connor was given Executive Council authority to raise an overseas loan of $US 4,000 million. Whitlam left Australia that day on a mission to the Soviet Union, Europe and Britain. As it seemed likely that sales of Australian uranium would be a major matter in his discussions, he took Sir Lenox Hewitt with him as an adviser. In Hewitt's absence, Treasury was able to reassert its opposition to the loan proposal. For a week after the signing of the Executive Council authority, Connor waited for the word from Khemlani that the money would be delivered. The Attorney-General's Department, closely involved at all stages of the negotiations, made a record of the meetings at which it was represented. The final entry in this document records:

Saturday, 21 December 1974

There was first a meeting of officers at the Department of Minerals and Energy and then officers were summoned to the Reserve Bank Building for a further meeting with Ministers. The following were present—the Acting Prime Minister/Treasurer Dr Cairns, the Minister for Minerals and Energy, the Attorney-General, Mr J. Menadue, Sir Frederick Wheeler, Mr R. Hardman (Treasury), Sir John Phillips (Reserve Bank), Mr Townsend and Mr Jennings (Minerals and Energy) and Mr Menzies and Mr Rose (Attorney-General's).

Mr Connor described the events leading up to that date. He said at 8 p.m. on Friday night he had sent a telex message to the Union Bank of Switzerland referring to a code given to him by Mr Khemlani. The Union Bank had replied that it was unaware of any proposed transaction of the kind mentioned. The amount involved exceeded by far any public or private proposals of which it was aware and it saw no validity in the proposal. The Union

Bank had further said that in view of past negotiations with Treasury on other matters it would be pleased to talk to Treasury about matters that might be available to the Commonwealth through ordinary channels. The Acting Prime Minister expressed the view that the Prime Minister should be told immediately and that there should be no more dealings with Mr Khemlani. *Mr Connor said that there would be none by him* (Author's italics). Ministers noted Mr Connor's statement that he did not propose to take any action pursuant to the Executive Council authority. Attorney-General's Department assisted in the preparation of draft telexes to Mr Khemlani and the Union Bank of Switzerland indicating that the Government did not intend to proceed with the transaction.

Of all the millions of words put forth about the loans proposal, there are none so poignant as these: 'The Acting Prime Minister expressed the view that the Prime Minister should be told immediately and that there should be no more dealings with Mr Khemlani. Mr Connor said that there would be none by him.' On 7 January 1975, Connor's authority was revoked. On 28 January, his authority—this time to the more modest limit of $US 2,000 million—was renewed by the Executive Council.

*

Why did Connor change his mind? The answer is that he didn't, because in his heart he was still convinced that he would pull off the deal; he would show those little-minded men, clerks without vision, what bigness and courage and faith in Australia could do.

In the following weeks he lived for, almost literally lived with, his dream, waiting for the message that never came. His young friend and closest Caucus colleague, Paul Keating (destined to fill briefly the Cabinet vacancy left by Connor in October), found him morning after morning, in March and April, asleep in his office, unshaven and dishevelled, waiting for the ring and rattle of the telex machine, which would herald Khemlani's message that he had at last delivered the goods. It never came. Connor had learnt how to operate his own machine. Keating pleaded with him, 'For God's sake, Rex, this is no way for a Minister to behave.' The strain was telling visibly. Connor's giant frame seemed to shrink. And still he held on, in pursuit of his dream.

*

The first victim of the Government's loan-raising activities was to be James Ford Cairns. From the high point of his prestige as Acting Prime Minister in December and January, his position had deteriorated rapidly. His office was demoralized; his relations with Treasury bitter. At the Terrigal Conference he had secured approval for a proposal to establish a Department of Economic Planning. Treasury saw this as another threat to its power and bitterly resisted it. Treasury's hostility became acute when it became known that Whitlam was considering Hewitt, their arch-foe, as its chief.

For some time, Whitlam had been considering a large-scale Cabinet reshuffle. The opportunity came with Barnard's decision to cash in the promise of an ambassadorial post that Whitlam had made him after his defeat for the deputy leadership in June 1974. The prospect of a by-election for Barnard's seat of Bass in Tasmania deeply alarmed most members of the Party, especially the President, Bob Hawke, and the National Secretary, David Combe. Barnard assured Whitlam that, given the right candidate, the seat could be held. In any case, Whitlam felt that he was obliged to carry out his undertaking to his old comrade-in-arms. Cairns was overseas when the rumours of Barnard's impending resignation surfaced. Arriving home, he issued a statement deploring any Cabinet changes and the creation of a by-election. This was designed to raise the standard of revolt in Caucus. It is a measure of Cairns' loss of support in Caucus that nobody rallied. Whitlam offered him a straight exchange — Hayden to take Treasury, Cairns to have Social Security. Cairns preferred Environment.

A much less smooth transfer involved Clyde Cameron, Minister for Labour and Immigration. Since the November measures, the linchpin of the Government's economic policy was wage restraint. This was formalized by the policy of wage-indexation whereby the unions would agree to forgo negotiated wage increases outside the Arbitration Commission, which in return would grant quarterly increases in line with the increase in the cost of living. Cameron himself pioneered the policy and worked hard to secure union agreement. However, when Australia's largest union, formed through the amalgamation of the metal trades unions, the Amalgamated Metal Workers' Union, sought an increase outside indexation, Cameron urged in Cabinet that it should not be opposed. From then on, Whitlam became increasingly convinced that Cameron should leave his old Department, though he did not express his feelings as bluntly as Science Minister, William Morrison: 'We don't trust you Clyde.' In any case, Whitlam believed that Cameron had exhausted his creativity in the portfolio which meant so much to him. Cameron flatly refused to go. Whitlam told him he had no choice—he would advise the Governor-General to withdraw his commission. Cameron became Minister for Science and Consumer Affairs; Morrison moved to Defence; Senator James McClelland became Minister for Labour and Immigration; and Moss Cass became Minister for the Media.

The end for Cairns came devastatingly. On 6 March, he had been authorized by Executive Council Minute to raise up to $US 500 million overseas. On 7 March, he signed a letter to a Melbourne businessman, George Harris, stating:

> The Australian Government is interested in exploring available loan funds overseas. In the event of a successful negotiation which may be introduced or arranged by you, and provided the interest rate for a term loan does not exceed 8 per cent per annum in total, we would be prepared to pay a once only brokerage fee of 2½ per cent . . .

Cairns was later to deny existence of this letter and to this day does not remember having signed it. This claim is perfectly explicable in the context of an office fallen into chaos. Such was the state of Cairns' private office in 1975. Cairns signed, and has always acknowledged signing, other letters to Harris expressing general interest in loan raising. The significance of the 7 March letter is that it established an agency between Harris and the Australian Government and, unlike all the Khemlani documents, committed the Australian Government to payments to its agent. On 4 June, Cairns replied to a question from Lynch denying the existence of the Harris letter of 7 March.

In the last week of the Bass by-election campaign (the result, on 28 June, was a massive 17 per cent swing against Labor), Whitlam was given a copy of the 7 March letter. On Monday 30 June, the *Age* published documents alleging unauthorized activities in a land development project in Melbourne on the part of Cairns' stepson, Philip Cairns, whom Cairns had put on his staff. On 1 July, Whitlam told Cairns he could not defend him in the House and that he would have to resign. In a letter to Whitlam on 2 July, Cairns wrote:

> I answered the question in Parliament on June 4th, 1975, consistently with my recollection of these events. I answered the question as I did, believing that I spoke the truth. I have no recollection at all of having signed the letter quoted in your press statement and I have a clear recollection that I rejected it because I found it unacceptable.
>
> I do not intend to resign from any position I hold because I answered the question in Parliament in the clear and sincere belief that what I said was true.

Whitlam replied immediately:

> I do not regard these documents as satisfactory explanations of the two issues which I put to you.
>
> Since you write that you do not intend to resign from any position you hold, I must advise the Governor-General to terminate your commission. He will receive me at 8 o'clock this evening.

And so it was done, and one of the authentically great Australians of his time was ruined. Cairns had been brought to such a pitch that he was without any support in Caucus. 'The rebuke had broken his heart; he was full of heaviness; he looked for someone to have pity on him; but there was no man; neither found he any to comfort him.'

This tremendous crisis coming into the wake of the Bass disaster determined Whitlam to bring the whole matter of the loans affair to the open forum of Parliament. He called Parliament to a Special Sitting on 9 July. Only in the days immediately before this Special Sitting were the Prime Minister's personal staff brought into the business, and then only for the matter of preparing his speech. He was urged, particularly by his Press Secretary, Evan Williams, to cut the Gordian knot and appoint a Royal Commission. Williams urged this with all his natural eloquence and instinctive sense of the decent, straightforward thing to do, but with that diffidence which sometimes masked his deepest passions and his profound compassion. Whitlam replied that a Royal Commission was appropriate only in cases where there were specific charges of illegality or impropriety; in this case there were none. He spurned a Royal Commission as a 'cheat's device' — yet it might have saved his government, as events were to turn out.

In his speech, Whitlam said:

> This house has been recalled so that once and for all the people of Australia may hear and judge any allegations of impropriety, illegality, malpractice or malfeasance against the Government or any Minister. In all the welter of information and misinformation of the past two weeks, in all the orgy of trivia drummed up as investigative reporting, only one clear fact emerges: no responsible person has expressly or directly made any specific charge of impropriety, of illegal or corrupt conduct on the part of my Government, myself or any individual Minister. Innuendo, insinuation, the sneer, the smear, yes; the Leader of the Opposition and his Deputy, willing to wound but afraid to strike, have proved adept at it; but no specific charges. If such charges are to be made, this is the place to make them. Now is the time to make them. Now is the time to put up or shut up.
>
> The privileges of this Parliament fully protect members who believe that their information, even partial information, even suspect information, would warrant making specific charges of impropriety against Ministers which the laws of libel might render dangerous if made outside. Here, allegations can be made, persons named and documents produced with impunity and immunity. It is not only a privilege which Parliament bestows but a responsibility which it imposes. Through this Parliament, Ministers are responsible to the people, but an equal responsibility rests with the Opposition, to produce their information, if they have any, to formulate their charges, if they have any. This is the place to make them; this the place where they shall be answered. The

inquisition by innuendo is over. This is the tribunal in which the Opposition as much as the Government, will be judged — in the highest court, by the jury of the people. We are all on trial now.

There is a special and overriding reason why this Parliament is the proper place. For it is upon the very question of proper Parliamentary conduct that the one authentic event in a week of squalid intrigue turned — not the pseudo-events of the media, but the one definite event, tremendous in the life of a party, of a nation, supremely tragic in the life of a man.

In the case of Dr Cairns' dismissal, the issue was very precise; the standards which this Government sets and upon which I insist, were not upheld. The personal integrity of my colleague was not an issue and is not in doubt. The fault was grievous, but it lies not in his integrity or reputation as a man of honour. He has rendered remarkable service to his Party and the nation. But the lapse from the standards which this Prime Minister at least, and this Government at least, insists upon, left me with no choice. And in this lies the supreme irony of the present occasion. The dismissal of a Deputy Prime Minister, for whatever reason, cannot but be damaging to any Government, the dismissal of a Deputy Leader, from the Ministry, particularly one held in the regard — affection — of his Party in and out of Parliament, as is the member for Lalor, is a tragedy for all the Party, not least its leader. The course was taken because it had to be taken, because of standards, because of Parliamentary propriety. Yet the Government and Party which has set for itself these standards and which has shown in action how swift, complete and condign is the penalty for any lapse from them, is at the same time being smeared for improper conduct . . .

Let me bring all these points (relating to the charges against the Government) together, in one summary of all the issues involved.

At no time was Parliament misled about the nature, size or source of our loan proposals. Their purpose was to pursue policies in the interests of Australian control of Australian resources. The proposal was economically sound. Legal advice was obtained from the Government's legal advisers before the Minister for Minerals and Energy was given authority by the Executive Council to proceed with negotiations for the loan. Proper care was exercised with the assistance of the Government's legal advisers to ensure that the intermediary in this matter would have no claim against Australia and the Australian Government.

Australia and the Australian Government was protected at all points in connection with the proposed borrowing. Proper checks were made on the bona fides of the gentleman involved.

Not a cent has been paid or will be paid or is liable to be paid to the intermediary. There has been no impropriety on the part of the Australian Government. There have been no breaches of the law. Australia's international reputation as a Government with prime borrowing status remains unchanged, remains as high as it has ever been.

These are the real matters at issue. These are the important issues.

It all boils down to this: was there impropriety in the Australian Government seeking a loan which would help promote Australia's immediate and urgent well-being and promote great long-term Australian objectives? Was there any impropriety in the manner in which the Australian Government sought to achieve those objectives? The objectives were proper; the means were proper.

Let those who say otherwise now specify their allegations of impropriety or illegality—specify the precise charge, the precise breach.

The purposes for which we sought the loan command the clear support of the Australian people. The purposes, the means were perfectly proper. The Australian people have shown their clear support for our objectives in promoting Australian control of Australian resources, which in the final analysis is nothing short of Australian control over Australia's destiny.

In the perspective of these great objectives, the Australian people will despise the pettiness, the self-serving, of those who have chosen this issue to undermine the fulfilment of those objectives. They will quickly see that those who would use any weapon to damage this Government, do not give a damn if in the process, they damage Australia herself.

Nothing which emerged from the Special Sitting, or from efforts in the Senate to call all the officials involved before the Senate bar, seemed to create new problems for the Government. It seemed for a time that the loans affair was at last at an end. But if the Government thought that its loan activities were over, others did not.

*

On the very day of the Special Sitting, 9 July, Tirath Khemlani telexed his Adelaide contact, Mr G. Karidis, from London:

> I thank you for your support for the past few days . . .
> This is what we do, that is, we use codes for the loan and we shall talk on the phone using code names and this way we are free of any leakage. Loan: raw sugar. Ten million dollars: one thousand tons. Mr Connor: Uncle. Prime Minister: Big Boss.

Connor had failed in the most crucial point: he had failed to tell Khemlani, simply, shortly, sharply, that it was all over. Perhaps Connor still held to the vestiges of his dream; but Khemlani was still living in a dream-world.

Long before the Special Sitting of 9 July, Whitlam had become anxious about his lack of full knowledge of Connor's activities. John Menadue was particularly worried about this. On 23 May —as it turned out, a fatal day for Connor— Menadue, on behalf of the Prime Minister, asked Clarence Harders, Secretary of the Attorney-General's Department, to obtain from Connor 'possible communications which could result in financial or other embarrassment to the Government'. Connor provided documents concerning activities only up to the end 1974. On 29 May, Whitlam wrote to Connor requiring 'all papers on recent loan raising matters'. On 30 May, another batch on documents was delivered to Harders. On 13 June, Whitlam again wrote to Connor pointing out that 'you would be responsible for any matters which may emerge which are referred to in documents in your possession and which you have not referred to me . . .' On 18 June, Connor replied that he had 'produced all appropriate documents'. In the days before the Special Sitting, Michael Delaney, Whitlam's Private Secretary, was given the task of winkling from Connor and Hewitt the documents to be tabled. Because of this incredible coyness, Whitlam decided that he could not take responsibility before Parliament of tabling the loan documents; Connor must do that. On 7 July, Whitlam again wrote to Connor listing the documents which must be tabled. On 8 July, he again wrote to Connor, enclosing copies of documents received that day, on the very eve of the Special Sitting, from the *Age*.

On 9 July, the morning of the Sitting, Whitlam demanded

further documents for tabling in Parliament. On 10 July, Whitlam summoned Connor and Hewitt to his office and directed that all papers, including files, documents and records, relating to Khemlani be delivered to Harders by 5 p.m. the following day. On 11 July, Hewitt delivered to Harders a further set of papers up to 14 May. On 13 July, Whitlam wrote to Connor saying that papers were still clearly being withheld. The next day, Hewitt handed to Harders a further set of notes, files and other material.

The concern of Whitlam, Menadue, Harders and Delaney was about one matter: whether anything had been said to Khemlani that might give him a legal claim on the Australian Government. On that basis, Connor and Hewitt undoubtedly believed that they were providing all 'relevant' material. But in his obsessive secrecy, Connor had set himself on a course of self-destruction. It needed only one significant document to surface and his position would become untenable. In the political climate of Spring 1975, with Fraser hunting for his 'extraordinary and reprehensible circumstance', with some Treasury officials acting as informers, with newspapers prepared to pay money for documents genuine or otherwise, it was inevitable that something would turn up. There was no need for such a document to be intrinsically improper or irregular. The most innocent document—dated after 20 May—could destroy Connor by its mere existence, an existence which he had denied.

There was a lull. For a few weeks it seemed that the 'loans affair' had at last run its course, damaging in the extreme but not fatal. On 8 October, the Melbourne *Herald* published a story under the by-line of Peter Game, who had been sent overseas to dig into the affair. The story alleged that Connor had had dealings with Khemlani after 20 May. Connor, from his sick-bed, denied the story. Next day, Lynch asked Whitlam:

> Will the Honourable Gentleman assure the House that all communications between the Minister for Minerals and Energy and loan-raising intermediaries were tabled by the Minister and himself on 9 July. If he is not prepared to give that assurance will he now table the documents that have been withheld?

As with other questions asked by the Opposition that day, based on reports appearing the previous day in the Melbourne *Herald*, Whitlam asked that the question be placed on notice. That evening he answered Lynch in writing:

> I am assured by the Minister for Minerals and Energy that all communications of substance between him and Mr Khemlani were tabled by him on 9 July 1975.

On 13 October, Whitlam received, under cover of a letter from Corr and Corr, Solicitors, of Melbourne, acting for the Herald and Weekly Times Limited, a statutory declaration signed by Mr Tirath Hassaram Khemlani and copies of a number of telex messages. The messages comprised a series of communications between Khemlani's office in London and Connor's office between 23 and 26 May 1975 claiming that funds were still available. There were a score of messages from Khemlani to Connor; only one was a reply from Connor. Connor replied on 23 May—the day Harders had visited him and three days after his loan authority had been revoked:

> I await further specific communication from your principals for consideration.

In the *Herald*'s 'revelations' only ten words came from Connor. Those ten words were a noose around his neck.

Carpeted by Whitlam, Connor said that he did not regard that telex as a 'matter of substance'. Perhaps it was not, but these things had become plain: Connor had not complied with Whitlam's repeated and insistent requests, rising to demands, for all relevant information; he had obviously failed to inform Khemlani that his authority to raise a loan had been revoked; his public denial could not be sustained; and, most important of all, he had let Whitlam make a statement, on record in Hansard, which was untrue. It was the end of the line.

In a statement on 14 October, Whitlam said:

> In Mr Connor's judgement, these are not communications of substance. I have already said that in my judgement, they are. But further, it was for me both as head of Government and as the Minister accepting responsibility for the answer I gave to the

Deputy Leader of the Opposition, to be in a position to make that judgement. I regret I was not in that position.

In these circumstances therefore I today advised the Governor-General to terminate the Commission of the Minister for Minerals and Energy.

I wish to make it absolutely clear that the reason for my decision rests solely and wholly on the matter of the answer I was led to give last Thursday. It is a precise, clearly defined issue. I have made it clear throughout the life of this Government that there is the one standard which, once departed from, must carry the heaviest penalty. It is a principle on which the integrity of Parliament itself depends.

There was I believe a departure caused from that principle. The principle and my insistence upon it was made amply clear, tragically clear, only three months ago in the case of the Honourable Member for Lalor. In that case and in this, there was and is, no reflection whatsoever on the personal integrity of the Minister involved. The error of judgement in both cases was the same; the consequences of that error of judgement, the same. But let me emphasize, those consequences spring not from misconduct, but from standards in relation to a narrow, precise but fundamentally important principle which this Government and I as head of this Government must insist upon.

I want to make the clearest possible distinction between the conduct of the Member for Cunningham in his capacity as Minister, the manner in which he fulfilled his duties, including all his involvement in overseas loan raisings and this one precise principle. In particular, his communications with Mr Khemlani after 20 May were not of themselves improper or irregular. It is not the contents of those telex messages which constitutes any breach of propriety. It is the discrepancy between those communications and the assurances the Member gave me last Thursday.

Up to 20 May, the Minister for Minerals and Energy had Executive Council authority to conclude a special loan-borrowing overseas. That authority was revoked.

On 20 May, that authority was revoked. The revocation of that authority did not mean that the Minister for Minerals and Energy or indeed any other Minister should not even discuss possibilities of loan borrowings with any intermediaries nor did it mean that the Australian Government would not receive information about the possibility of loan raisings. When this whole matter was debated on 9 July, I said specifically (page 3595) 'Executive Council authorities are usually sought only when a loan matter approaches finality. This does not mean that negotiations or discussions about possibilities are precluded without Executive Council authority.' Nothing in the communications after 20 May between the Minister for Minerals and Energy and Mr Khemlani or any of his associates indicate that the Member for Cunningham went beyond these guidelines. There was no impropriety involved in his receiving messages and there was no impropriety involved in his response to one of them. The error was in his failure to inform either the Treasurer or myself about these activities. This was an error of procedure, not an error of propriety. It is because that error of procedure led to a breach of Parliamentary standards last week that I asked for the resignation of the Minister.

The basic facts remain as they have always been and have always been revealed in this Parliament. They are as I stated them on 9 July and nothing has changed them. Since 14 December 1974, the Minister for Minerals and Energy (Mr Connor) had Executive Council authority to borrow up to US$ 4,000 million and to determine on behalf of the Australian Government the terms and conditions of the borrowing. The Minister was also authorized to sign and deliver promissory notes for the purposes of the borrowing, or to authorize any other person in writing to sign and deliver the promissory notes. That authority was revoked on 7 January 1975 since it had not been used and it conflicted with a Deutschmark loan then pending. On 28 January 1975, the Executive Council authorized the Minister to raise a loan not exceeding US$ 2,000 million. The authority was revoked on 20 May 1975 since negotiations were in train for a borrowing of US$ 100 million through Morgan, Stanley and Company Incorporated, of New York.

In all the scrutiny to which this matter has been subjected, for all the money offered to informers, for all the travels of the Deputy Leader of the Opposition there is still not one allegation of impropriety, illegality, malpractice or malfeasance charged against this Government or any Minister. There has been no specific charge of impropriety in negotiations of an illegal or corrupt conduct on any member of the Government involved in these transactions at any time from 14 December to the present day. A specific charge, a specific allegation can be investigated; no such charges no such allegations have been made. In two cases breaches of the Parliamentary principle have been alleged, investigated and punished. But in no case has there been an allegation of improper conduct, of dishonest conduct, of reprehensible conduct of an illegal or corrupt nature. No such allegation or charges were made at the special sittings to discuss this matter on 9 July. No such charges were made in the Senate for all their abuse of Parliamentary privilege. No such charges have been made by the Deputy Leader of the Opposition even after his foray into the highways of New York and the byways of Europe. No such charges have been made by any of the newspapers who have investigated this matter. Never has so much been paid by so few for so little.

The wisdom of the Government's effort to raise overseas loans on behalf of Australia and for the good of Australia is a perfectly legitimate subject for political debate. We defended and continue to defend it. That particular effort failed and so, while Australia has received nothing, Australia has paid nothing and is under no obligation to pay anything.

Whitlam's declaration was to prove altogether wrong, for the price was to be immense. Fraser had found his 'extraordinary and reprehensible circumstance'.

24
Ambush

'One of the main responsibilities of the Governor-General as President of the Executive Council', Sir Paul Hasluck stated in a lecture, the Queale Memorial, in 1973, 'is to make sure that all actions of the Government are constitutionally correct and lawful.' This was not, Sir Paul said, a matter of the Governor-General's own opinion; if in doubt the Governor-General 'refers the matter to the Attorney-General for advice'. Even if Sir Paul's definition is accepted, it should be emphasized, as a matter of record, that at no time did Sir Paul's successor, Sir John Kerr, raise the question of the dismissal of the Whitlam Government with the Prime Minister who appointed him. Some writers and commentators have claimed to see a direct connection between the loans affair and the dismissal of the Whitlam Government. All that can be said is that if Sir John Kerr had doubts about the propriety of the Executive Council authorities given to Connor on 14 December 1974 and 28 January 1975, he never raised them with the Prime Minister, with the Attorney-General, Lionel Murphy, or with Murphy's successor, Keppel Enderby.

When Sir Paul Hasluck's term as Governor-General was due to expire, he declined Whitlam's offer of a further term. At Whitlam's request, he wrote down a list of seven possible candidates, including the Melbourne businessman, Kenneth Myer, Crean, Barnard and Sir John Kerr. Myer, head of the great Melbourne-based chain of department stores, was Whitlam's own first choice but he declared himself unavailable for private reasons. Of those remaining on Hasluck's list, Sir John Kerr, Chief Justice of New South Wales, was clearly the most suitable in terms of distinction, presence and, above all, public acceptability.

John Kerr, born in 1914, son of a boilermaker from Balmain, an inner industrial suburb of Sydney, had been a brilliant scholarship boy at Fort Street High School. On the advice of Dr Evatt, a family friend, he studied law. In 1942, he was summoned by Dr Alfred Conlon to join the Directorate of Research and Civil Affairs. This war-time outfit was as enigmatic as its director, Conlon. Its functions and activities were, and are, cloaked in mystery. Conlon, described by his contemporaries as 'the third most powerful man in Australia' is now remembered only by his contemporaries. On all who knew him, he made an indelible impression; on formal, written history, the impression is scratchy indeed. In his work with the Directorate, Kerr was apparently involved in intelligence work which first brought him into contact with the international intelligence network. Perhaps the most significant aspect of his war work was that it opened up to him a world in which real power could be exercised in ways other than through elective office. He could never be persuaded to run for public office. In 1966, he was appointed to the bench of the Commonwealth Industrial Court. In 1969, he provoked what seemed certain to prove Australia's worst industrial confrontation since 1949 when he jailed the Secretary of the Victorian Tramways Union, the communist Clarence O'Shea, for contempt of court and refusal to pay fines. The crisis was averted when a lottery winner paid O'Shea's fines. In 1972, the New South Wales Liberal Government appointed Kerr Chief Justice of New South Wales.

For six months Kerr hedged before definitely accepting Whitlam's offer. He made three conditions: a 50 per cent increase in the Governor-General's salary to $30,000 before he was sworn in (Section 3 of the Constitution forbids an increase during the Governor-General's term); the same pension as a Chief Justice of Australia; and an undertaking that a Labor Government would renew his term for a second five years. He pointed out, fairly enough, that in 1974 he still had ten years until his retirement as Chief Justice at the age of 70. Whitlam readily agreed to these conditions. Kerr then asked Whitlam to seek a similar undertaking from Snedden, then Leader of the Opposition. Whitlam refused to try to so bind any future Prime Minister. However, as soon as the Queen accepted the appointment (at Government House, Canberra, during her Australian visit in February 1974, Kerr thus becoming the first Australian Governor-General whose appointment was approved by the Queen of Australia in Australia) Whitlam did two things: he telephoned Sir Robert Askin, the New South Wales Premier, to tell him of the appointment; and he telephoned Snedden to urge him to give Kerr a second term should he ever have the opportunity to do so as Prime Minister.

Kerr later told Whitlam that one of the reasons both for his hesitation and his final acceptance was the critical health of his first wife. This gracious woman was terminally ill. After her death a few months later, Kerr told Whitlam: 'Peggy was desperately ill . . . couldn't look after the basic things around the house . . . but she wanted to do it . . . so I didn't have the heart to get in help . . . at least at Government House, there were servants and I could have a valet'. Whitlam, much moved, told this story to intimates as evidence of Kerr's compassion and sensitivity.

Whitlam was delighted when Kerr confided in April 1975 his intention to re-marry. On the day of the marriage, Whitlam was in Kingston, Jamaica, for the Commonwealth Heads of Government meeting. With a contrived air of mystery, he instructed his Press Secretary, Evan Williams, to assemble a press conference for an important announcement: 'I can now tell you, gentlemen, that the Governor-General was married in Sydney today.' And then he broke out the champagne.

The personal relationship which developed between Whitlam and Kerr became the most informal and apparently close which has ever existed between an Australian Prime Minister and an Australian Governor-General. Kerr, with his Sydney associations, used the official residence at Admiralty House more than any of his predecessors. Whitlam was the first Prime Minister to make the neighbouring house, Kirribilli House, his Sydney residence. Both superbly sited on Kirribilli Point, commanding harbour views unequalled in the world, these two houses share a garden; Kerr invited Whitlam to continue to use the Admiralty

House swimming pool built for the common use of both houses. Kerr frequently strolled down to the pool to discuss work and issues with Whitlam. The emphasis was always on intimacy and informality between these two neighbours who happened to be Prime Minister and Governor-General. This is basic to any understanding of Whitlam's approach to procedural matters involving Sir John Kerr.

Whitlam was pleased that Kerr abolished the curtsy at Government House. He failed to observe that this gesture to egalitarianism was compatible with an assertion of a greater role and powers for the Governor-General. In the pre-appointment months, there was much more discussion about the pay and conditions of the office of Governor-General than about the role and powers of the Governor-General. It scarcely occurred to Whitlam that such discussions were necessary.

Whitlam was negligent — with hindsight culpably so — in never having opened up with Kerr in any profound way Kerr's view of the constitutional role of the Governor-General. It can now be seen that three successive Australian-born Governors-General — Casey, Hasluck and Kerr — have each sought to define that role in terms of its increasing importance. Once Menzies, through Casey's appointment, confirmed that an Australian would always be appointed by Liberal Governments as well as Labor Governments, an attempt has been made to reverse the convention, the belief held unquestioningly by the general public and most constitutional writers, that the Governor-General had a purely ceremonial role and that his independent role had diminished to total insignificance. This at least was the merit of decayed peers and has-been English politicians. The instability of Australian politics after Menzies, particularly because of Holt's disappearance, lent a kind of relevance to the search for a new role. Even before Holt's death, Casey had intervened privately to express to leading members of the coalition government his anxiety at the government's lack of stability. Hasluck was the first Governor-General to represent Australia as its Head of State when he attended Iran's 2,500th anniversary. Kerr took this precedent as evidence of a real change in the Governor-General's role, claiming in a speech to the Indian Law Institute in New

Delhi in February 1975:

> The Governor-General nowadays, in acting on behalf of the Australian Head of State, the Queen, has a very different task from that of the Australian Governor-General years ago. Until recently, the Governor-General did not as such travel abroad with the Head of State status.

Whitlam encouraged Kerr in this role—with qualifications. He told Kerr that he believed that it was proper for Australia to have a person who could act overseas as Head of State in those countries which understood our system—the Commonwealth countries, Ireland, Belgium, The Netherlands and even France and West Germany. Whitlam specifically excluded Japan where the Emperor was at least residually more than a figurehead. This did not stop Kerr from approaching the Japanese Embassy for an invitation for an official visit to Japan, after he had met the son of the Crown Prince of Japan. The Japanese politely but firmly refused to be enticed.

Whitlam did not perceive—nor did any of his advisers—that Kerr sought to widen the dimensions of his role. In his speech, in New Delhi, Kerr said:

> I am very interested in foreign policy and defence policy and both as a new Governor-General and Commander-in-Chief, I have been seeking to come to grips with our policy in these fields . . . I am the Commander-in-Chief . . . As such, I maintain a constant interest in defence matters . . .

This literal interpretation of the formal words of the Constitution was a portent. Its implications were lost upon the Prime Minister. He never saw the Delhi speech. But, as he later acknowledged: 'I don't think I would have read it, even if he had sent it to me.'

*

After the trauma of Cairns' dismissal, the Government entered a period of relative calm. The survival of the Dunstan Government in South Australia by the narrowest margin served, in the perverse manner of the moods of parliamentary politics, as a rallying-point. Dunstan had called a snap election for 12 July when his Upper House rejected an agreement with the

Australian Government to transfer the State's railway system to the Commonwealth on terms very favourable to South Australia. This was one Federal issue on which Dunstan was eager to fight. But his campaign, begun so confidently, was stopped short by the Cairns crisis and the loans debate. Dunstan abruptly changed tactics to distance himself as far as possible from the Labor Government in Canberra. Whitlam accepted the change of tactics, knowing that the Labor Party would not easily forgive the loss of its most popular and successful State Government.

For the next three months, the theme for the Government was consolidation — consolidation of the team, consolidation of the programme, consolidation of the reconstructed Government. The new team, with Crean as Deputy Prime Minister, Hayden as Treasurer and Senator James McClelland as Minister for Labour settled down well. In search of Budget restraint, Hayden showed much the same skill, patience and perseverance with his Cabinet colleagues that he had shown with the States, the doctors and the health funds in establishing Medibank. For the first time since the oil crisis of 1973, the Government appeared to be having some success with the public in placing Australia's economic difficulties in the world context. On 17 July, the OECD Economic Outlook reported:

> The present recession in OECD countries is the most serious since the war. It is remarkable not only for its length and depth — a third consecutive half-year of negative growth has now been recorded for the area as a whole — but also for its widespread nature: virtually every OECD country grew by less than its medium-term average rate in 1974, and no economy is expected to take up slack in 1975. The margin of idle resources in the OECD area is now the largest in the post-war period, with unemployment at record levels. The forecasts presented in the December Economic Outlook, and to a greater extent those being made at that time by national authorities, proved to be too optimistic. Industrial output in the major countries fell very sharply in the last quarter of 1974 and the first quarter of this year. The extent and simultaneous nature of the decline was unlike anything recorded in the post-war period. The combined GNP of the major countries, which was thought at the time to have increased marginally in the second half of 1974, is

now estimated to have fallen, at an annual rate, by over one per cent. Output was expected to continue stagnating in the first half of this year; it may in fact have fallen at an annual rate of about five per cent. The December Economic Outlook gave reasons for supposing that the balance of uncertainties attaching to the forecasts was on the downside; but the extent to which this proved to be the case is astonishing.

In his Chifley Memorial Lecture at Melbourne University on 14 August, Whitlam claimed:

> What is not widely recognized is that Australia has fared comparatively well among the industrialized countries during this period of economic turmoil. Our unemployment has been a great deal lower than that suffered by a number of industrialized countries; our loss of production has been less; and many industrialized countries have had much more severe inflation. Unlike most industrialized countries, we have moreover been entirely free of balance of payments problems.
>
> In spite of our economic problems the Australian employee has on balance done well over the last two years. Average minimum award rates have risen by 55 per cent and average earnings by 48 per cent — while the cost of living as measured by the CPI has increased by 32 per cent. There have, therefore, been substantial real gains to wage and salary earners, supplemented by improved education, health, and other benefits by this Government.

The Budget, Whitlam said, would consolidate these gains.

Before the Parliament assembled for the Budget, an event occurred which was to transform the situation. A Queensland Labor senator, Bert Milliner, died. The Queensland Premier, Joh Bjelke-Petersen, seized the chance to deal a devastating blow to the hated Labor Government in Canberra. Since 1950, all State Governments, Liberal and Labor, had abided by the convention that a Senate casual vacancy should be filled by a member of the deceased or retiring senator's party. The New South Wales Premier, Tom Lewis, had broken this convention by refusing to appoint a Labor senator to replace Mr Justice Murphy. Bjelke-Petersen refused to accept the Labor nominee and appointed a person of outstanding obscurity, one Albert

Field, who said that he would vote against the Labor Government on all occasions.

This was the situation: the 1974 Double Dissolution had given Labor 29 senators, the coalition 29, with two independents. After Murphy's resignation, Labor had 28, the coalition 29, with three independents. The Tasmanian Independent, Michael Townley, rejoined the Liberal Party. With the arrival of Albert Field, the Opposition could muster 31 against Labor's 27 with two independents voting with the Government on vital matters — this in a Senate in which the people had given the Labor Government parity with the Opposition scarcely a year before. The Government challenged Field's appointment and for the crucial votes in October and November he was not in his place. Even so, the line-up was 30–29 against the Government.

It was this fatal equation that made possible the events of October–November 1975. If Senator Milliner had lived, or if he had been replaced by a Labor senator, the line-up on the deferral of Supply would at worst have been 30–30 — and thus the vote would have been 'carried in the negative'. It would have been lost. That is what Senator Steele Hall meant when he said on 16 October that the Opposition had 'marched on the sleazy road to power . . . over a dead man's corpse'.

Introducing his budget on 19 August, the new Treasurer, Bill Hayden, said: 'Our reforms are enduring; they will not be reversed. Now we propose to take stock of the achievements . . . we have exercised the utmost restraint on government spending . . . the key-note of this Budget is consolidation and restraint rather than further expansion of the public sector.' The Hayden Budget estimated spending of $21,915 million with a deficit of $2,068 million. Its centre-piece was a 'radical new personal tax system' to free 500,000 lower income earners from tax, reduce marginal rates of tax for most middle income earners and reform the tax rebate system.

Fraser's immediate reaction to the Budget was cautious: 'We'll be following normal procedure in the Senate and with the knowledge we have at the moment, at this stage, it would be our intention to allow it a passage through the Senate.' The Budget itself was generally well-received. In a mood of wish-fulfilment, the Government was inclined to take Fraser's statements on their favourable interpretation, ignoring his escape clause 'with the knowledge we have at the moment, at this stage'. The Connor resignation on 14 October rendered the escape clause fully operative. On 15 October, the Senate carried, 29–28, a motion to defer the Loan Bill 1975. On 16 October, the Senate carried a motion to defer the Appropriation Bills—the Budget. Both motions called on the Government to hold an election for the House of Representatives. Senator Ivor Greenwood said: 'Let it be clear that the Opposition's purpose is to secure an election.'

On 10 June 1976, the Chief Justice of Australia, Sir Garfield Barwick, told a National Press Club luncheon in Canberra:

> The only time I got any idea that the Governor-General was troubled was when I chatted at dinner on 20 September with the members of the Order of St Michael and St George. The Governor-General and the Governor of the State were both guests. During the meal I became aware, from something that was said, that he was troubled very much in his mind about what was occurring. I had no more contact with him until he rang me on the Sunday evening (9 November) and I had made up my mind to advise him.

Until Sir John Kerr tells his own story it will be impossible to say what precisely was troubling him on 20 September. There was at that time no real constitutional crisis. Neither side to the political dispute had taken an immovable position. While Fraser was under continuing pressure to force an election, particularly from Anthony, he had not advanced irreversibly from his declared position of 21 August: 'At this stage it's our intention to allow the Budget passage through the Senate.' He had not yet found a plausible 'extraordinary and reprehensible circumstance'. On 15 September, five days before Sir John Kerr unburdened himself to his fellow-companion, Fraser said on a television interview: 'Nobody knows what's going to happen. There could be another loans affair . . .' But having opened the door a little further, he pushed it back: 'You've got to balance that against the constitutional practices which have generally determined how long governments will last.'

Whitlam himself, on 12 September, had emphasized the fluidity of the situation. A rather mystified assembly of teachers and students at the opening of a College of Advanced Education at Goulburn, New South Wales, heard him say: 'There are no laws applying to a situation where Supply is refused by an Upper House, no laws at all ... one can only say that there is no obligation by law, by rule, by precedent or convention for a Prime Minister in the circumstances which are threatened to advise the Governor-General to dissolve the House of Representatives and have an election for it.'

Perhaps Sir John Kerr was troubled by the general instability of the political situation. Perhaps he was troubled by the implications of Whitlam's Goulburn doctrine. We cannot say. Only one thing is known for certain: the nature of the anxiety he felt on 20 September, which he vouchsafed to the Chief Justice, was never raised with his Prime Minister, in any shape or form, between then and 11 November.

This break down in communications at the top is at the centre of the events which led to the dismissal of the Whitlam Government. To the last hour, this failure to communicate was one-sided. Whitlam believed he was keeping Kerr fully informed of all developments. To the last hour, Whitlam had complete confidence in Sir John Kerr and had every reason to believe that he had Kerr's confidence. The 'trouble in his mind' was never made known to Whitlam directly or through any number of intermediaries available to Kerr — old friends and associates like Senator James McClelland, Minister for Labour and Immigration, Joseph Riordan, the Minister for Housing or Kep Enderby, the Attorney-General. And to the end, Whitlam believed that Kerr's appointment was one of the best his Government had ever made.

On 16 October, Robert Ellicott, Q.C., Liberal member for Wentworth and former Australian Solicitor-General, issued a press statement in which he expressed the opinion that the Governor-General should dismiss the Government if the Prime Minister would not advise him to dissolve the House of Representatives alone or both Houses of the Parliament together as a means of resolving the disagreement between the two Houses. Ellicott suggested that the

Governor-General should not accept the device of a half-Senate election as a means of resolving the disagreement unless he was satisfied that the new Senate would pass the appropriation bills. He did not say whose advice the Governor-General should accept as to the likely outcome of a half-Senate election. The Ellicott opinion became the scenario for the events of 11 November.

On 6 November 1975, the Attorney-General, Kep Enderby, Q.C., handed to Sir John Kerr an opinion signed by the Solicitor-General, Maurice Byers, Q.C., and concurred in by Mr Enderby. In nearly 10,000 words the opinion canvassed the matters raised by Ellicott, concluding: 'We have found ourselves for the reasons we have stated firmly of the opinion that Mr Ellicott's expressed views are wrong.' It rejected his basic contention, stating: 'Nor do we agree with the suggestion that were the Prime Minister unable to suggest means which would solve the disagreement between the Houses and left the Government without funds to carry on, it would be His Excellency's duty to dismiss his Ministers.' It concluded: 'It seems to us, if we may respectfully say so, that assumptions underlie Mr Ellicott's press statement which present dangers to the orderly working of Government. Those dangers are significant ones. That the possibility of their existence is a disquieting one cannot, we venture to think, be seriously doubted. For they may be indefinitely repeated and may involve deleterious consequences to the working of the constitutional provisions. That that working requires restraint on the part of both Houses is hardly open to doubt. A view which looking only to the existence of the legal power disregards or ignores constitutional practices hitherto apparently governing the exercise of those powers, requires, we venture to think, the gravest consideration before its adoption could even be contemplated.'

Thus, in most considered and deliberate terms, the Law Officers of the Crown put their views to Sir John Kerr. At no time did he express doubts about its arguments or its conclusions. Between 6 November and 11 November, he did not raise the matter again with Whitlam or Enderby. Yet, in his statement of reasons for the dismissal he asserted: 'I should be surprised if the Law Officers expressed the view that there is no reserve power in the Governor-General to dismiss a Ministry which has been refused supply by the Parliament and to commission a Ministry as a caretaker ministry which will secure supply and recommend a dissolution, including where appropriate a double dissolution.' This was technically correct and quite misleading. The Law Officers had not asserted that there was no reserve power; they had argued fully and carefully that the use of those powers, in the way urged by Ellicott in the circumstances of November 1975, would be wrong.

As well as providing his written opinion, Maurice Byers collaborated closely with the Prime Minister's Department and this author in preparing material for Whitlam's parliamentary *tour de force* in a series of speeches between 16 October and 11 November. Byer's drafts provided detailed arguments against the Ellicott memorandum. While we gratefully incorporated a great deal of Byer's material, none of Whitlam's speeches gave emphasis to Ellicott's argument on the reserve powers of the Governor-General. In Whitlam's office, there was a distinct feeling that, on this point, Byers was more concerned to argue down his predecessor as Solicitor-General than to deal with the political realities of the situation. Such was our invincible blindness! Nonetheless, the substance of Byer's arguments was included in Whitlam's speeches. At no time did Sir John Kerr ever suggest that Whitlam's assertions about the role of the Governor-General — that he was bound to act on the advice of his Ministers — were wrong or even open to argument. At no time did he make the slightest comment which might have changed the focus of Whitlam's attention from the role and powers of the Senate to the role and powers of the Governor-General.

Even if it had occurred to Whitlam or his advisers that the Governor-General was taking the Ellicott opinion seriously, the Attorney-General's Department advised him to play it down. In a memorandum sent to Whitlam, through John Menadue, the Secretary of the Attorney-General's Department, (now Sir Clarence) Harders, wrote on 20 October:

> It would not be wise for the Prime Minister to simply assert that the Governor-General must at all times and in all circumstances act on his advice. It would be much better to rely upon the combination of the several strong points that the Prime Minister has in his favour and that he has already been making publicly. There could be a risk that the strong points will become obscured if the public debate is

confined to the single issue of whether the Governor-General must, all the time, act on the Prime Minister's advice.

Ellicott's analysis is incomplete. It stops short of considering the place occupied by certain important facts. He ignores the breach of the convention that the Senate does not reject Supply and he ignores the fact that the Prime Minister and his Government continue to have a majority in the House of Representatives.

The vital issue has been stated again and again by the Prime Minister. He should continue to state it. There is a dispute between the two Houses. A convention of fundamental importance to the future of parliamentary government would be breached by the rejection of Supply by the Senate. It is not the short-term interest of any of the political parties that is important. What is important, and what must be maintained in the interest of parliamentary government in Australia, is the maintenance of the convention. The convention *must* prevail.

So long as, in the end, the Governor-General is thoroughly seized of what is at stake, and acts accordingly, it does not really matter whether, in point of fine analysis, he has acted on the Prime Minister's advice or in the exercise of a discretion reposed in the Queen's representative. Let the historians of the future work it out. Indeed, some 'blurring' may be desirable if, in the last resort, *the Governor-General were to be minded to take the step of addressing both Houses of the Parliament* (Author's italics). It would, of course, be hoped that in doing so the Governor-General would take the view that the convention *must* prevail.

To sum up, the Ellicott analysis falls short of 'biting the real bullet' — namely, the breach of the convention. By the same token it would be best if the Prime Minister were not simply to reduce the constitutional debate to the question whether the Governor-General is, or is not, obliged to take the Prime Minister's advice at all times and in all circumstances.

As Harders' memorandum shows, Whitlam's advisers did suggest one form of intervention by Sir John Kerr: a message requesting the Senate to stop deferring the Budget and vote directly upon it. Whitlam's response to the suggestion was: 'It's certainly something to keep in mind but only as a last resort.' His reservation was, of course, that such action would involve the Governor-General in a party political struggle.

Only twice did Sir John Kerr raise directly with Whitlam the question of his own role. And both times he raised it in a way and in a sense exactly contrary to the course he actually took. Both times he left Whitlam with the belief that he resented outside attempts to force him into an active role.

On 19 October, Kerr told Whitlam he was worried by press reports, particularly in the Murdoch papers, urging him to intervene in the crisis. The author was with Whitlam at Kirribilli House preparing a speech. 'He says it is intimidation,' Whitlam told me after his conversation with Kerr. As a result of this conversation, the following paragraph was included in Whitlam's speech in the House of Representatives on 21 October:

> The conditioning which the Leader of the Opposition and the (press) proprietors sought to impose upon the public last week, and indeed for weeks before that, they now seek to impose upon the Governor-General himself. There have been those long months of conditioning by the political, the business and media interests who have never been prepared to accept the legitimacy of an Australian Labor Government. Now we are seeing a fresh phase in this exercise. Now we have the headlines, 'Will Sir John Kerr act?' and 'Fraser says Kerr must sack PM'. Where will this intimidation stop . . . We now have the extraordinary spectacle of the Opposition, apparently recognizing that it has failed to blackmail me into an election for the House of Representatives, seeking to bring reprehensible pressure to bear on the Governor-General, the representative of the Queen of Australia, to achieve that very thing, and to do so by dismissing me as Prime Minister.

This was the first time Whitlam had raised in Parliament the spectre of a vice-regal dismissal — and he did it because Sir John Kerr had complained of press 'intimidation'.

Kerr's second mention was in response to the Ellicott memorandum itself. In a telephone conversation with Whitlam at Parliament House, he volunteered: 'This Ellicott thing. It's all bullshit isn't it?' Whitlam agreed. Kerr then asked Whitlam for the Solicitor-General's opinion. Whitlam reported this conversation to the Attorney-General, Kep Enderby. On 6 November, Enderby gave the Byers opinion to Kerr at Government House. He told Kerr he had decided not to sign it because he was even stronger than Byers against the idea of the use of the Crown's reserve power. 'I must say', Enderby told Kerr, 'that in this I am a Whig. I don't believe in 1975 the powers exist.' Kerr then asked Enderby what he thought would happen. 'It will be over in a week. The Senate will crack,' Enderby replied. Kerr said: 'That's not what Malcolm tells me.'

Despite the intimacy and mutual confidence he believed he had established with Kerr, Whitlam was never contemptuous or cavalier towards the office of Governor-General. On the contrary, throughout the constitutional crisis, he was protective towards both the office and the person of the Governor-General. This is one of the reasons why he was slow to come to a decision to use his option of a half-Senate election. Most of his colleagues were in favour of a half-Senate election. The Federal Executive urged it. Whitlam hesitated. He was worried that the State Governors in non-Labor States would not issue the writs for a half-Senate election on the request of the Governor-General. (As late as the night of 10 November, he discussed this very matter with the Governor of Victoria, Sir Henry Winneke, at the Lord Mayor of Melbourne's banquet. Sir Henry said that whatever the propriety of non-issuance, he would be bound to act on the advice of the Premier. 'Of course you're right,' said Whitlam. 'That's the only proper course.') The Liberal Party Council had called for such a refusal. Whitlam thought that it was likely that the Governor of Queensland, Sir Colin Hannah — 'the Premier's palindromic pro-consul' as Whitlam once described him — would certainly refuse. He had recently secured from the Queen the revoking of Hannah's 'dormant commission' to act as Australian Administrator in the absence of the Governor-General. This had been done because Hannah had made a blatant political attack on the Government on 16 October 1975.

In a note to the author explaining his hesitation over a half-Senate election, Whitlam wrote: 'Am not going to allow the present Governor-General to be the first in our history to be rebuffed by a State Governor not only wrongly advised but instinctively willing to break an unfailing and unchallenged tradition of three-quarters of a century.' And to explain why he was reluctant to ask Sir John Kerr to send a message requesting the Senate to vote on the Budget, Whitlam wrote in the same note: 'It is not the function of the Crown to bail out the Opposition.'

*

There were two political campaigns in 1975 — the campaign against the Senate before 11 November and the election campaign after 11 November. The first was one of the most brilliant and successful in Australian history.

From 15 October to 11 November, Whitlam waged one of the great battles in the history of Parliament. Clyde Cameron, still smarting under his demotion and still determined on revenge, said in Parliament on 16 October:

> How could any lover of democracy fail to admire the fighting qualities, the resilience and the intellectual qualities of the present Prime Minister? Why is it that the Federal Parliamentary Labor Party is now being seen at its magnificent best? Why is it that the Labor movement outside Parliament is now more solidly united than ever before? It is because of the inspiration given by the Prime Minister in this present crisis, the man who in this Parliament stands out like a giant against the intellectual and moral pygmies who sit opposite him. It is because the Prime Minister has thrown down the gauntlet in defence of parliamentary government that I stand proudly beside him. That is why I and my ministerial colleagues stand solidly behind the Prime Minister in this, the most important fight of his life, the most important fight that the Australian people have ever fought in their lives.

From Cameron, it may have been tongue-in-cheek; but it caught exactly the deepest feelings within the Labor Party. In Caucus, Senators Wriedt and Wheeldon argued in favour of an election. Overwhelmingly, Caucus urged Whitlam to 'tough it out' — a phrase he himself did not use and which did not really express his strategy.

Whitlam's strategy had a single, clear aim: to force the Senate to vote directly on the Budget. Everything else was secondary to this essential aim. To get such a vote was the key to victory.

There were good reasons to believe that Fraser could not keep all his followers in the Senate in line if they were asked to vote directly to reject the Budget. At least four—Senators Donald Jessop, Eric Bessell, Alan Missen and John Marriott—were known to object on principle to the use of the Senate power to deny Supply. After Connor's resignation, they were willing to go along with deferral. But outright rejection was another question altogether. Fraser asserted in Parliament that his reason for choosing the tactic of deferral rather than outright rejection was to allow the Budget to be revived by a new government. But the tactic was really chosen because it was the only way he could hold the support of all his followers in the Senate.

The pressure was applied by Whitlam at all points—in Parliament, by mass meetings in all capitals, through the press, on television. Robert Hawke immersed himself in the campaign to ensure complete union involvement. The result was a remarkable turnaround in public opinion. From a low of 37 per cent support in August, the Government's support rose to 46 per cent. More significantly, a poll published on 30 October showed 70 per cent support for passing the Budget. From then on, the mood in Parliament was transformed. Whitlam's ascendancy became perhaps more complete than at any previous time. For the first time, a crack appeared in Fraser's confidence. In an interview with the Governor-General, he proposed a compromise—he would allow the Budget to pass on condition that a House of Representatives election be held within six months. Whitlam rejected the proposal out of hand. 'I would like to warn Honourable Members', he told Parliament on 4 November, 'that the proposals derive no greater sanctity from the fact that they were divulged to the Governor-General at the request of their author before they were made public.' Sir John Kerr had not discussed the Fraser proposal with Whitlam, who felt entitled to say: 'I have no doubt that His Excellency was amused. One of the proposals is, I gather, that I should seek an audience of His Excellency and advise him that in six months I intend to advise him to issue writs for an election of the House of Representatives. This is a most diverting proposal. When I advise the Governor-General to issue writs for the House of Representatives, I shall do so at a time of my own choosing. It is in this

House that such decisions are made . . .' Whitlam then reopened the possibility of holding an immediate election for half the Senate. He said he was not yet persuaded to do so. 'Nevertheless,' he said, 'I am influenced by the advice of the editorials in the *Age* and the Australian *Financial Review* to consider an earlier Senate election . . . I am not over-euphoric because of the overwhelming results of the opinion polls . . . I still want to look at the matter.' At this stage, Whitlam's chief purpose in discussing his options was to keep up the pressure. The threat of a half-Senate election in December was just another turn of the screw.

The critical date was 30 November, when the Government's supply would run out. In swinging public opinion around, equally as important as Whitlam's *tour de force* was the threat of disruption caused by the actions of the Senate. On 16 October, Hayden hammered the theme of impending chaos:

> The present course the Opposition is taking will grind this nation to a halt . . . There will be a major economic collapse . . . a substantial number of enterprises in the corporate sector will fail, there will be an upsurge in unemployment . . . aged persons hostels, aged and disabled persons homes . . . will not obtain money necessary to pay the people who provide the services . . . hospital services will grind down. Medical research will have to be stopped . . . education in the States will be short of some $360 million. I cannot see how the defence forces will operate at all . . . people throughout this country will find that they will not be able to obtain the Medibank medical benefits . . . companies will find that taxation will continue at a much higher rate than we propose.

Hayden further pointed out that although pensions were paid automatically and were therefore not dependant on the passing of the Budget, pensioners would be deprived of the increase proposed in the Budget. As the days passed, as the deadline of 30 November approached, the vague menace of October became the grim reality of November. It was this prospect, more than any concern over the constitutional issue, which caused the swing back to the Government. Business grew anxious. Newspapers, including the *Age*, the Melbourne *Herald* and the Adelaide *Advertiser*, began to urge that the Budget be allowed to pass. And within the Parliamentary Liberal Party there were daily increasing signs of nervousness. Each weekday in Parliament they saw their leader diminished, in a way disturbingly reminiscent of the Snedden days; each weekend in their electorates they heard complaints and resentment against their leader's actions.

The two essential elements of Whitlam's strategy were to keep up the pressure and to buy time. The Government's arrangements to provide emergency finance after 30 November were designed to do both. The centre-piece of the plan was a proposal to pay public servants and essential contractors by a voucher system, in effect, to allow banks to give credit on the basis of a government guarantee. There was never any suggestion that such arrangements could be more than a stop-gap measure. Their purpose was not, as Fraser alleged, to allow the Government to govern without Supply. Their purpose was to buy time — to allow even more time for public pressure to force the Senate to vote one way or the other. The proposals were closely scrutinized by the Law Officers to ensure their legality. Sir John Kerr did not raise with the Attorney-General or the Treasurer any question of their legality. He did, on 6 November, question Hayden on their workability. In his statement dismissing the Government on 11 November, Sir John Kerr said merely: 'The announced proposals about financing public servants, suppliers and contractors and others do not amount to a satisfactory alternative to supply.' They were not intended to be. They were part of the political effort to force a political solution to a political problem.

The object of Fraser's strategy was as clear-cut as Whitlam's; his need was the mirror-image to Whitlam's. Whitlam needed to force the Senate to vote directly — to pass the Budget or reject it. Fraser needed to prevent such a vote — the vote he could not win. So he had to hold his senators to the line agreed upon on 15 October — to vote for deferral, to delay but not reject Supply. In a battle of nerves, Fraser had to worry about not only his own nerves but his followers' nerves, particularly the nerves of the senators opposed to the rejection of the Budget. In that sense, 'toughing it out' describes Fraser's stance much more accurately than Whitlam's. Fraser had two handicaps: the Connor resignation, his ostensible excuse for blocking the Budget, was swamped in the massive political crisis; and public opinion was clearly turning against him.

To overcome these handicaps, Fraser astutely tried to keep alive marginal issues—by a number of little matters to keep building up a picture of dishonesty and shadiness on the part of the Whitlam Government. An attempt was made to portray a briefing on the Budget given to Hawke a few hours before its presentation as a Budget leak of scandalous proportions—another 'extraordinary and reprehensible circumstance'.

The centre-piece of all these efforts was the apparition of Tirath Khemlani himself in Canberra on 27 and 28 October. The purpose of this visit has never really been disclosed, nor who paid for it. But there he was. All that is known for certain is that from the time he reached Canberra to the time he left Canberra he was in the custody of two members of Lynch's staff and was visited by two Liberal members, Robert Ellicott and John Howard, (Bennelong N.S.W.). Khemlani arrived with a pile of documents in suitcases. After Ellicott had gone through the documents, he stated (in Sydney): 'Nothing I have seen involves the Prime Minister.' A few hours later, after consultation with Fraser (in Canberra) he revised his statement and his view.

The bizarre nature of Khemlani's descent on Canberra was summed up with perfect accuracy by Fred Daly in answer to a parliamentary question:

> I understand that yesterday afternoon a Commonwealth ministerial car was booked by the Deputy Leader of the Opposition to meet Ansett flight 361, 2.10 p.m. from Sydney, and that of course commenced a drama in Canberra yesterday that has rarely been equalled. The car was to meet a person named Mr Khemlani. I understand that the gentleman approached the Commonwealth car dressed in a safari suit and wearing dark glasses. He was met by bearded investigators who hustled him into the VIP room while the Commonwealth car backed into the normally restricted luggage area and his 8 bulging briefcases were loaded into it.
>
> Mr Khemlani was then pushed into the Commonwealth car along with 2 sinister bearded staff members and taken on a high speed car chase through the back streets of Fyshwick reaching

speeds of 100 kilometres per hour, turning down side streets and doing sudden U turns before coming to a sudden stop at his destination—a $23 a night room at the Hotel Wellington. Mr Khemlani, still using the car, and the men then disappeared into room 49—the room adjoining the motel shoe-shine box. Lemonade, potato chips and 2 Sydney afternoon papers were pushed through the breakfast hatch. He stayed locked in his room while the staff members stayed huddled in a corner sifting through his 8 suitcases of documents. Later in the afternoon Mr Khemlani was taken on another high speed car chase. This time, as a taxi pulled up at the front of the motel, Mr Khemlani disappeared out the back door and sped off in a late model gold Torana with the manager of the Wellington Hotel at the wheel. That is service. It raced through the peak hour traffic, went one and a half times around State Circle, and reached speeds of up to 120 km along Commonwealth Avenue before swinging around and returning to the hotel. Then Mr Khemlani disappeared.

An hour later his brief cases were lugged into a lift at the $33 a night Lakeside Hotel where Mr Khemlani usually stays. But he was not booked in there last night. Last night Mr Khemlani was locked up with two Opposition front benchers, Mr Bob Ellicott and Mr John Howard, going through suitcases full of documents. As if he were not in enough trouble without being locked up with them! I come back again to the Commonwealth car. Poor Mr Khemlani: he had come all the way from Singapore, at his own cost and without a visa, to clear his name and he had all that excess baggage with him. What must he think of Australia—his life was endangered by high speed car chases in Commonwealth cars; his bags were searched by bearded investigators, and as far as we know they were not false beards; he was booked into a $23 a night room next to a shoe shine box yet his bags were booked into a $33 a night international hotel; he was locked up all afternoon with bearded men and then all night with 2 members of the Opposition.

Mr Whitlam: And fed with peanuts.

Mr Daly: And then fed with peanuts. He must also be wondering why the Opposition would pay out all that money for his bags but was too lousy to pay for a taxi fare for him to go from the airport to the hotel. That brings me back to the original point about the misuse of a Commonwealth car . . .

Equally bizarre were the circumstances surrounding an incident which the Defence Department chief, Sir Arthur Tange, was to describe as 'the gravest risk to the nation's security there has ever been'. At Port Augusta on 2 November, Whitlam made an off-the-cuff reference to allegations that the CIA had funded the National Country Party and that its leader, Doug Anthony, had friends in the CIA. The reaction was remarkable. The *Financial Review* revealed in a story by Brian Toohey that Anthony had rented his Canberra house to one Richard Stallings, a CIA officer at the Pine Gap installation. The real significance of this revelation was not any link between Anthony and the CIA (Anthony was unaware of Stallings' CIA role) but the link between Pine Gap and the CIA, an involvement of which the Government had been unaware. Against the specific request of the Defence Minister, William Morrison, Anthony insisted on pursuing Whitlam's Port Augusta generalization in Parliament. Thus, at Anthony's insistence, not by Whitlam's conduct, and against the pleas of the responsible minister, the role of the CIA and the name and activities of one of its operatives were dragged into the parliamentary debate at a time of supreme political tension. According to an extraordinary cable from the ASIO officer in Washington to ASIO Headquarters, dated 10 November, the CIA believed that Whitlam was about to reveal the names of current CIA operatives in Australia; the CIA was threatening to break its intelligence links with Australia. On 8 November, a Defence Department official briefed Sir John Kerr on these developments. Whitlam was unaware of the briefing and he was unaware of the alleged 'crisis' in relations between the intelligence networks of Australia and the United States.

Until Sir John Kerr tells his story, it is impossible to say what part, if any, this matter played in his decision to dismiss the Whitlam Government. All that is now known is that on 6 November he congratulated Whitlam for his handling of the constitutional crisis and left the Attorney-General, Kep Enderby, and the Minister for Labour, Senator James McClelland, an old friend and associate, with the impression that he had no intention of intervening; that on 8 November he received, unknown to the Government, the Defence Department briefing;

and that on 9 November he telephoned the Chief Justice, Sir Garfield Barwick (again without Whitlam's knowledge and contrary to Whitlam's expressed advice), to ask him for advice on the exercise of his powers to dismiss a government — advice on, as Sir Garfield wrote to Sir John on 10 November, 'a course on which you had determined'.

Whitlam, also, had determined upon a course on 9 November. He had at last decided to request an election for half the Senate. It was to be the last turn of the screw in the battle of tactics which he was waging and which he and most observers believed he was winning. The threat of a half-Senate election gained force from the fact that the High Court had recently ruled that legislation to provide for two senators each from the Australian Capital Territory and the Northern Territory was valid. There was every chance that the Government could obtain at least a temporary majority in a new Senate. When, on the afternoon of 9 November, his public relations officer, David Solomon, informed Whitlam that Fraser had requested a meeting to discuss the deadlock Whitlam agreed to a meeting at 9 a.m. on Tuesday 11 November at which he intended to proffer a half-Senate election as a last bargaining counter. The real target was not Fraser but the Liberal Senators Missen and Jessop, who were on record as saying that they would not vote outright to reject the Budget.

On Sunday 9 November, Whitlam made what was designed to be a final appeal to the people as their Prime Minister. It was made with absolute confidence that the crisis was about to end. It was a message of reassurance:

> There has been this paradox: that on an issue containing the seeds of deep division and disruption there has been a display of national unity and common purpose from the vast majority of ordinary Australians and from all sections of the community such as we have not had on any issue since the war. I have drawn great reassurance from it and I believe all Australians, whatever their political persuasion and indeed whatever their opinion on this specific issue may be, can take reassurance from it — confidence about the fundamental stability and maturity of our country . . . The second ground for reassurance is the loyalty and solidarity already shown by the public service throughout Australia. Clearly

the basic business of the nation cannot continue without their co-operation. For instance, while money is certainly available to pay all pensions, the actual delivery of payment requires that officers can continue to do their job. It's because of this that the Government is determined to do all within its constitutional and legal power to ensure that pensions are paid, that public servants can do their job and that the armed forces can continue at their posts ... The third ground for confidence and reassurance is the attitude of the industrial movement and increasingly, as the implications of the Senate's conduct become clearer, the attitude of the business community. I acknowledge that before the Budget, most businessmen would have preferred a change of Government. But it's been very clear for the past month that what most businessmen now want is for our Budget to pass and for that Budget to be given a real chance to work ... The people of Australia are now very properly asking 'Where will it end?' It will end as soon as the Senate stops this nonsense, abides by the Constitution and passes the Budget and allows the duly elected government to govern.

This speech had been prepared at Kirribilli House, Sydney. Unknown to Whitlam, events taking place a hundred yards away at Admiralty House had made it irrelevant. In late October, Kerr had asked Whitlam if he should consult the Chief Justice of Australia, Sir Garfield Barwick, about the crisis. Whitlam advised that it would be inappropriate; the Chief Justice was not under the Constitution a constitutional adviser to the Crown, this role being reserved for the Law Officers; the title of 'Chief Justice' did not confer an authority upon the holder higher than that of his brother-judges; and there were matters arising from the crisis which could become matters for litigation before the High Court. Whitlam further told Kerr that while a previous Governor-General, Munro-Ferguson, had sought an opinion from the Chief Justice on the question of a double dissolution in 1914, this precedent had been superseded by the Statute of Westminster defining the nature of Australia's Dominion status. This specific advice notwithstanding, Sir John Kerr, late on the evening of 9 November, rang Sir Garfield Barwick at his Sydney home. According to Barwick's statement at the National Press Club on 10 June 1976, Kerr told him that 'he had decided on a course of action' and 'he asked me to give him some advice. I said it depends what it is, and that I would call in and see him, which I did'.

Barwick saw Kerr at Admiralty House next day and later wrote to him: 'in response to Your Excellency's request for my legal advice as to whether *a course on which you had determined* (Author's italics) was consistent with your constitutional authority and duty' that it was his opinion 'that if Your Excellency is satisfied in the current situation that the present government *is unable to secure supply* (Author's italics), the course upon which Your Excellency has determined is consistent with your constitutional authority and duty.'

The Government's ability to secure Supply before the deadline of 30 November was a political question, not a constitutional one. Whatever the propriety of Kerr's taking legal advice from Barwick against Whitlam's express advice, there can be no question about the proper source of advice on political questions. Yet, between 9 November, when Kerr had 'determined' to dismiss the Whitlam Government, and 11 November, when he did so, he made no attempt to seek advice from his Prime Minister or any other of his constitutional political advisers. In particular, he sought no advice on the ability of the Government to secure Supply. The well-grounded advice of the Prime Minister would have been that the crisis would end that week and certainly before the deadline when Supply would be exhausted. If that advice had been sought and tendered, Kerr would then have been in the position of accepting or rejecting the advice of his political advisers. He avoided that course by the simple expedient of not seeking the advice of the only person constitutionally competent to give it—the Prime Minister.

Soon after 9 a.m. on 11 November, Whitlam, Frank Crean and Fred Daly met with Fraser, Phillip Lynch and Doug Anthony for a final confrontation in the Prime Minister's suite at Parliament House. Fraser said that nothing had changed; unless the House of Representatives election or a double dissolution was announced the Budget would not pass. Whitlam said he would that day wait on the Governor-General to advise an election for half the Senate. The meeting ended quickly. Crean said to Whitlam,

'They seem awfully cocky. Are you sure Kerr's O.K.?' Whitlam said, 'Of course. He understands the Constitution and he knows his duty — to accept the advice of the elected Government.' Daly expressed similar misgivings, for he too was disturbed by the apparent confidence and intransigence of Fraser and Anthony. Yet it must be emphasized that the doubts of Crean and Daly were exclusively about whether Sir John Kerr would accept the advice of the Prime Minister for a half-Senate election. Dismissal was beyond their wildest fears or imaginings.

Returning to his own Party room, Fraser did not report on his meeting with Whitlam. He did not inform his supporters that Whitlam intended to seek a half-Senate election. Instead, he asked for 'support and trust'. And, at a press briefing after the meeting, his deputy, Phillip Lynch, said: 'We believe events will work themselves out . . . We believe the present course is sound *for reasons which will become apparent to you later*' (Author's italics).

Caucus applauded as Whitlam announced his decision to request an election for half the Senate. Whitlam, confident and ebullient, suggested to Enderby that he himself might, after all, attend the Remembrance Day wreath-laying ceremony at the Australian War Memorial where Enderby was to represent Whitlam and the Government. Enderby however insisted on going. At the War Memorial, he escorted the Governor-General and Lady Kerr to and from their car. Kerr spoke not a word. As Enderby farewelled the vice-regal couple, only Lady Kerr spoke, with stony finality: 'Goodbye Mr Attorney.'

It was a crystalline spring day in Canberra. From his office window at Parliament House, Whitlam witnessed a strange tableau. On the road beneath him, two people were being photographed. Jim Cairns and Junie Morosi were posing for pictures to promote a forthcoming book by Morosi. Arm in arm, under a flawless blue sky, they posed for the cameraman — now beneath a tree, now in the rose-garden. It was Arcadia before Armageddon.

The scheduled parliamentary event of 11 November was to be a debate on a motion of want of confidence in the Government of which Fraser had given notice the previous Thursday. Yet the debate — the first full dress debate on so solemn a question in his period as Leader of the Opposition—was curiously low-key. For so important a question, theoretically at least, involving the survival or fall of the Government, it was marked by a strange lack of tension and attention. With hindsight, it can be seen that the only significant part of Fraser's speech was his reference to the Governor-General. Fraser said: 'The Prime Minister has not said that he would accept the Governor-General's decision taken in accordance with his constitutional prerogative. There are circumstances, as I have said repeatedly, where a Governor-General may have to act as the ultimate protector of the Constitution. He ignores that prerogative.' The point is that Whitlam had not said, or not 'not said' anything of the sort; such a question had never been raised with him—not in the Parliament, not by the press, not by Fraser and above all, never by Sir John Kerr. No such use of the prerogative had ever been raised. For the rest, Fraser's speech was a rehash of the old allegations about the loans affair.

Whitlam replied in what, unknowingly, was to be his last speech as Prime Minister. It was the speech about which he had commented a few hours earlier, 'Are we being too hard on Malcolm?'

> I said three weeks ago that it would be my duty to make the Australian people aware that the Leader of the Opposition was, in his own description of the Right Honourable Member for Higgins (Mr Gorton), 'Unfit to be Prime Minister'. The people seem to be getting the message very quickly. But when one considers the lengths he is apparently prepared to go to damage Australia, to cause harm to innocent people, to delay or even destroy the chance of economic recovery and in particular, the role he is playing in attempting to undermine the rights and powers of this House—this House of Representatives, this people's House—he is unfit to be Leader of the Opposition.

At Government House, soon after 1 p.m. on 11 November 1975, the Governor-General, Sir John Kerr, handed the Prime Minister of Australia, Edward Gough Whitlam, a letter dismissing him and his Government. Ten minutes later, he commissioned John Malcolm Fraser to form a government.

Whitlam (on 13 November) made the following record of the events of 11 November as they appeared to him at the time:

> My appointment was for 1.00, made at 10 a.m. The appointment was not for immediately after I finished speaking because on a no-confidence motion I would not leave the House. I left immediately the House adjourned, i.e. 12.55. Fraser received (his) invitation while in the House, presumably while I was speaking, i.e. 12.9 to 12.35. He did not remain in the House to hear me speak, which would have been not only the courtesy but the norm.
>
> In his eagerness he left for Government House and arrived there before I did. If I had seen his car at the entrance, I would have been alerted. If I had seen him before I entered the Governor-General's study, I would have been alerted. Fraser was, I believe, in a reception room opposite the study. The doors were closed. They were closed because, in the aide's words, a successor was being interviewed in there right now. The aide meant me to assume it was a successor to himself, because he had told me he was leaving. It was, of course, a successor to me.
>
> 1. The Melbourne *Herald* yesterday had a story about my asking the Queen to withdraw the Governor-General's commission. No such story came from me. It is clear that the Governor-General set up this scenario to prevent my contacting the Queen.
>
> 2. Before concluding his statement announcing that he had been commissioned to form a caretaker government Fraser said that double dissolution papers were being prepared. He sat down at 2.47. Before adjourning at 3.15 the House had expressed its want of confidence in him and its confidence in me. Any double dissolution papers therefore should have been presented to the Governor-General not by Fraser but by me.
>
> 3. The Governor-General refused to see the Speaker for one and a half hours. He used the intervening period to draft and sign (the) double dissolution proclamation.
>
> 4. There may be differences of opinion on my right to contact the Queen, on my Governor-General's right to dissolve the House of Representatives or both Houses, but there can be no doubt about the Governor-General's duty to receive The Speaker on a message from the House concerning the Prime Minister he had just appointed and about the Governor-General's duty to have as his Prime Minister during the election campaign a person in whom the dissolved House had confidence.

The Governor-General at all stages set aside and ignored and avoided the views of the House of Representatives which is the very issue on which the issues arose and the election must be held. The Governor-General may or may not have been right in bringing about an election. He had no right whatsoever to continue Mr Fraser's appointment as Prime Minister during the campaign for the election. The Governor-General never brought his influence on Fraser to have his Senators vote on the Budget before Tuesday. The Governor-General in the end ignored the rights of the House of Representatives. Whatever I said or did at 1 p.m., whatever the House of Representatives or Senate thereafter did, the Governor-General was determined that Fraser should go to the people as Prime Minister.

On returning to the Lodge from Government House, Whitlam's first action was to telephone Margaret Whitlam at Kirribilli House, Sydney. It is a measure of the lack of any sense of crisis on that day that she was not in Canberra—where she has always been in any serious crisis—at her husband's side. Whitlam's Principal Private Secretary, John Mant, and the author were then summoned to the Lodge at 1.15 p.m. We thought Whitlam wished to discuss dates and venues for the campaign for the Senate election which he had gone to Government House to request. Whitlam was sitting alone in the small, glazed breakfast room beside the Lodge dining-room. Whitlam was eating his customary steak. There is perhaps something to be said about a man who can eat a steak minutes after the greatest shock of his life. As we entered, Whitlam stated quietly but bluntly: 'I've been sacked.' One's first reaction was that this was one of his more heavy-handed sallies. 'No. I'm serious. Here's the letter.' And he passed the letter across the table:

Dear Mr Whitlam,
  In accordance with section 64 of the Constitution I hereby determine your appointment as my Chief Adviser and Head of the Government. It follows that I also hereby determine the appointments of all of the Ministers in your Government.
  You have previously told me that you would never resign or

advise an election of the House of Representatives or a double dissolution and that the only way in which such an election could be obtained would be by my dismissal of you and your ministerial colleagues. As it appeared likely that you would today persist in this attitude I decided that, if you did, I would determine your commission and state my reasons for doing so. You have persisted in your attitude and I have accordingly acted as indicated. I attach a statement of my reasons which I intend to publish immediately.

It is with a great deal of regret that I have taken this step both in respect of yourself and your colleagues.

I propose to send for the Leader of the Opposition and to commission him to form a new caretaker Government until an election can be held.

Yours sincerely,

(signed)

John R. Kerr

The exchange was repeated as each of Whitlam's colleagues arrived — Frank Crean, Fred Daly, Kep Enderby, David Combe, John Menadue, except that in the case of his Cabinet colleagues he announced: '*We've* been sacked.' To Enderby, he said: 'He's done a Game on us' — referring to the dismissal of Premier John Thomas Lang and the New South Wales Labor Government by the Governor, Sir Philip Game, in 1932. In each case there was the same reaction — puzzlement followed by disbelief, outrage quickly subsiding into shocked silence. It is a measure of the degree of shock that the irrepressible Fred Daly could find nothing funny to say. The difficulty was to find anything at all to say.

With his colleagues in this catatonic state (better described by Daly as 'stunned mullets') Whitlam tried to work out a plan of action. His preoccupation was to frame a resolution for the House of Representatives, where his majority remained unbroken, which could circumvent Sir John Kerr's action. Whitlam was later to be harshly criticized for failing to include his Senate colleagues, particularly Ken Wriedt, in these discussions and keeping them in ignorance of their dismissal. This caused abiding resentment. As a result, Wriedt, in total ignorance, was delighted when Senator Withers told him the Budget would at last be allowed to pass. Wriedt thought that the crisis was over and that the expected had at last happened and the Opposition had caved in. While the final outcome could not have been changed (Fraser had the commission, he was already Prime Minister, and the coalition controlled Senate business), Whitlam's failure had closed off a tactical option. The omission is difficult to excuse but easy to explain: nobody around that table at the Lodge thought of the Senate! Whitlam's preoccupation was with the House of Representatives. And the strategy, the last desperate throw forming in his mind required that the Budget should be passed. Once the Budget passed, the crisis was at an end; the House of Representatives could resume its normal role as the House determining who was the government of Australia.

The proposed resolution Whitlam wrote at The Lodge in his own hand stated: 'That this House declares that it has confidence in the Whitlam Government and that this House informs Her Majesty the Queen that if His Excellency the Governor-General purports to commission the honourable member for Wannon as Prime Minister the House does not have confidence in him or in any government he forms.'

The debate on Fraser's no-confidence motion continued. Frank Crean had the call to speak after the luncheon adjournment. He rose to deliver at 2 p.m. what must be one of the most remarkable efforts in restraint in the annals of the Australian Parliament or any parliament. Whitlam, still seeking time to work out tactics, had asked him not to reveal directly what had happened an hour before. Crean's speech covered the familiar ground of Senate obstruction. His only reference to the momentous, incredible event to which he was privy was at the beginning of his speech (Hansard still referring to him incorrectly as the Minister for Overseas Trade):

> I want to say to begin with that this issue today is a constitutional issue of great importance. It is about who has the right to govern in

this country. It is about the rights of another place with respect to this House in what are called financial matters. It was said earlier this morning that the Queen, the House of Representatives and the Senate are what constitute the Parliament as a functioning body. What needs to be spelt out is that the Queen's representative in Australia, the Governor-General, does not act on his own initiative but acts on the advice of his Ministers. Who the Ministers are is conditioned by who has the majority in the House of Representatives. I would hope that everybody, in this House at least, would assert that as a fundamental ground rule of the Australian parliamentary system.

When Crean finished, the House of Representatives by a majority of seven voted to reject Fraser's motion of no-confidence in the Whitlam Government and voted, by a majority of seven, its confidence in the Whitlam Government. At 2.34 p.m., Fraser arose on the call of Speaker Scholes to announce that he had been commissioned as Prime Minister, that the Budget had already passed in the Senate and that the Governor-General had accepted his advice as Prime Minister for a double dissolution. He then moved the adjournment of the House. His first motion as Prime Minister was defeated by ten.

Whitlam scrapped the resolution he had drafted at The Lodge and moved: 'That this House expresses its want of confidence in the Prime Minister and requests Mr Speaker forthwith to advise His Excellency, the Governor-General, to call the Honourable Member for Werriwa to form a government.' Whitlam then said:

The Governor-General's views have been read at sufficient length to show that the circumstances upon which he relied no longer apply. There is no longer a deadlock on the Budget between the House of Representatives and the Senate. The Budget Bills have been passed. Accordingly, the Government which twice has been elected by the people is able to govern. Furthermore, as has been demonstrated this afternoon, the parties which the Prime Minister leads do not have a majority in the House of Representatives. The party I lead has a majority in the House of Representatives. It has never been defeated in the year and a half since the last election and in those circumstances it is appropriate, I believe, that you, Mr Speaker, should forthwith advise the Governor-General—

> waiting upon him forthwith to advise him — that the party I lead has the confidence of the House of Representatives, and you should apprise His Excellency of the view of the House that I have the confidence of the House and should be called to form His Excellency's Government.

Whitlam's motion was carried by ten votes. In five votes that afternoon the majority of the House of Representatives reaffirmed the Whitlam Government. Subsequent argument has naturally focused on Sir John Kerr's refusal to see Speaker Scholes until the Parliament had been dissolved. An equally serious aspect of his conduct in relation to the House of Representatives has been ignored. He dismissed the Whitlam Government at the very time the House of Representatives was debating a motion of no-confidence in the Government. The full extent of Kerr's contempt for the House of Representatives can only be grasped if one recalls that the Government was dismissed in the middle of debate upon a 'no-confidence' motion — the highest and most serious of all parliamentary motions. Whatever the role and powers of the Governor-General may be, one thing is beyond doubt: it is not the role of the Governor-General — or anybody — to predict the outcome or pre-empt the House of Representatives in its highest function.

Fraser had first raised the possibility of Sir John Kerr's intervention as early as 17 October. In Hobart, he said that he expected Sir John Kerr to act 'quite soon'. At the first of his public meetings on the crisis, he said: 'It quite clearly is not just a question of what Mr Whitlam does or what the Opposition does. There comes a time that if it has been demonstrated the Government can't govern the Governor-General himself has a role to play.' The *Sydney Morning Herald* next day had as a headline: 'Leaders clash: Will Sir John Kerr act?' The *Australian* headline was: 'Governor-General will act soon, says Fraser'. It was this statement and its press coverage which caused Kerr to complain to Whitlam of 'intimidation' and the inclusion in Whitlam's speech of 21 October of a demand for an end to the 'intimidation' of the Governor-General.

In the absence of any statement about what happened and

who said what to whom and when, by either Kerr or Fraser, dates become the crucial reference point. This debate of 21 October is significant because it was on that night that Fraser had his first interview with Kerr on the constitutional crisis, with Whitlam's knowledge and consent. It was the first of four such interviews before 11 November. The known dates of such conversations are 21 October, 30 October, 3 November and 6 November.

The critical day was 6 November. That was the day on which Kerr saw Whitlam, Fraser, Hayden and Enderby at Government House. It was the last time he spoke to Whitlam before he dismissed him. It was the last time in his life he lunched, drank and joked with one of his oldest friends and associates, Senator James McClelland, then a Minister of the Crown. For the last time, he congratulated Whitlam on his handling of the crisis. And perhaps for the first time he told Fraser exactly what he had in mind—that unless the crisis was resolved in the following week, he would dismiss the Whitlam Government.

*

At his first press conference as the newly-installed Prime Minister, Fraser was asked: 'Mr Fraser, on that note—you've said on 'This Day Tonight' last Friday night that the Governor-General would speak for himself sometime before you could say in confidence that he would sack Mr Whitlam. But you added to your statement that the Governor-General *can* speak for himself you said: "And he will." Why did you make such a firm and definite statement? What information did you have?' He replied: 'Only because I had a proper understanding that the Parliament of Australia is comprised of the Queen, in her case represented by the Governor-General, and the Senate and the House of Representatives.' Next day, 12 November, he repeated the formula when he was asked: 'Mr Fraser, in your radio interview with us on Thursday afternoon last, you said that you wouldn't back down and you thought Mr Whitlam wouldn't back down. But you said "There is another alternative" that you wouldn't expand. Would you say that what has happened is the alternative you talked about?' He replied: 'What I'd pointed out, going back over three or four weeks, that the Parliament of Australia is composed of the Queen and her representative, the Senate and the House of Representatives.' The truth is that this formula — 'the Parliament is the Queen, and in Australia's case, the Governor-General, the Senate and the lower House' — was only used by Fraser at precisely 11.50 a.m. on 11 November in his last speech in Parliament as Leader of the Opposition, made scarcely an hour before he was appointed Prime Minister.

This was only the second time, in thousands upon thousands of words over the previous twenty-four days of crisis, that Fraser had used this formula. Yet he was to use it repeatedly in the next twenty-four hours. Indeed, in all those previous speeches and statements, Fraser had only once joined the Crown, the Senate and the House of Representatives in a single sentence. Speaking on 30 October, he had said: 'The essential principle of the money power is this: it should be wholly within the fully-elected Parliament controlled by the people's vote free from the influence of the Crown. The Senate is as much the people's House as is the House of Representatives.' Whatever that may be as a statement of the Constitution, it was certainly not a statement about the Parliament which existed on 30 October 1975.

In a letter to the Queen after the elections, Whitlam wrote of 'the serious implications' of Sir John Kerr's action for the 'future of the Crown in Australia'. Sir John Kerr had, he wrote, 'put in jeopardy the future of the Crown in Australia and gravely undermined the respect and regard attaching to the office of the representative of the Crown and therefore, to the Crown itself'. In the draft of this letter, Whitlam continued:

> The very clear result of the elections convincingly settles Australia's immediate political future. The elections in no way, however, resolve the legal and constitutional questions raised by the conduct of the Crown's representative on and before 11 November. Nor could the election result of itself legitimise that conduct.
>
> It is not my present purpose to canvass the legality, or even the propriety, of the Governor-General's actions. I enclose however relevant documents expressing the opinions of eminent jurists on these points.

My immediate concern and contention is that the manner in which the Governor-General chose to invoke and exercise the reserve powers of the Crown has put in jeopardy the future of the Crown in Australia and has already gravely undermined the respect and regard attaching to the office of the representative of the Crown, and therefore, to the Crown itself.

I assert that the Crown can have no enduring future in Australia except by the continuing consensus and assured assent of the overwhelming majority of the people. I further assert that that majority must transcend traditional political allegiances and temporary political attitudes. I finally assert that these conditions can apply only if the Crown continues to avoid any intervention, or appearance of intervention, on behalf of any of the contending political parties. I fear that these conditions no longer apply in Australia.

Sir John Kerr used the reserve powers of the Crown to make at least five political decisions. All these decisions favoured one political combination against the other, which happened to be the party with an assured majority in the Lower House.

At no time did he inform me as Prime Minister of the resolution he had formed to dismiss my Government. He refused not merely to accept but even receive my advice recommending steps to bring about an election for half the Australian Senate. He rejected the opinion of the Crown Law Officers and accepted the contrary opinion of a private member of Parliament, albeit a former Solicitor General. Against my express advice, and contrary to all proper practice, he consulted the Chief Justice on a matter that could well have become a matter for judgement by the High Court itself. He refused to receive the Speaker of the House of Representatives, acting on the express instructions of the House, until Parliament had been dissolved.

The events leading up to 11 November were essentially a political crisis, a political deadlock between the two Houses and capable of political solution. The Governor-General chose to make a political judgement to the effect that his constitutional advisers had exhausted all political means to solve this political crisis by procedures legally and constitutionally open to them. He refused to receive the advice which would have shown that such a conclusion was unwarranted . . .

Far from resolving the constitutional issues, the recent political crisis in Australia has only obscured them. They must await future clarification. I regret to say, but am in duty bound to say it, that the actions of Sir John Kerr, as representative of the Crown in Australia, have been such as to call into question on the part of many millions of Australians, particularly the younger majority, not merely the limits of the powers of the Crown, but its whole future role in Australia.

*

What happened on 11 November 1975? The Australian writer, Donald Horne, answered the question in his book, *Death of the Lucky Country*, struck off in three weeks of passion and pain:

> This: The Governor-General secretly made a decision, the effect of which was to support the political plans of the Liberal and National Country Parties.
>
> Against all contemporary practice he did not discuss that decision with the government that was then in power. But having contemplated the decision secretly he secretly got for it the support of the Chief Justice, a person of no more constitutional significance in this matter than you or me, but one whose respected office could seem to give extra authority to what the Governor-General had decided. The Governor-General then mounted a time-tabled operation, for which the phrase 'constitutional coup d'etat' seems a useful description. It was an operation which had the general effect of leaving the Prime Minister with a false sense of security, then, without discussing any alternatives, kicking him out of office, installing the minority leader as Prime Minister, then dissolving Parliament. It all happened so quickly that no preventive action could be taken.

In its deepest sense, however, what happened on 11 November was simply the crowning point of all that had gone before: the denial of the legitimacy of a Labor Government. The dismissal of that Government by the Governor-General, the Queen's representative, the Crown in Australia, was merely the ultimate expression of that denial of legitimacy. The denial began first with the assertion in April 1973 that the election of 1972 was just an aberration, 'an act of lunacy by the larger states', as Senator Withers had put it. Then began the process of Senate obstruction, even on matters like health insurance and the Schools Commission, so clearly within the mandate, however narrowly or widely the doctrine of the mandate is interpreted. This led to the Double Dissolution of 1974. Yet, even by its re-election and the strengthening of its position in the Senate, the Labor Government was still unable to establish its legitimacy. It remained a Government under siege, under question, under doubt as to its legitimacy. By November 1975, a climate had been created in which the Crown, the fount of legitimacy, the fount of honour, the fount of authority, could at one stroke declare that the Labor Government was indeed illegitimate.

After the proclamation dissolving the Parliament was read out on the steps of Parliament House, Whitlam declared, 'Well may they say "God Save the Queen" for nothing will save the Governor-General'. In a second appearance before a crowd which had assembled in anger in front of Parliament House, he said, 'Maintain your rage'. Throughout the campaign which followed, these were the last and only direct references to Sir John Kerr and what Sir John Kerr had done to him and his Government. One of the great myths of the 1975 election campaign is that Whitlam campaigned against Sir John Kerr and that the election result was a public endorsement of what he had done. Whitlam knew and accepted the advice that the campaign could not be about or against the Governor-General or his actions.

By keeping his Prime Minister so utterly in the dark about his intentions, Kerr had closed off every option and every weapon available to an elected Prime Minister. By forcing him to an election as Opposition Leader, he deprived the Labor Party of its major remaining asset — the authority of a Government and the information available to it. Even his terms for the conduct of the 'caretaker' Government were couched in the most damaging way; Fraser's undertaking that there would be no Royal Commission into the Labor Government's activities merely ensured that the public was left with the impression that the Governor-General believed that its loan-raising activities were fitting subjects for a Royal Commission at some later time.

One illustration of how seriously deprived Labor was because of Kerr's action will suffice. During October, Treasury had revised its Budget forecasts, indicating an improvement in the economic situation. The Treasury now believed that inflation would fall to 10 per cent by the end of 1976, that the wage splurge had eased and that recovery would be well underway by June 1976. Hayden was naturally anxious to publish these new, optimistic forecasts. Treasury resisted publication — its standard line of resistance — and brought a new argument to bear. 'These forecasts', wrote First Assistant Secretary, R. W. Cole, in a Treasury minute dated 30 October, 'take no account of the current funds crisis and could not be put to any practical use until that crisis is resolved. If there is significant delay in resolving it, the forecasts would have to be revised.' Next day, Sir Frederick Wheeler wrote to Hayden endorsing Cole's opinion: 'I also emphasize that I share Cole's view that the question of whether to publish, or not to publish, is quite academic until we have a settlement of the funds crisis and a reappraisal of the situation at that time.' By the time the crisis was over, the Treasury — with its favourable forecasts — was in other hands. During the campaign, Fraser accused Hayden of 'stealing Treasury documents'. The truth was of course that he had been deprived not only of his documents but of the authority as Treasurer which would have made his release of the Treasury forecasts a formidable campaign weapon. Nothing had contributed so deeply to the Labor Government's decline as the fear of the middle class that inflation had taken an unbreakable hold and was beyond control. It was this fear which made the removal of Labor by any means acceptable to the conservative middle class. At the time of the greatest fear, the Government was deprived of its best means of allaying the fear.

The opinion polls showed exactly what had happened. As long as the constitutional crisis lasted, the Labor Government had improved its rating massively. As soon as the Governor-General delivered his 'verdict', the polls reverted to their pre-crisis levels. Surveys conducted by the Labor Party's agency told exactly the same story. During the crisis, public support for the Government's stand, and the Government's general standing, rose rapidly; as soon as the crisis was resolved, interest in the constitutional issues fell away and, with it, support for Labor.

The public meetings were the biggest and most enthusiastic in memory—except for those who could remember the meetings of J. T. Lang after his dismissal in 1932. It has been asserted, by Reid among others, that Whitlam, his colleagues and his staff were misled by this manifestation into disbelieving the opinion polls. This is only partly true. The whole truth is that we drew a mantle of unreality around us, some deliberately, others subconsciously, creating a world of make-believe. It was only that which allowed us to function and allowed us to keep at bay the effects of the shock of 11 November. But almost all Labor people were infected by the enthusiasm of Labor supporters. In Perth, the veteran of a dozen campaigns since 1943, Kim Beazley, predicted that Labor would keep all its Western Australian seats—and win Forrest. In the event, Beazley survived as the only member in the House of Representatives in Western Australia.

But the opinion polls had told it all. For a national vote of 43 per cent, Labor was reduced to a rump of 36 seats in a House of 127—a better vote than in 1966, but a worse result. The gains of 1969, 1972 and 1974 were swept away. The electoral land-marks of Whitlam's long road to power were torn up—for days it seemed that even Corio, the turning point in 1967, might be lost.

*

In the golden summer of 1972–73, Whitlam had confidently predicted eight years for himself as Prime Minister and at least three terms for the Labor Government. If his opponents regarded this as an example of Whitlam's arrogance, most at least conceded the likelihood of two terms—six years—for Labor. In the event, Labor under Whitlam did win two terms. Yet, it governed Australia for just under three years. The devastation of 13 December 1975 occurred on a day which might very well have been the chosen polling day for the Whitlam Government at the end of the three-year term for which it had been elected in December 1972. In the world economic climate of the time, probably no Government could have been re-elected. Across the Tasman, the New Zealand Labour Government went down in an even greater electoral avalanche. If mere days in office were any worthwhile measure of political achievement, then the Double Dissolution of 1974 and the dismissal of 1975 may have made little difference. But by the measure of effective power these events transformed the nature of the achievements and the reputation of the Whitlam Government.

Throughout its whole period, but particularly after 1974, the decisive factor was its lack of a majority in the Senate. More important than the rejected or stalled legislation was the psychological effect of the constant threat from the Senate. After 1974, the Labor Government was a government under siege. As Whitlam said in his 1975 Chifley Memorial Lecture: 'It may be true that hanging concentrates the mind wonderfully, but it is not generally regarded as good for the morale or health.'

It is not necessary to conjure up a conspiracy by the CIA, as allegations made in 1977 suggested, to find the real reason for the 'destabilization' of the Whitlam Government. The threat from the Senate and the certain knowledge after the 1974 Double Dissolution that Labor's lack of majority in the Senate would be used against it whenever it suited the Opposition is sufficient explanation. This uncertainty and instability extended beyond the Cabinet and the Caucus to the public service itself. Disaffection of many senior officials — and it was not from Treasury alone that leaks to the Opposition and the press came — was the reaction of men who wished to cultivate their likely new masters. The situation in the Senate eroded the authority of the Government at a time when, beset by unprecedented economic problems, it needed every ounce of authority that the position of Government should confer. And underlying all, from the beginning, was the denial of the legitimacy of a Labor National Government.

No Prime Minister has unashamedly enjoyed the exercise of power more than Whitlam yet no Prime Minister has cared less for the mere possession of the titles and appurtenances which come with power. The possession of the Prime Ministership was important to him mainly for the opportunity to carry out the programme which he had developed over twenty years. The key elements of Australian politics between 1972 and 1975 were Whitlam's determination to carry out the programme and

Labor's lack of a majority in the Senate. These are the two things which shaped the events of this turbulent time — from the duumvirate to the dismissal.

Nine months after the crisis, on 16 August 1976, Edward St John, Q.C., spoke to a Constitutional Seminar at the University of New South Wales and put this view:

> ... I believe the academics failed to appreciate the degree of concern for Australia's future sincerely felt by responsible people throughout Australia, many of whom, like myself, had voted for a change in 1972, and some of whom had still voted for the return of Labor, even in 1974. They were rightly concerned about the Gair affair and the loans affair, and all that had followed — the sacking of senior Ministers for the deception of Parliament, and so on. Side by side with that was the well-justified concern for the serious inflation, unemployment, and economic mismanagement, and the apparent trend towards nationalization without electoral mandate which was becoming evident in many ways.
>
> Add to this our apprehension that Australian democracy itself was increasingly endangered by this increasingly wild man, Gough Whitlam — more and more the demagogue, less and less the responsible statesman.
>
> From much of this the academic community was sheltered, but practising lawyers were necessarily very conscious of it. They had not the time, perhaps, nor the inclination, to write letters to the press about it, but they had a strong gut feeling that the Senate was at last justified in doing what it did, and that the strong action of Sir John Kerr, to match Mr Whitlam's intransigence, represented our last chance to pull Australia back from the brink before the processes set in train by the Whitlam Government became irreversible.
>
> Extraordinary measures (entirely within their legal powers, but never previously employed), were called for, both on the part of the Senate and the Governor-General, by an extraordinary phenomenon, Edward Gough Whitlam, and his extraordinary conduct.
>
> Historians of the future will need to appreciate this aspect of the matter also if they are to understand the reactions of the Opposition parties, and the majority of the Australian public, in the events of 1975 ...

This reveals more about St John and a whole class than it does about Whitlam and his Government. He mentions the Gair appointment of April 1974 as a cause for concern in November 1975, while conceding that many like him still voted Labor in May 1974. He links the loans affair with the dismissal of senior Ministers; yet those sackings occurred not because of loan-raising activities but because of Whitlam's highly conservative definition of parliamentary propriety. But St John's essential criticism and the source of his fear is that in the weeks of October and November 1975, Whitlam was 'increasingly wild . . . more and more the demagogue'. And this gets to the core of why the events of November happened. Whitlam aroused fear not because of any demagoguery — unless a successful appeal to public opinion at highly enthusiastic rallies is automatically demagoguery. The fear arose because he was so nearly successful.

If victory had been achieved in the circumstances of the 1975 crisis, it is as certain as anything can be in politics that the Senate would never again have attempted to block, much less reject, a Budget. The primacy of the House of Representatives would have been established beyond doubt for all time. In attempting to establish these two principles, Whitlam believed he was doing no more than establish forever what had always previously been taken for granted. In seeking to settle once and for all a principle that had not been challenged in seventy-five years of Federation, Whitlam believed he was acting as a constitutional conservative. But his opponents saw it as nothing short of revolution. Labor saw itself as confirming something long established; conservatives saw Whitlam's actions as a threat to the last bastion of conservatism, the Senate. It was because he was coming so close to success that Whitlam had to be destroyed.

If only by reason of its extraordinary end, there would always be great difficulty in making a final judgement on the Whitlam Government of 1972–1975. Friend and foe alike allow that it was never able to reach its full potential; those who hated it thank God and Sir John Kerr that it was not allowed to destroy Australia; those who loved it will forever feel a sense of loss, not the loss of office — for that was the least valuable part of it — but the loss of what might have been. For a St John who found it a time for fear, there is a Patrick White, a Manning Clark, who found it a time of liberation; and between them lie thousands who found it a time of hope and even happiness. Once Sir John Kerr had acted to force an election, there is no great difficulty in explaining why the Labor Party lost, even to the extent that it did. The denial of legitimacy, sealed by the action of the Queen's representative, economic conditions and the events of 1975, real and trumped-up, together sufficiently account for a swing of six per cent. There is no great mystery about that. It is not so easy to explain the hatred and fear — the fear about which St John wrote — and the reaction against it which allowed Sir John Kerr to destroy it so easily. The speed with which public opinion, recorded in the opinion polls, turned around in the last weeks of the constitutional crisis showed that the majority were by no means unrelenting and unforgiving about the Government's economic policies, much less matters like the loans affair. The majority still thought that the Budget should pass and that there should not be an election.

The most common and most shallow verdict on the Whitlam Government is that it tried to do 'too much too soon'. This is the preferred self-criticism within the ranks of the Labor Party itself. It is less common to hear from that source specific criticisms of particular decisions, just as criticism of each of Labor's budgets by economists ranged from 'too much' to 'too little'. It is probably true that in 1973 and 1974 the Labor Government could have got as much political mileage from slightly less new spending in the big areas — schools, welfare, cities and health, and might have weathered any criticism for cutting defence spending. But its decisions in these areas were in strictest accord with its mandate. The vast increase in education spending rested on the basic principle of accepting the recommendations of the Schools Commission and other education commissions; tying pensions to average weekly earnings meant that these increases were tied to a higher tax rate; it was the Senate, not the Government, which decided that Medibank should be financed from general revenue, rather than the proposed levy. In any case, the most restrained of Labor's Budgets, the Hayden Budget, was the rejected Budget. It is clearly absurd to say that the Whitlam Government was destroyed because its Budget deficits were too big. The cry 'too much too soon' is really from those who want to be good — but not yet. The most violent criticisms came over not what the Labor Government spent but what it cut — the withdrawal of the superphosphate bounty for example. On the other hand, trifles, such as the purchase of the Pollock painting 'Blue Poles' (by and on behalf of the National Gallery, not the Government) or a $500 grant for a country festival were used as examples of a recklessly extravagant government.

Why then was the Government so easily destroyed? Part of the explanation may lie in the Australian character itself. This was a highly visible government, distinctively Australian in style and character. John Grey Gorton's was a pallid Australianism compared with this. These Ministers — Connor, Cameron, Murphy, Cairns, Hayden, Daly, Uren, as much as, perhaps even more than Whitlam himself, personified, larger than life, the Australian style and character. Their strengths and faults were very visible and very visibly Australian — their independence, their turbulence, their touchiness, their irreverence, their impatience with 'proper channels'; yet, with all, a certain insecurity and self-doubt, by no means convinced of their right to rule, diverse as men of common loyalties could be, but each unmistakably Australian.

Is it possible that the Australians so easily allowed the destruction of the Whitlam Government, and then endorsed those who had procured its destruction, because they saw a government so palpably reflecting themselves — and, after the shock of self-recognition, were rather relieved to see the image smashed?

In his 1976 Boyer Lectures for the ABC, Australia's poet-historian, Manning Clark, said:

> After watching men in high places closely in Canberra, I came to the same conclusion about mankind as the author of the Book of Ecclesiastes, as the Greek tragedians, as Shakespeare and many others — namely, that men suffered from a fatal flaw in the being which stood between them and what they wanted to achieve.

But is it possible that the fatal flaw lies not just in those we put to lead us, but in our dislike of ourselves as Australians? Sooner or later we shall have to come to terms with this question, for if we decide that we do not like ourselves and each other very much, it is certain that the rest of the world has no reason to like us at all.

At the height of the constitutional crisis, Whitlam delivered his Curtin Memorial Lecture at the Australian National University on 29 October. He spoke about the significance of the crisis to Australia's future and the future of the Australian Labor Party:

> A conservative Government survives essentially by dampening expectations and subduing hopes. Conservatism is basically pessimistic; reformism is basically optimistic. The great tradition which links the American and French revolutionaries of the Age of Reason with the modern parties of social reform is the tradition of optimism about the possibility of human improvement and human progress through the means of human reason. Yet inevitably there will be failures, and the higher expectations rise, the greater the likelihood of at least temporary failure to meet them.

Whitlam then spoke of the counter-view, which dismisses or downgrades government action as the best means for the betterment of the human condition:

> The argument is in fact based on a particular view about human nature and human motives. In the final analysis it predicates fear and greed as being the principal spurs to human action. It says in effect that if people know they are guaranteed an income in retirement they will be both lazy and improvident during their earning days; if people are not afraid of the price of sickness they will abuse and over-use the health services the community provides; if public schools are made as good as private schools then parents will not work so hard to earn the fees for their children's education; if the attempt is made to make underprivileged communities more decent places to live, people will lose the competitive urge.
>
> We are going to hear a lot more of these arguments, put in perhaps more subtle ways. The counter-argument is that the removal or reduction of basic fears and insecurities, far from being a limitation on individual incentive, represents a liberation for human creativity. The contest between the two opposing views of human nature and human society is still the essence of the philosophical debate between the Parties—to the extent that the political debate is conducted in terms of philosophy and to the extent to which either Party can be said to have a philosophy.

He then related these wider questions to the immediate crisis:

> It is now clear that behind the present constitutional struggle there is a wider political question the answer to which is central to the way in which Australia's whole political life will develop for the rest of this century and beyond. The question is not just whether this particular government, the Whitlam Government, will be allowed to govern for the term for which it was elected. The question is whether any duly elected reformist government will be allowed to govern in the future. What is at stake is whether the people who seek change and reform are ever again to have any confidence that it can be achieved through the normal parliamentary processes.
>
> I would not wish on any future leader of the Australian Labor Party the task of having to harness the radical forces to the restraints and constraints of the parliamentary system if I were now to succumb in the present crisis. It is clear that the basic attack which has been mounted against the Labor Government from April 1973 onward was not an attack on its competence or its effectiveness but on its very legitimacy — the legitimacy of any reform government now or in the future.

The essence of Whitlamism is optimism about the possibility of human progress through the application of human reason and in a democracy achieved through and applied by parliamentary means. In opening his campaign for the 1975 election — a confession of faith and a communion with the faithful more than a policy speech, he said:

> My whole public career has been dedicated to the proposition that reform and change needed in Australia can and must be achieved through democratic Parliamentary means. For fifteen years as Deputy Leader of my Party and Leader of my Party and Prime Minister of Australia, I have maintained that faith; and the Australian Labor Party has held to that faith with me.
>
> Now that faith is challenged in a way none of us would ever before have believed possible. I shall never abandon that faith — because my faith rests not just in Parliamentary democracy itself, but in the Australian people themselves — in their common sense, their intelligence, their decency, their instinctive sense of fair play.

At those elections, the victory went to a leader who had developed a philosophy far removed from the Whitlam doctrine of optimism. In his Deakin Memorial Lecture of 1971, a few months after the fall of Gorton, Malcolm Fraser, referring approvingly to the historian of civilizations, Arnold Toynbee, said:

> His thesis can be condensed to a sentence and is simply stated — that through history nations are confronted by a series of challenges and whether they survive or whether they fall to the wayside depends on the manner and character of their response. Simple, and perhaps one of the few things that is self-evident. It involves a conclusion about the past that life has not been easy for people or for nations and an assumption for the future that that condition will not alter. There is within me some part of the metaphysic and thus I would add that life is not meant to be easy.

Seldom have the two rival political chiefs been so fundamentally opposed in their personal philosophy as now — the metaphysic pessimist, John Malcolm Fraser, and the rational optimist, Edward Gough Whitlam. Since he is the victor in possession at the end of the tumultuous events I have tried to describe, I have left Malcolm Fraser with the last word. But on none of these events — from Vietnam in 1965 to the coup of November 1975 — has the last word yet been spoken, or the final verdict of history been given.

*

Afterword

On the morning of anguish of 14 December 1975, Whitlam, still in residence at the Lodge, made two attempts to divest himself of the leadership of the Australian Labor Party. He telephoned Bill Hayden in Ipswich, Queensland, offering to stand aside and support Hayden's candidature. A Labor leader can go no further to secure the succession for a colleague. Hayden declined to run for the leadership. Bob Hawke came to the Lodge in the afternoon. He and Whitlam discussed the possibility of Hawke entering Parliament and becoming a candidate for the leadership. In the days following, however, it became clear that none of the survivors of the great crash was prepared to give up his seat to provide Hawke a passage into Parliament. The two chances Whitlam offered Caucus for a fresh start under a new leader passed. On 27 January 1976, Caucus narrowly re-elected him as its leader.

Two years later, on the night of 10 December 1977, Whitlam acted alone and at once. At 11 p.m., as soon as the extent of the débâcle became clear, in many ways more shattering than 1975, he declared briefly, with memorable grace: 'I shall not myself renominate for the leadership of the Parliamentary Labor Party.' He ended his record term as Labor leader on 22 December 1977, after nearly eleven years. He had surpassed Curtin's term in the job on 11 November 1976, the first anniversary of his dismissal by Sir John Kerr. He had been the first Labor leader to win office from Opposition since Scullin in 1929; he had been the only Labor leader to win two successive elections; he had led the Labor Party into its two heaviest defeats, in terms of seats if not in terms of aggregate vote.

The intervening two years swung violently between bitterness and hope. It could not be denied that Whitlam was a seriously wounded leader and there were those who lost no chance to rub salt into the wounds. At the very moment of confirming him as leader in 1976, Caucus damaged and diminished the leadership itself. It decided that leadership positions should be declared vacant and recontested in mid-term, thus ensuring instability at the leadership level for the whole Parliamentary term. It was this new rule which forced Hayden reluctantly and prematurely into a challenge to Whitlam in June 1977. At least on the part of the two contenders, the contest was a remarkably civilized affair; but their mutual civility could not overcome the mutual damage done by a challenge which Whitlam survived by one vote.

Yet quite early in 1976, hope was rekindled. Labor's victory in New South Wales under the leadership of Neville Wran in the State elections on 1 May vastly encouraged the Labor Party throughout Australia. Obviously it was premature to write off a party which could win government of the leading State only six months after its massive national defeat. The resignation of a former Liberal minister, Don Chipp, to form a new party, the Australian Democrats, raised in a new form the warning given by John Gorton in 1975, after Fraser's demolition of Snedden, that Fraser would split the Liberal Party. Contrary to his election promise to 'maintain Medibank in letter and spirit', Fraser restored the private health funds as the dominant force in the provision of health insurance. A $17\frac{1}{2}$ per cent devaluation in November 1976 (subsequently modified) cast doubts on the Fraser Government's ability to curb inflation. Throughout 1977, key indicators showed a deepening recession and unemployment continued to rise. A redistribution of seats seemed marginally to favour Labor. The opinion polls showed a partial recovery in Labor's support; their steady trend in 1977 was to show the two great parties approaching equality in public support, or lack of it. Elections in South Australia and the Northern Territory indicated a Labor resurgence.

Thus when Fraser announced on 27 October that he proposed to advise an election of the House of Representatives and half the Senate on 10 December, Labor looked forward to the contest with an optimism inconceivable twelve months before. Fraser had thrown away one of the three years he had demanded in 1975 to right the economy; the message seemed plain: he was taking a huge gamble to save his government before unemployment deepened even further in 1978. The expert predictions differed only in the degree of gloom, between 400–500,000 unemployed by early 1978. Then, results in the Queensland State elections and a Victorian by-election, showing swings of 7–10 per cent to Labor (with the new Democrats' preferences slightly favouring Labor), pointed to a huge and perhaps fatal miscalculation by Fraser. At the very least, Labor could look forward to big gains, if only through a normal recovery from the 1975 low, and Whitlam could look forward to leaving a rehabilitated party within striking distance of victory in 1980.

The actual outcome on 10 December 1977 showed the skilfulness of Fraser's timing—that same 'impeccable' skill he had shown in moving against Gorton in 1971, against Snedden in 1975 and against the Whitlam Government in 1975. He had not, after all, risked one year; he had gained at least six years. He had made at least two successful calculations: that, as in 1974, the people would not be prepared to change a government elected less than two years before; and that it was too close to the traumatic events of 1975 for considerable numbers to be willing to change their vote or radically reassess their attitude towards the Whitlam Government or their perception of the Labor Party under Whitlam.

It is not true that on 10 December 1977 the Australian people themselves moved sharply towards the 'right'. They had re-elected a government whose leader was driving the nation sharply to the 'right' and whose economic policies were designed to achieve a marked redistribution of wealth back in favour of the higher income-earners. Nobody, however, who has lived through this period of our history or followed the thread of this book should be surprised that 10 December 1977 confirmed the conservatism of the Australian electorate rather than signalling any radical lurch to the 'right'.

In the aftermath of a second devastation, bitterness was inevitable. It is important for the future of the Labor Party that it

Malcolm Mackerras, *Elections 1975*, Angus & Robertson, Sydney, 1975 (despite its title, it is the indispensable guide to all recent elections except 1975).

I have also found invaluable the Political Review of the Australian Quarterly. This quarterly survey varies in scope, quality and tone according to the authors, but it remains an informed, accessible and topical account.

Introduction
page
xii   For Whitlam on Theodore, see his foreword to Irwin Young, *Theodore, His Life and Times*, Alpha Books, Sydney, 1971.
xii   The Scullin Government was defeated in December 1931; it retained its commission until 16 January 1932.

Chapter 1   The Menzies Inheritance
page
5   Patrick White in a speech at the Sydney Opera House, 13 May 1974.
6   Robert Murray, *The Split: Australian Labor in the Fifties*, Cheshire Publishing Pty Ltd, 1970.
9   A. A. Calwell, *Be Just And Fear Not*, Lloyd O'Neil Pty Ltd, 1972, p. 146.
9   'In sympathy with Savonarola': A. A. Calwell, *Labor's Role in Modern Society*, Lansdowne Press, Melbourne, 1963, p. 8.
10   A. A. Calwell, *Be Just and Fear Not*, p. 154.
11   'He was a thorough Celt': Justin McCarthy, *A Short History of Our Own Times*, Chatto & Windus, London, 1905, p. 57.
13   B. A. Santamaria writing in the *Australian*, 1977. For an account of Ward, see E. Spratt, *Eddie Ward, Firebrand of East Sydney*, Rigby Ltd, Adelaide, 1965.
15   For an account of Australia's approach to West Irian, see Bruce Grant, *The Crisis of Loyalty — A Study of Australian Foreign Policy*, Australian Institute of Foreign Affairs, Angus & Robertson Publishers, Sydney, 1972.
23   A. A. Calwell, *Be Just And Fear Not*, p. 227.

Chapter 2   'Is the Tumbril Ready?'
page
24   'a system of Satanic delusion': see Manning Clark (ed), *Sources of Australian History*, Oxford University Press, London, 1957, p. 358. For the early history of the State aid controversy in Australia, also see A. G. Austin, *Australian Education, 1788–1900*, Sir Isaac Pitman & Sons Ltd, Melbourne, 1961.

Chapter 3 How Australia got into Vietnam

page

40–50   In 1975, Whitlam instructed the Department of Foreign Affairs to make available to the Archivist, Professor R. Neale, the documents surrounding the decision to involve Australia militarily in Vietnam. A version was published as a White Paper in August 1975. The background to the Quat letter was revealed in Parliament in August 1971, during the controversy following the publication of the Pentagon Papers in the *New York Times* in July 1971. A series of articles in the *National Times* in February–March 1975 by Evan Whitton also contains valuable information.

51–4   See *Commonwealth Parliamentary Debates*, 4 May 1965, p. 1102 ff.

58   R. G. Casey, *Australian Foreign Minister: The Diaries of R. G. Casey 1951–60*, Collins, London, 1972, p. 133.

60   A. A. Calwell, *Be Just And Fear Not*, p. 6.

Chapter 4 The Leader

page

66   W. Denning, *Caucus Crisis: The Rise and Fall of the Scullin Government*, Cumberland Argus, Parramatta, 1937, p. 106.

70   Whitlam's papers and addresses on the Constitution have been brought conveniently together in E. G. Whitlam, *On Australia's Constitution*, Widescope International Publishers, Melbourne, 1977.

71   'the didactic impulse': W. Bagehot, *Historical Essays*, Anchor Books (Doubleday & Co. Inc.), New York, 1965, p. 243.

71   Chifley Memorial Lecture 1975.

72   An address to the Victorian ALP Conference, June 1967.

72   'we forget Section 96': E. G. Whitlam, 'Socialism within the Constitution' in *On Australia's Constitution*, p. 65.

73   Curtin Memorial Lecture 1961.

74   Lenin, *The Labor Government in Australia 1913*. Lenin's famous put-down of the Australian Labor Party has been quoted in support of their views by authors as diverse as A. A. Calwell (*Labor's Role in Modern Society*, p. 59) and the new Marxists. (See R. Catley and B. McFarlane, *From Tweedledum to Tweedledee — The New Labor Government in Australia*, Australian and New Zealand Book Company, Sydney, 1974. The standard Marxist critique of Australian radicalism and nationalism is Humphrey McQueen, *A New Britannia*, Penguin Books, Melbourne, 1970.)

75   Chifley Memorial Lecture 1975.

78   H. Fairlie, *The Kennedy Promise: The Politics of Expectation*, Eyre Methuen, London, p. 32.

79   W. Bagehot, *The English Constitution*, Oxford University Press, London (World's Classics 1963 ed. pp. 115–9).

81   For the debate between Whitlam and Bland, see *Liberty in Australia*,

Australian Institute of Political Science Seminar 1955, Angus & Robertson, Sydney, 1955, p. 159 and p. 171.

Chapter 7 The Coming of John Grey Gorton
page
119 Paul Hasluck, *A Time For Building: Australian Administration in Papua and New Guinea 1951–53*, Melbourne University Press, 1976, p. 6.
119 Alan Reid, *The Power Struggle*, Shakespeare Head Press, 1969 and Tartan Press, Sydney, 1972, p. 157.
120 Alan Reid, *The Power Struggle*, p. 194.

Chapter 9 'I did it my way'
page
147 Edward St John, *A Time To Speak*, Sun Books, Melbourne, 1969, pp. 179–215. The fullest account of the Gorton period remains Alan Reid's *The Gorton Experiment*.

Chapter 13 Whirlwinds of Change
page
190 For Hasluck on Calwell, see *A Time For Building*, p. 215.
194 Don Woolford, *Papua-New Guinea: Initiation or Independence*, University of Queensland Press, 1976, p. 109.
195 Speech at the Chief Minister's dinner, 18 February 1973.
203 Laurie Oakes, *Whitlam PM*, Angus & Robertson, Sydney, 1973, p. 216.
205 The transcript of the Chou En-lai–Whitlam interview was made by David Barnett, the representative of Australian Associated Press, and later press secretary to Malcolm Fraser.

Chapter 14 'Tiberius with a telephone'
page
217 Menzies' appeal quoted in the political review of 'The Australian Quarterly', December quarter 1971.

Chapter 16 The Duumvirate and After
page
248 C. J. Lloyd and G. S. Reid, *Out of the Wilderness: The Return of Labor*, Cassell Australia, Melbourne, 1974, p. 32.
251 For Whitlam on Hughes, see his foreword to Irwin Young, *Theodore, His Life and Times*, Alpha Books, Sydney, 1971.
255 For Hasluck on his reactions to Menzies' offer, see *A Time For Building*, p. 5.
258 The description of the Christmas bombing of North Vietnam is taken from Kalb, Marvin and Kalb, Bernard, *Kissinger*, Hutchinson of London, 1974, p. 415. The reaction of Australian Ministers was moderate compared with that of the Prime Minister of Sweden, Olaf Palme, who said the bombing was 'a form of torture and an outrage similar to those

linked to names like Guernica, Lidice, Babi Yar, Sharpeville and Treblinka.' Pope Paul warned that 'the unforeseen worsening of events has intensified bitterness and anxiety in world opinion.' (quoted in Kalb, p. 416.)

Chapter 17 Opposition
page
268 A. A. Calwell, *Labor's Role in Modern Society*, Cheshire, Melbourne, p. 74.

Chapter 18 The Economy v. The Programme
page
282 For Whitlam on the 1974 credit squeeze, see his address at the National Press Club, Canberra, 20 July 1977.

Chapter 19 Double Dissolution
page
298 Laurie Oakes and David Solomon, *Grab for Power: Election '74*, Cheshire, Melbourne, 1974, p. 307.

Chapter 20 A Question of Legitimacy
page
304 *The Australian Government — Second Whitlam Ministry*. Prime Minister: E. G. Whitlam, Q.C.; Deputy Prime Minister and Minister for Overseas Trade: J. F. Cairns; Minister for Minerals and Energy: R. F. X. Connor; Minister for Social Security: W. G. Hayden; Government Leader in the Senate, Attorney-General and Minister for Customs and Excise: Senator L. K. Murphy, Q.C.; Minister for Foreign Affairs: Senator D. R. Willesee; Treasurer: F. Crean; Minister for Services and Property and Leader of the House: F. M. Daly; Minister for the Media and Manager of Government Business in the Senate: Senator D. McClelland; Minister for Defence: L. H. Barnard; Minister for Agriculture: Senator K. S. Wriedt; Minister for Northern Development and Minister for the Northern Territory: R. A. Patterson; Minister for Labour and Immigration: C. R. Cameron; Minister for Education: K. E. Beazley; Special Minister of State and Minister Assisting the Prime Minister in Matters Relating to the Public Service: L. F. Bowen; Minister for Repatriation and Compensation: Senator J. M. Wheeldon; Minister for Urban and Regional Development: T. Uren; Postmaster-General: Senator R. Bishop; Minister for Housing and Construction: L. R. Johnson; Minister for Transport: C. K. Jones; Minister for Health: D. N. Everingham; Minister for Manufacturing Industry: K. E. Enderby, Q.C.; Minister for the Capital Territory: G. M. Bryant; Minister for the Environment and Conservation: M. H. Cass; Minister for Aboriginal Affairs: Senator J. L. Cavanagh; Minister for Science, Minister Assisting the Minister for

Foreign Affairs in Matters Relating to Papua New Guinea and Minister Assisting the Minister for Defence: W. L. Morrison; Minister for Tourism and Recreation and Vice-President of the Executive Council: F. E. Stewart.

Chapter 22   The Last Casualty

page
338   The *Age*, 30 April 1975.

Chapter 23   Loans

page
344   Geoffrey Blainey, *The Rush That Never Ended — A History of Australian Mining*, Melbourne University Press, 1963, p. 340.

Chapter 24   Ambush

page
402   Donald Horne, *Death of The Lucky Country*, Penguin, Melbourne, 1976, p. 12.

Appendix

This book has been republished in its original form. Some inaccuracies have been drawn to my attention.

(i)

In his own book, Mr Whitlam gives a version, more accurate than mine in chapter 2, of the NSW Government's 1963 education proposal.

> The NSW Government resolved to act upon the decision of the State Conference in formulating its 1963 Budget. Proposals were developed to fund the construction of science laboratories in private schools and pay a means tested allowance of £21 per year to the parents of children enrolled in the third and subsequent years of non-State secondary schools and to the parents of those pupils enrolled in the third and subsequent years in State secondary schools who were required to live away from home. Heffron took the precaution of informing Calwell of these proposals to obtain his endorsement in the light of the July Federal Conference decision on State aid. Calwell vetoed the science block proposal but allowed the bursaries scheme to remain in the Budget when it was announced in September 1963.
>
> *The Whitlam Government* Penguin (1985) p. 298

(ii)

In chapter 21, Dr Cairns made his avowal of 'a kind of love' in the *Sydney Sun*. However the description of the character of the *Daily*

Mirror is historically accurate for 1975 and, indeed, accurate enough as a description of the *Sydney Sun* in 1987.

(iii)

The most comprehensive textual analysis of this book was that undertaken by Sir William McMahon. His Personal Explanation in the House of Representatives on 25 October 1977 deserves to be quoted in full:

> **Sir WILLIAM McMAHON** (Lowe)—Mr Speaker, I seek your approval to make a personal explanation.
>
> **Mr SPEAKER**—Does the right honourable gentleman claim to have been misrepresented?
>
> **Sir WILLIAM McMAHON**—Yes. In Graham Freudenberg's book *A Certain Grandeur* there are several statements which misrepresent me. Firstly, on page 221 of the book, Mr Freudenberg says:
>
>> His last prop was removed in 1972 when Sir Frank Packer sold the *Daily Telegraph* to Rupert Murdoch. Murdoch had committed his papers to a change of government. He telephoned McMahon in London to tell his of his deal with Packer:—
>
> He is supposed to have said this:
>
> **Mr Young**—I raise a point of order. I ask you to rule on this. Previously members have helped authors to sell their books by raising these questions. Is the right honourable member for Lowe again trying to help the book publishing—
>
> **Mr SPEAKER**—There is no point of order. The honourable gentleman will resume his seat.
>
> **Sir WILLIAM McMAHON**—Mr Freudenberg says:
>
>> He telephoned McMahon in London to tell him of his deal with Packer:—
>
> And this is supposed to be an actual quote—
>
>> 'I can promise, Prime Minister, that we will be as fair to you as you

deserve.' In the background, Packer warned: 'If you do that, you will murder him.'

Mr Murdoch did not telephone me in London or in any other overseas city to tell me of his deal with Sir Frank Packer, or for any other purpose. It is a lie, Sir. The final episode in the signing of the contract between Sir Frank and Mr Murdoch was in Sydney. I was present with my wife at Sir Frank's home. He informed me of the agreement. I remarked: 'Well, that just about ends our prospects in New South Wales.' Mr Murdoch assured me that he would, except in editorials, give me equal treatment with Mr Whitlam. I was not able to make proper use of this facility until Mr Phil Davis came on to my staff late in 1972. From then on 'repair jobs' were always published. Looked at in retrospect the whole story is amusing and worth while telling, one day.

Secondly, as to the reference to the December 1971 United States devaluation, Mr Freudenberg says:

> Nixon devalued the American dollar by 8.5 per cent. On 20 December, Cabinet met to settle whether Australia should devalue with the United States dollar or revalue with sterling. The Country Party members insisted on devaluation; Treasury and Snedden were as vehement in favour of revaluation. By 4 a.m. on 21 December, Cabinet reached the decision to revalue by less than the full percentage required to keep the dollar in line with sterling. By the afternoon, Treasury returned, with new arguments to its original line of full revaluation with sterling. Anthony thereupon threatened to take the Country Party out of the coalition. McMahon caved in.

As the House will remember, on 4 October I set out the facts in the House and made it clear that there was no caving in on my part. The facts had been made known publicly on at least two previous occasions. The problems associated with the two days between the first meeting of Cabinet and the last at which the decision was confirmed have never been stated. If the statement is made, I am sure it will show that I subordinated my interests to those of my Party. Thirdly, on pages 217 and 218 he says:

> There is a tape extant which records McMahon's speech at the White House in October 1971. He had approved a speech drafted by his

> foreign affairs adviser, Richard Woolcott, and Dr H. C. Coombs, whom he had publicly designated as his 'Guru'. As Richard Nixon made, or appeared to make, a neat off-the-cuff speech, McMahon decided to reply in kind and pocketed his prepared speech. On and on he warbled.

This is an awful mix-up. If a speech was drafted it was not given to me. I went to the White House without any notes. I could not have made a speech of the same dull quality as the Press Club draft I will mention later even assuming one had been given to me. Mr Freudenberg was not in the Australian Press team in Washington. He might have been referring to a speech at the Washington Press Club. Whilst on the platform I was handed a speech. I read it through. It seemed to me to be too trite and superficial for such an audience. I dropped the last part completely and ad libbed considerably. When I sat down I received a note from Mr Woolcott informing me that the part deleted had already been distributed to the Press. So I had to explain the matter to the Press Club Chairman. We agreed that it would be better if I gave a complete explanation. I did so and read out the part which had been dropped. I doubt whether Dr Coombs could have been associated with the writing of this speech. His writings were always elegantly expressed. In any event I did not see much of Dr Coombs in Washington or London except for an unrehearsed game of squash in San Francisco. He did stay at Blair House in Washington while I was there. Fourthly, as background to the United States visit I mention that Alan Ramsey reported in the *Australian* on 8 November 1971 relative to the mission to the United States:

> The point is that on the official level Mr McMahon's Washington visit was as effective as he could wish.

Even so I was plagued daily by reports by the Australian Press that had little or nothing to do with the facts. For example, it was reported that I did not, as expected, turn up to a church service in Washington and that I kept the Minister waiting. No arrangements were made for me to attend the service. The Minister did not expect me. I also mention the clamour in Washington from the Australian pressmen about an alleged breach of protocol in an answer to a question concerning Senator Muskie, an aspirant for presidential honours. I cleared the answer with the Press Club Chairman as to its propriety and was subsequently informed by the Australian Embassy Information Attache, Mr Roger Henning, that Senator Muskie had been pestered by the Australian Press but could see nothing in my speech to object to.

Fifthly, there are other parts of the book which are incomplete and misrepresent the facts, such as his allegation on page 185 relating to the appointment of Mr Gorton to the Defence Ministry and the influence of Sir Frank Packer. Mr Gorton, as Deputy Leader, was entitled to ask for the ministry of his choice. Sir Frank Packer did not speak to me about this matter as is suggested. The reference on page 237 to the speech by Archbishop James Carroll, with which the Leader of the Opposition (Mr E. G. Whitlam) was very much involved, is but an extract of what actually occurred. Obviously Mr Freudenberg did not read the book of Oakes and Solomon *The Making of an Australian Prime Minister*. Nor does Mr Freudenberg's contribution cover the relevant facts, particularly the letters passing between myself and His Grace. His Grace informed me that he had been seriously misrepresented but would not correct the misrepresentations.

The next point relates to the story on page 111 in which Mr Freudenberg says that 'William McMahon influenced by poor advice and worse champagne spoke at a Liberal Party rally in Rockhampton on the danger of atheistic communism'. At that time, as you, Mr Speaker, will know and the Leader of the Opposition ought to know, it was unusual for me to drink alcohol. This statement is slick but of course recklessly untrue. There is one person in the House who was there who wrote to the *Sydney Morning Herald* subsequently denying the accuracy of what had been said and setting out the facts. I should add that during the time I have been in Parliament no one has seen me affected by alcohol. I wonder whether that can be said about many others. Atheistic communism was not the subject of my speech. In any event the one word which was criticised was added to a copy of my speech and was distributed without my knowledge.

I have no wish to mention all the misrepresentations. I contacted Mr Freudenberg's office on Friday and asked whether he could be informed that I had called. He did not respond. In each of his statements Mr Freudenberg is guilty of the journalistic sin of plagiarism. He was not connected with the reporting at the relevant time nor was he present when any of the events took place. It is fair that I should say this about Mr Freudenberg: Even with all its mistakes the book is pleasantly written and easy to read, but it is neither 1971 nor 1972 vintage Freudenberg. What a pity Freudenberg did not consult me before publication. I have considerable respect for him and, even though he is back working for Whitlam again, I would have been happy to have cleared up the misunderstanding. I will send a copy of this statement to him and to the publishers, the Macmillan Co., because I do not believe that history should be massacred by faulty research. It is worth mentioning that a large number of members of the Canberra Press Gallery of 1971-72 became Press officers for Labor Ministers in 1973 and later but nearly all lost their jobs in 1975.

(iv)

I now restore to this edition the concluding words of the 1977 edition, which I omitted from the 1978 edition:

> On 14 July 1977, Sir John Kerr announced that he would resign as Governor-General. Whitlam commented to the author: 'How fitting that the last Bourbon should bow out on Bastille Day.'

Index

Aborigines, 25, 236, 253, 286
Abortion, 237
Admiralty House, 368-9, 389-90
Allsop, Jack, 143
Anderson, H. D., 47
Angleton, James, 264-5
Anthony, Douglas, 201, 210, 219-20, 226, 272, 274, 278, 292-3, 300, 374, 387, 390
A.N.Z.U.S., 41, 139, 205-8
Arbitration Commission, 280, 354-5
Army, 183-4
Arts, 236
Askin, Robert, 61, 180, 250, 300, 320, 323, 368
Aston, William, 146
Australia Party, 60, 272
Australian Association of Social Workers (Vic. Branch), 103
Australian Broadcasting Commission, 182, 337
Australian Council of Trade Unions, 98, 176-7
Australian Industry Development Corporation, 289, 307
Australian Institute of Political Science, 80-2, 191-2
Australian Labor Party,
    Caucus (Federal Parliamentary Labor Party), 13, 19-20, 22-3, 26, 33, 35, 51, 55, 83-7, 91, 106, 115-6, 129-30, 132-7, 168-73, 225, 227, 240-1, 243, 250-5, 277-8, 284, 304-7, 319, 322, 343-4, 354, 381, 391, 406
    Committees, 30, 90
    Commonwealth Labor Advisory Council, 284
    and Democratic Socialism, xi, 73-75
    Federal Conference, 20, 31-3, 36, 51, 89-91, 99-100, 102, 133, 137, 151-3, 155-8, 168, 170-1, 205, 225-6, 243, 261, 310, 322, 354
    Federal Executive, 7, 20-1, 28-36, 59, 62, 89-91, 100, 112, 129-33, 135-7, 152-3, 170, 174-6, 202, 227-8, 241, 380
    Government, 101-2, 242, 248-52, 278, 281, 302-4, 318, 375, 402-11; dismissal of, 375-80, 390-410
    leadership, 22-3, 61-4, 71, 75, 83-9, 106, 129, 131-7, 172, 228, 241-2, 250-2, 304
    legitimacy, 264, 266, 302, 308-9, 379, 402, 406, 409, 412
    and Marxist critics, xi, 74-5
    Mid-term campaign, (1971), 224-5
    Ministerial staff, 256, 258-60, 282-3, 307, 312, 318-22, 348, 354-7, 360-1, 405
    Ministry, 240-3, 250-7, 261, 284, 304-7, 319, 322-4, 354-8, 362-5, 406-7, 410
    and nationalization, 17, 72-4
    New South Wales Branch, xi, 21-2, 26, 28-30, 34, 76, 93, 98-9, 134-5, 152, 174, 221-2
    Opposition staff, 50-1, 85, 88-9, 100-3, 109, 203, 211, 225-7, 229, 247
    Platform, 30, 51-3, 71-6, 101, 257
    Department of Economic Planning, 354
    Queensland Branch, 14, 26, 34-5, 76
    rank and file, 33, 71, 89-92, 94, 175-7, 381, 405
    and reorganisation, 89-100, 105
    South Australian Branch, 34, 76, 93, 97, 370-1
    splits, 6, 25
    1955, 6, 9-10, 12, 19, 26-7, 37, 58, 71, 92, 163, 173, 175, 177, 201, 261, 293.
    state machines, 90-1, 94, 135, 172

Tasmanian Branch, 34, 76, 87, 99, 130-3
and trade unions, 90-2, 95
Victorian Branch, 6, 14, 32-4, 36, 59, 76, 91-6, 98-9, 129-31, 133, 136-7, 168-78, 238, 243
Western Australian Branch, 34, 93-4, 99, 166-8
Australian Medical Association, 182
Australian Security Intelligence Organization, 260-6, 387

Barbour, Peter, 261-3
Barnard, Lance, 86-7, 93-4, 99-100, 129, 131-3, 136, 146, 155, 171-2, 210, 225, 241-3, 247, 253-6, 304, 318-9, 354, 366
Barnes, Allan, 143
Barnes, C. E., 190-2
Barton, Gordon, 272
Barwick, Garfield, 15, 41-2, 374-5, 388-9, 400
Beale, Howard, 40-1
Beaton, Noel, 150
Beazley, Kim, 55, 73, 80, 86, 195-6, 242, 254-5, 257, 274, 405
Benson, Sam, 59, 162, 170
Berinson, Joe, 90, 253, 309
Bessell, Eric, 382
Bevitt, George, 105
Bijedic, Dzemal, 262-3
Bishop, Reginald, 254-5
Bjelke-Petersen, Johannes, 250, 291-3, 310, 372, 380
Blainey, Geoffrey, 344
Bland, F. A., 81-2
Bland, Henry, 81
Blue Poles, 410
Bolte, Henry, 171, 180, 272
Border, L. H., 125
Bowen, Lionel, 254-5, 274
Bowen, Nigel, 214, 273
Bowers, Peter, 34
Britain, 81, 139
Conservative Party, 271
Labour Party, 90
Broken Hill Proprietary Company Limited, 279
Brown, Wallace, 143
Brown, William, 174-5
Browne, Frank, 145-6
Bruce Government, 66
Bryant, Gordon, 254-5, 320
Budgets, 284
1967 and Postal Charges Increase, 115-6
1971, 221
1972, 278, 282
1973, 273, 280, 283
1974, 305-8
1975, ix, 371-85, 388-91, 394, 396-7, 408-9
Bunker, Ellsworth, 125
Bunting, John, 50, 239, 245, 260
Burchett, Wilfred, 246
Burgmann, Bishop, 56
Burns, Tom, 35, 152, 200
Bury, Leslie, 118-20, 216
Byers, Maurice, 348-50, 376-7, 379-80

Cabban, Peter, 107-8
Cabinet, 216-7, 220, 237, 249-52, 255, 258, 260, 305, 357, 406
Cahill, Geoffrey, 288
Cairns, James, 8, 51, 56, 86, 134-6, 155, 169, 218, 242, 254-5, 258-9, 282, 304-7, 318-24, 336, 342, 346-9, 352-8, 363, 370-1, 391, 410
Cairns, Philip, 356
Calder, Samuel, 281
Calwell, Arthur, 7-23, 28-36, 39, 45, 49-57, 60-2, 64, 73, 83-6, 89, 94, 96, 112, 116, 126, 131, 134, 146, 149, 171, 190, 192, 228, 236-7, 253, 268-9, 293
Cameron, Clyde, 83, 85, 99-100, 130-1, 136, 138, 146, 149, 155-6, 170, 173-7, 225, 227, 242, 249, 254-5, 258, 311, 342, 346-7, 354-5, 381
Canberra, 66-7
Carrington, Lord, 247
Carroll, Archbishop, 30-1, 237
Casey, Lord, 57-8, 114, 369
Cass, Moss, 102-3, 254-5, 355
Catholic Church, 6, 8-13, 21, 24-7, 30-1, 56, 58, 154-8, 237-8, 274
Cavanagh, James, 254-5
Central Intelligence Agency, 145, 260, 264-5, 328, 336, 387, 406
Chalfont, Lord, 64-5, 71
Chamberlain, F. E. (Joe), 27-36, 93-4, 99-100, 151-2, 156-7, 166-70, 175
Chief Justice, 374-5, 388-9, 401-2
Chifley, Ben, 14, 64, 73, 83-4, 159, 163, 253
Chifley Government, 71, 98, 263
China, 189-90, 200-14, 221, 245-6, 273, 287, 340
Chou En-Lai, 202-14, 273
Cities, 5, 14, 67, 70, 75, 101, 157, 221, 224, 229, 300

431

Clark, Manning, 56, 409, 410
Cohen, Sam, 86, 100, 129
Colbourne, William, 134
Cole, R. W., 404
Combe, David, 274, 285, 288, 322, 354, 395
Commonwealth Banking Corporation, 307
Commonwealth Police, 262-3
Commonwealth Prime Ministers' Conference, 19, 339
Communism, 20, 41, 96, 164
Conlon, Alf, 367
Connor, Rex, 133-4, 136, 249, 254-5, 304, 308, 342-54, 358, 360-4, 366, 374, 382, 384, 410
Connor, Xavier, 175
Conscription, 8, 45, 51-2, 114, 225, 232, 242, 244-5, 247
Constitution, 3, 29-30, 70, 72-4, 82, 288, 295, 391, 394, 400
Constitutional Crisis, ix-xi, 375-412
Cook, Joseph, 239
Cooley, Alan, 240, 245, 260
Coombs, H. C. (Nugget), 217-8, 236, 280, 283
  Task Force, 280-1
Cope, James, 309, 311-3
Country Party, See National Country Party
Cox, Harold, 143
Crawford, George, 168, 170, 172, 175
Crean, Frank, 86, 155, 227, 242, 254-5, 305-7, 366, 371, 390-1, 395
Croatian Migrants, 261-3
Cross, Manfred, 253
Crown, 392, 399-403
Cullen, Peter, 109
Curtin, John, xii, 8-9, 14, 17, 64-5, 70, 77, 87, 163, 229, 251, 253, 296-8
Curtin Government, 98

Daly, Fred, 86, 112, 146, 188, 225, 254-5, 385, 390, 395, 410
Daly, Lt. General T., 183-4
Darling and Company, 348
Darwin Cyclone, 322
Davies, Ron, 86
Deakin, Alfred, 11, 62, 264
Dedman, John, 83, 98
Deeble, John, 103-4
Defence, 224-5, 370
  F-111, 42, 45-6, 113
  'Fortress Australia', 139, 164

Indian Ocean, 159, 245
Delaney, Michael, 337, 360-1
Democratic Labor Party, 6, 9-10, 12, 26, 92, 96, 109, 115-6, 122-3, 125-6, 138-40, 150, 159, 164-5, 178, 209, 272, 274, 289-90, 293
Denning, Warren, 66
Departments:
  Air, 113
  Attorney-General, 241, 256, 262, 348-53, 360, 377-8
  Defence, 42, 46, 219, 246, 387; Joint Intelligence Organization, 260, 334
  Foreign Affairs, 16, 42-43, 46-7, 139, 159, 201, 203-4, 209-10, 214, 217, 245-7, 259, 262, 331-8
  Minerals and Energy, 308, 349-52
  Prime Minister and Cabinet, 50, 112-3, 122, 256, 262, 308, 334, 350, 377
  Treasury, 141, 220, 256, 281-3, 304-8, 322-3, 342, 346, 348-50, 352, 361, 403-4, 406
Dougherty, Tom, 173-5
Dovey, Margaret, see Whitlam, Margaret
Dovey, Wilfred, 63
Drewe, Robert, 265
Ducker, John, 174
Dulles, John Foster, 190, 206-7, 234
Dunstan, Don, 97-8, 171-4, 370-1
Duumvirate, 239-50, 252, 263
Dwyer, Lorraine, 344

Economy, 4, 16, 95, 158-9, 218-20, 371-2, 405
  business, 249-50, 285
  credit squeeze, 282
  currency, 220, 278-9, 282, 305
  Full Employment White Paper, 98
  Government spending, 280
  growth, 4, 77-8
  inflation, 222, 231-2, 273-6, 278-9, 282, 296, 304-5, 350-1, 404
  interest rates, 274
  tariffs, 242, 278, 282-3, 305
  unemployment, 222, 276-9, 283, 287, 305-6, 350-1
  wages, 273, 283, 354-5, 404
Education, 70, 74-5, 101, 224, 231, 242, 285-6
  Schools Commission, 153-4, 156, 274, 402, 409
  State Aid, 21, 24-36, 151, 153-9, 169-75, 237, 274
  Universities, 26-7, 72

432

Egerton, John, 283
Elections (Federal)
  1961, 3, 13-17, 41, 73-4, 92, 163
  1963, 6, 21-22, 29, 43
  1964 Senate, 22, 46
  1966, 22-3, 28, 36, 59-61, 85, 93
  1967 Senate, 94, 115-6, 129
  1969, 152, 158-65, 170, 189
  1970 Senate, 169, 178
  1972, 163, 189, 222-38, 240, 260, 271
  1974, 230, 271, 296-303, 373
  1975, 163, 228, 403
  By-elections; Denison (1964), 22; Dawson (1966), 31-5, 109; Corio (1967), 96, 98-9, 109-10, 177; Capricornia 109-11; Curtin (1968), 150; Bendigo (1969), 109, 150; Gwydir, 150; Parramatta (1973), 273-4; Bass (1975), 319, 354, 356
  Half Senate option (1975), 380-3, 388, 391, 401
  Policy speeches, 160, 226-36, 243-4, 257, 296-300
Elections (State),
  1965 N.S.W., 49
  1970 S.A., 171-2; Vic. 171-2, 174
  1974 Qld., 309-10, 324
  1975 S.A., 370-1
  1976 N.S.W., 49
Electoral Redistributions, 151-2, 298, 303
Ellicott, Robert, 375-9, 385-6, 401
Enderby, Kep, 254-5, 366, 375-6, 379-80, 387, 391, 395, 399
Environment, 75, 285-6
Equality, 75, 78, 80-3, 95, 286
Erwin, Dudley, 119, 146, 179
Evatt, H. V., (Bert), 6, 13, 15, 50, 64, 74, 83-4, 90-1, 203, 228, 261, 293, 343, 367
Everingham, Douglas, 110-1, 253-5
Executive Council, 343, 348, 352-3, 355, 358, 363-4, 366

Fairbairn, David, 180-1, 185, 219
Fairfax, Warwick, 16
Fairhall, Alan, 148, 159
Federal—State Relations, 70, 76-8, 81, 97-8, 122, 157, 164-5, 179-81, 220-1, 224, 250, 350-2, 371
Field, Albert, 372-3
Fisher, Andrew, xii, 64, 74, 84
Fitchett, Ian, 143
Fitzgerald, C. P., 56
Fitzgerald, Stephen, 203-4, 210, 273
Forbes, James, 45, 311

Forde, Frank, 83, 87
Foreign Investment, 141, 232, 279, 282
Foreign Policy, 74, 101, 124, 218, 234-5, 370
France, 234, 264
Fraser, Allan, 35, 55, 80
Fraser, Malcolm, ix-x, 81, 119, 126, 146, 148-9, 153-4, 159, 181-8, 211-4, 258n, 272, 274, 282, 308-10, 312-6, 318-9, 325, 337-9, 351, 361, 365, 373-4, 379-85, 388, 390-400, 403-4, 413
Freeth, Gordon, 159, 202
Freudenberg, Graham, 50-1, 88, 109, 203, 211, 225, 227, 229, 257, 260, 292, 312, 377, 379-80, 394

Gair, Vincent, 139, 209, 273, 288-94, 407
Galbally, John, 174
Game, Peter, 361
Game, Philip, 395
Gaul, Jonathon, 143
Gilroy, Cardinal, 10
Gorton, John, 4, 106-7, 110, 113, 115, 118-28, 138-50, 153, 158-62, 166-7, 178-88, 195-7, 199, 215-7, 220-2, 237, 250, 270, 300, 316, 321, 345, 392, 410
Gotto, Ainsley, 122, 179, 321
Governor-General, 257-8, 295, 356, 363, 366-70, 374-80, 382, 384, 387-409, 413
  and reserve powers, 376-80, 392, 400-402
Grassby, Al, 254-5, 301, 304, 319
Greenwood, Ivor, 261-2, 374
Grey, George, 110-1
Gruen, Fred, 276
Guise, John, 194

Hall, Richard, 202, 259-60
Hall, Steele, 303-4, 373
Hancock, Lang, 344
Hannah, Colin, 291, 380
Hansen, Rubensohn-McCann, Erickson Pty Ltd, 225
Harders, Clarence, 240, 348-9, 360-2, 377-8
Harradine, R. W. B., 130, 133, 152-3
Harris, George, 355-6
Hartley, William, 32-3, 112, 171, 173, 175
Hasluck, Paul, 43, 118-20, 179, 188, 190-2, 240-1, 244-5, 255, 257-8, 290, 366, 369
Hawke, Robert, 101, 168, 176-7, 227, 274,

433

283-4, 307, 322, 354, 382
Hayden, Bill, 105, 195-6, 244-5, 354, 371, 373, 383-4, 399, 404, 410
Haylen, Leslie, 19
Health, 70, 72-3, 75, 97-8, 100-5, 157, 231, 253, 285-6, 402
Heffron, R. J., 28
Henderson, R. A. G., 17
Hewitt, Lenox, 345-54, 360-1
Higgins, Henry Bourne, 251
High Court, 30, 264, 303, 310, 317, 324, 387, 401
Hill, E. F. (Ted), 202
Hogg, Robert, 177-8
Holding, Clyde, 171-2
Holt, Harold, xiii, 3-4, 7, 31, 33, 61-2, 86, 106-20, 122-3, 125-6, 138-40, 142, 188, 201, 236, 269-70, 321, 369
Holt, Zara, 4
Horne, Donald, 401
Howard, John, 385-6
Howson, Peter, 113
Hughes, Alan, 150
Hughes, Thomas, 146, 149
Hughes, William Morris, 8, 64, 75, 84, 251

Immigration, 11, 224-5, 232
Inall, Ruth, 102
Indonesia, 42, 45-7
  East Timor, 18
  West Irian, 15-19
Industrial Relations, 224-5, 232, 242
Innes, Ted, x
Institute of Applied Economic and Social Research, Melbourne University, 103, 306
Insurance Lobby, 299
Inter-Departmental Committees, 262

James, Bert, 144-6, 149
James, Francis, 213-4
Japan, 189, 205-7, 211-2, 370
Jess, John, 107
Jessop, Donald, 382, 388
Johnson, Leslie, 199
Johnson, Leslie Royston, 254-5
Johnson, Lyndon Baines, 60-1, 113-4, 124, 142, 145, 207
Jones, Clem, 90
Journals
  Australian Quarterly, 150, 276

Karidis, G., 346-7, 360
Keating, Paul, 353
Keefe, James, 129

Kelly, C. R. (Bert), 127
Kennedy, David, 150, 237
Kennelly, Patrick, 13, 293
Kent Hughes, Wilfred, 17
Kerr, John, ix, xi, 258n, 366-70, 374-80, 382, 384, 387-409, 413
Kerr, Lady
  Peggy, 368
  Anne, 391
Khemlani, Tirath, 325, 347-8, 352-3, 355, 359-64, 385-6
Killen, James, 15, 146, 148-9
Kirribilli House, 282, 368-9, 379, 389, 394
Kissinger, Henry, 208, 210, 259, 327
Kocan, Peter, 60
Korean War, 200

Lang, John Thomas, 395, 405
Law Reform, 263-4, 286, 324
Lazzarini, H. P., 63
Lee Kuan Yew, 176
Lewis, Tom, 310, 324, 372
Ley, Frank, 240, 302
Liberal-National Country Party Governments, 7, 117, 158-9, 182, 220, 278
  'Caretaker Government', 403
Liberal-National Country Party Opposition, x, 264, 266-74, 282, 307, 309, 315, 335, 349, 357-8, 373-4, 379-98
Liberal Party, 4-6, 22, 60, 65, 92, 106-7, 110-1, 118-27, 138-42, 147-9, 164-6, 178-88, 209-10, 215-23, 268-73, 292, 294, 300, 303, 309-16, 382, 383-4, 388
Liberal Reform Group, 60
Lloyd, Clement, 226, 247-8
Loan Council, 350-2
Lombard, John, 291
Luchetti, Anthony, 51
Lynch, Phillip, x, 127, 268, 270, 308, 313, 343, 355-6, 361-5, 390-1
Lyons, Joseph, xii, 106, 239

MacCallum, Mungo, 143, 312
McClelland, Douglas, 254-5
McClelland, James, 355, 371, 375, 387, 399
McEwen, John, 4, 80, 108, 117-8, 129, 147, 155, 160, 181-2, 201, 284, 344
McFarlane, A. B., 113
McGarvie, Richard, 175
McKell, William, 65
McKenna, Nicholas, 13, 199
McLeay, John, 320
McMahon, William, ix, 4, 49, 111, 114-5,

434

117–8, 129, 141, 146, 148, 155, 159–61, 164–5, 179, 182, 185–6, 188, 203–4, 207, 209–11, 215–23, 225–6, 234, 236–7, 239–40, 260, 270, 279, 300, 321
McManus, Frank, 293–4
Macmillan, Harold, 191
Malaysia, 139–40
Mandate, 243–5, 278, 283, 287, 295, 302, 403, 406, 409
Mannix, Archbishop, 8, 10, 12
Mansfield, Senator Mike, 114, 341
Mant, John, 394
Maori Kiki Albert, 194
Marriott, John, 382
Mathews, Race, 103, 226
Medibank, 104–5, 289, 303, 371, 409
Menadue, John, 51, 88–9, 100–3, 279, 308, 318–9, 349, 352–3, 360–1, 377, 395
Menzies, Robert, xiii, 4–5, 7–8, 14–18, 20–2, 25–7, 29, 39–41, 44–50, 53–5, 67, 72, 77, 79, 86, 88–9, 98, 106, 113, 119, 125, 138, 140–3, 158–9, 161, 163, 179–80, 191, 200–1, 216–7, 223, 234, 252, 255, 261, 268–70, 284, 310, 369
Middle Classes, 4, 6, 11, 164, 404
Milliner, Bert, 372–3
Milte, Kerry, 262
Millar, Robert, 105
Minogue, Daniel, 343
Missen, Alan, 382, 388
Molotov Letter, 203
Monk, Albert, 98, 177
Moore, John, 279
Moratorium, *see* Vietnam
Morosi, Junie, 317–27, 391
Morrison, William, 187–8, 254–5, 355, 387
'Mr Williams', 307–8
Murdoch, Rupert, 221, 236, 279, 379
Murphy, Lionel, 86, 100, 113, 116, 129, 131, 155, 178, 241–2, 249, 253–6, 261–5, 290, 295, 310, 317–21, 324, 342, 348–53, 366, 372, 410
Murray, Robert, 6
Myer, Kenneth, 366

National Anthem, 236
National Civic Council, 6, 9, 92
National Country Party, 4, 93, 117–8, 220, 226, 272–4, 278, 281, 387
National Press Club Luncheons, 236, 283, 374, 389
Navy,
 H.M.A.S. *Melbourne*, 107
 H.M.A.S. *Voyager*, 107

Newsletters,
 Incentive, 143
 Inside Canberra 143
 Things I Hear, 145
Newspapers, 180–1, 259, 325, 365
 Advertiser, 383
 Age, 47, 56, 143–4, 154, 325, 336, 338, 356, 360, 383
 Australian, 143, 183, 208–9, 265, 398;
  Sunday, 216
 Bulletin, ix–x, 11, 143, 182
 Canberra Times, 56, 143, 182, 260
 Courier Mail, 143
 Daily Mirror, 143, 322
 Daily Telegraph, 143, 179, 182, 221
 Financial Review, 143, 312–3, 383, 387
 Herald (Melbourne), 143, 262, 361–2, 383, 393
 Murdoch Group, 379
 Nation Review, 312–3
 National Times, 314–5, 320–1
 Private Eye, 145
 Sun (Sydney), 143
 Sun-News Pictorial (Melbourne), 143–4, 160–1, 291
 Sydney Morning Herald, 16–17, 21, 34, 46, 127, 143–4, 208–10, 228, 230–1, 237, 325, 331, 336, 338–9, 398
Newton, Maxwell, 16, 143
New Zealand, 81, 139, 405
Nicholls, Martin, 129
Nixon, Richard, 202, 207–12, 218–20, 234, 258–9, 277, 327
Nuclear Weapons, 20, 234

Oakes, Laurie, 143, 160–1, 203, 291, 293, 297–8
O'Byrne, Justin, 289–90
O'Donnell, Joan, 85
Off-shore Legislation, 180–2, 220
Oil Prices, 300, 345–6
Oliver, Charles, 174
Opinion Polls, 85, 220, 222, 383, 404, 409
Opperman, Hubert, 98
Order of St Michael and St George, 374
O'Reilly, Neil, 143
Organization of Economic Co-operation and Development, 276, 371–2
O'Shea, Clarrie, 367
Overseas Loans, 308, 342–66, 371, 403, 407

Packer, Frank, 179, 186, 216, 221
Paltridge, Shane, 43, 46
Papua-New Guinea, 15–17, 184, 189–200,

435

221, 250
Pangu Party, 193-4
Parliament, ix-xii, 14, 47-50, 70, 78-83, 87-8, 94-5, 108, 120, 127, 144-7, 168-9, 179-81, 219, 248-9, 257, 273, 338-9, 362-5, 381-2, 391-403
 and conventions, 375-8
 Double Dissolution (1974) of, 267, 294-5, 303, 403, 406
 1975; 377, 393-4, 397, 400-2
 House of Representatives, 146, 243, 266-7, 295, 303, 309-15, 374-5, 382-3, 408
 Joint Sitting, 303
 Relations between Houses, x, 294, 375-8
 Senate, ix, 6, 108-9, 113, 115-6, 241, 243-4, 264, 266, 272-4, 288-92, 302-3, 311, 319, 351, 396-7, 403, 406-7, 409; casual vacancy, 310, 324, 372-3; and deferral of Supply, 1975, 374-84, 388-9; and refusal of supply, 116, 268, 292-5, 303, 314-5; and Territorial representation, 388
 Special Sitting, 356-60
Patterson, Rex, 31-5, 202-4, 254-6
Peacock, Andrew, x, 183, 221, 313, 335-6
Petrov, Vladimir, 50, 261
Petty, Bruce, 222
Phillips, John, 282, 348, 352-3
Pickering, Larry, 222
Power, Kevin, 143
Press Gallery, 11, 143-6, 221, 259, 300, 307, 312, 320
Price, G. J., 336-7
Prices Justification Tribunal, 231, 279-80
Primary Industry, 255-6, 281
 superphosphate subsidy, 281
Protestant Churches, 11, 25-7
Prowse, Russell, 282
Public Service, 66, 122-3, 138, 143, 256-8, 263, 348-9, 388, 406
 Public Service Board, 256

Queensland Government, 291, 343, 372

R.A.A.F., 178
Ramsey, Alan, 183-6
Referendums,
 1944, 65, 69-70
 1951, Communist Party Dissolution, 3
 1967, Aborigines, 108-9
 1967, Nexus, 108-9
 1973, Prices and Incomes, 272, 286
Refugees, 336-7, 339

Reid, Alan, 89, 119-20, 143, 155, 186, 216, 405
Reid, Elizabeth, 306
Reid, George, 75
Reid, Gordon, 248
Renouf, Alan, 246, 331
Reserve Bank, 282, 348-50, 352
Rhodesian Information Centre, 250
Rice, Walter, 259-60
Richardson, Michael, 336-7
Riordan, Joseph, 375
Robertson, Captain R. J., 107
Rose, Dennis, 348-9, 352-3
Royal Commissions;
 Petrov, 203
 Voyager, 107-8, 127

Samuel, Peter, 143, 182
Santamaria, B. A., 6, 12-13, 92, 136, 139, 209
Scholes, Gordon, 98, 110, 309, 397-8
Scott, Malcolm, 119
Scotton, Richard, 103-4
Scullin, James, xii, 84, 227, 239
Scullin Government, 66
S.E.A.T.O., 46, 54, 139, 201, 206
'Secret Cables Affair', 331-8
Singapore, 139-40
Snedden, Bill, 106, 118-20, 185, 220, 230, 268-73, 278, 292-4, 300, 302, 309-16, 319, 324-5, 368, 384
Social Welfare, 157, 159, 221, 231, 285-6, 409
Solomon, David, 387
Somare, Michael, 194-5, 199
Speaker, 309, 311, 397, 401
Spender, Percy, 192
Spigelman, James, 256, 260, 282, 318-9
Staley, Anthony, 312
Stallings, Richard, 387
State Governors, 291, 374, 380, 395
Stevens, Captain D., 107
Stewart, Bishop, 150, 237
Stewart, Frank, 51, 254-5
St John, Edward, 107-8, 127, 147-50, 159, 161, 407-8
Stuart, Barbara, 109
Summerhayes, Carol, 109
Sweeney, John, 175
Sydney Airport (Galston), 273

Taiwan, 201, 221, 234, 245
Tange, Arthur, 239, 246-7, 260, 387
Tariff Board, 182
Tasmanian Bush Fires, 86

Taxation, 227, 283
Taylor, Bruce, 109
Television, 119-20, 142, 222, 228-9, 300
Terrill, Ross, 202
Theodore, Edward, xii
Toohey, Brian, 387
Townley, Athol, 41
Townley, Michael, 303, 373
Townsend, R. N., 352
Trade Practices Legislation, 289, 324
Trade Unions, 284-6, 382
  Amalgamated Metal Workers Union, 355
  Australian Workers Union, 11, 173-4
  Miscellaneous Workers Union, 11, 173-4
Treasury, *see* Departments
Turnbull, R., 112

United Nations, 51-3, 240, 245
  Southern African Resolutions, 245
U.S.A., 12, 15, 23, 30, 39-61, 93, 113-4, 123-5, 138-9, 145, 147, 164, 167, 201, 206-12, 217-20, 234, 258-60, 339-41
U.S. Bases, 20, 89, 243, 277, 387
Uren, Thomas, 39, 254-5, 258-9, 410
U.S.S.R., 50, 159, 164, 203, 207, 261

Victorian Council of Social Service, 103
Vietnam, 4, 5, 7, 39-59, 93, 94, 109, 110, 113, 114, 115, 124, 125, 126, 127, 139, 142, 145, 164, 166-9, 182, 189, 192, 200, 205, 206, 218, 230, 247, 248, 258, 259, 327-41
  Origins of Australian involvement, 40-5
  Economic consequences of, 77, 276, 277
V.I.P. Flights, 111-3

Waller, Keith, 239, 245-6, 259-60
Walsh, Eric, 143, 237, 260
Walsh, Maximilian, 143
Ward, Edward, 8-9, 13-14, 19, 73, 79, 83, 253
Warner, Denis, 337
Watson, J. C., 11, 64, 74, 84, 251
Webb, C. H., 129
Wentworth, W. C., 57, 143, 213-4
Westmoreland, General, 125
Wheat Sales (China), 201-2
Wheeldon, John, 304, 381
Wheeler, Frederick, 282, 296, 349, 352-3, 404
White, David, 260
White, Patrick, 5, 409
Whitington, Don, 143

Whitlam, E. G.,
  Family and school background, 63-5
  Influence of father on, 66-8
  At Sydney University, 68-9
  War service, 63, 69
  Early interest in ALP, 69, 70
  Development of political philosophy, 69-73, 74, 79, 80
  On equality, 75, 76, 79, 80, 160
  Elected Deputy Leader, 13
  Relations with Calwell, 7-11, 18, 22, 23, 31, 85, 89, 116, 126
  Expulsion attempt against, 25-34
  Relations with Chamberlain, 33, 94, 167-70
  Campaigns in Dawson by-election, 34
  Attitude to State Aid, 32, 34, 155
  Vietnam, 50, 56-9, 114, 124, 125, 126, 167-8, 205, 230, 258, 259, 326-41
  Elected Leader, 86
  Relations with Barnard, 86-8, 131
  Plans Party re-organisation, 91, 92-6
  Corio by-election, 98, 99, 109
  1967 Federal Conference, 99, 100
  Health reform, 101-5
  Relations with Holt, 108, 110
  Capricornia by-election, 110, 111
  V.I.P. disclosures, 111-3
  1967 Senate election, 115, 116
  Relations with Gorton, 118, 119, 120, 121, 122, 149, 181, 186, 187, 195, 196, 198
  Voyager debates, 127
  Resigns and recontests leadership, 129-36
  Re-elected, 136
  1969 Federal Conference, 151-8
  1969 Federal election, 161-5
  Reconstruction of VCE, 170-8
  Relations with McMahon, 188, 215
  Visits Papua New Guinea, 190, 192, 193, 195, 196-9, 200, 221
  China visit, 201-14
  Interviews Chou En-lai, 204-9
  Intervenes on behalf of Francis James, 213-4
  Japan visit, 211
  1972 elections, 224-38
  Decides to form 'two-man' government, 239-42
  Doctrine of the mandate, 243-5
  ASIO, 261, 266
  Attitude to Senate, 267
  Parramatta by-election, 273, 274
  Dollar revalued, 278, 279

Prices Justification Tribunal, 279
Coombs task force, 280
1974 credit squeeze, 282
Taxation, 283
Reaction of business, 285
Reviews achievements of 1973, 286, 287
Gair appointment, 288, 290–2
Advises double dissolution, 295
1974 election campaign, 296–301
Relations with Treasury, 305, 307, 308
Appoints Cairns Treasurer, 306–7
European visit, 310
Attacks Snedden, 310, 311, 312
Resignation of Speaker Cope, 309, 311
Attitude to Morosi appointment, 324
Writes to Vietnam leaders, 329, 330
Defends Vietnam policy, 339–41
Loans affair, 345, 346, 348, 349, 351, 352, 353, 357, 358, 359, 360, 385
Dismisses Cairns, 355–6
Connor resignation, 360–5
Relations with Kerr, 366–70, 374–81, 413
Defends Government's economic performance, 371–2
Constitutional crisis, 372–95, 407
Dismissed as Prime Minister, 392–3
1975 campaign, 404–5, 412
Lectures: Robert Garran (1973), 66; Curtin Memorial (1961), 69–70, 102, 346; Chifley Memorial (1957), 72, 102; A.I.P.S. Summer School (1955), 80; Chifley Memorial (1975), 243–4, 323, 372, 406; Curtin Memorial (1975), 411–12

Whitlam, H. F. E., 63, 66–8
Whitlam, Margaret, 63, 394
Whitlam, Nicholas, 22
Wilenski, Peter, 256, 259, 318–9
Willesee, Don, 86, 100, 241, 253–6, 290, 336
Williams, Evan, 357, 368
Williamson, David, 162
Wilson, Bruce, 337
Wilson, Roland, 348
Winneke, Henry, 380
Withers, Reginald, 266–7, 292, 295, 395
Women, 286, 306
Woodberry, B. W., 40
Woolcott, Richard, 217, 245
Wran, Neville, 49
Wriedt, Ken, 254–6, 381, 396
Wyndham, Cyril, 29, 32, 36, 55, 90–1, 99, 112, 131–2, 151–3

Young, Michael, 131, 151–2, 156, 170, 173–4, 176, 202–3, 210, 224, 274
Yugoslavia, 261–3